D1520080

TOXICOLOGY
biochemistry and pathology of mycotoxins

Contributors

Kageaki AIBARA, National Institute of Health, *Kamiosaki, Shinagawa-ku, Tokyo 141, Japan*

Makoto ENOMOTO, Sagamihara Kyodo Hospital, *Sagamihara-shi, Kanagawa-ken 229, Japan*

Hiroshi KURATA, National Institute of Hygienic Sciences, *Kamiyoga, Setagaya-ku, Tokyo 156, Japan*

Kiyoshi TERAO, Research Institute for Chemobiodynamics, *Chiba-shi, Chiba-ken 280, Japan*

Ikuko UENO, The Institute of Medical Science, University of Tokyo, *Shiroganedai, Minato-ku, Tokyo 108, Japan*

Yoshio UENO, Tokyo University of Science, *Ichigaya, Shinjuku-ku, Tokyo 162, Japan*

Kenji URAGUCHI, Tokyo University of Agriculture, *Sakuragaoka, Setagayaku, Tokyo 156, Japan*

Mikio YAMAZAKI, Research Institute for Chemobiodynamics, *Chibashi, Chiba-ken 280, Japan*

TOXICOLOGY
biochemistry and pathology of mycotoxins

Edited by
Kenji URAGUCHI
Mikio YAMAZAKI

A HALSTED PRESS BOOK

KODANSHA LTD.
Tokyo

JOHN WILEY & SONS
New York–London–Sydney–Toronto

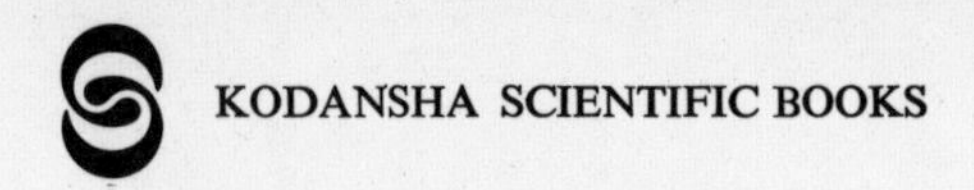

Library of Congress Cataloging in Publication Data

Main entry under title:
Toxicology, biochemistry, and pathology of mycotoxins.
"A Halsted Press book."
Includes bibliographical references and indexes.
1. Mycotoxicoses. 2. Mycotoxins. 3. Food contamination. I. Uraguchi, Kenji, 1906- II. Yamazaki, Mikio, 1931- [DNLM: 1. Mycotoxins. QW630 T755]
RA1242. M94T68 1978 615.9'5 78-8992
ISBN 0-470-26423-3

Published in Japan by
KODANSHA LTD.
12–21 Otowa 2-chome, Bunkyo-ku, Tokyo 112, Japan

Published by
HALSTED PRESS
a Division of John Wiley & Sons, Inc.
605 Third Avenue, New York, N.Y. 10016, U.S.A.

PRINTED IN JAPAN

Contents

Preface

Although certain mycotoxins have been identified as causative agents of distinct diseases, the concept of mycotoxins is not yet widely appreciated. Recent research has however promoted a wider recognition of the chemical and biological properties of mycotoxins.

As stated by Forgacs, mycotoxicosis is a "neglected disease". Mycotoxicoses or injuries attributable to mycotoxins have been demonstrated mostly in domestic animals and fowls: the human incidence is relatively low. This situation probably reflects the fact that, while farm animals and fowls are usually compelled to feed on an unchanging formula even after the feed has become polluted by fungi and mycotoxins, man is generally not subject to such passive feeding.

Geographic pathology and epidemiology have revealed high incidences of liver cancer and closely related diseases in certain areas of Asia and Africa. The moderate climate with high humidity in these areas constitutes and accelerative factor for fungal growth. Fungal contamination of foodstuffs is liable to occur during transport, storage and food processing. In fact, among the rice exported from the rice-producing countries of S.E. Asia and other regions, so-called yellowed rice has been detected. This is polluted with *Penicillium islandicum* Sopp, which produces two mycotoxins, luteoskyrin and cyclochlorotine, that are both simultaneously carcinogenic and hepatotoxic.

In work on the turkey "X" disease that occurred in Great Britain in 1960, another carcinogenic mycotoxin of surprisingly high potency, aflatoxin B_1, was discovered. The possible participation of mycotoxins in certain human liver cancers was thus emphasized, and the finding had a great impact in the field of public health throughout the world. Moreover, many unidentified diseases are presumed to have mycotoxins as the etiological agent. The suggestion that fungal toxins may thus be responsible for a broad spectrum of human diseases will undoubtedly contribute to a further concentration of research effort in this field.

Positive confirmation of mycotoxins as etiological agents is often difficult, due to the great complexity of causal factors involved in diseases. Studies on mycotoxins and mycotoxicoses thus demand organized cooperation among mycologists, chemists, phytopathologists, pathologists, pharmacologists, experimental toxicologists, and medical doctors. In Japan, such extensive collaboration has a comparatively long history, dating back to the early efforts to determine a causal link between beriberi and a neurotoxic mycotoxin. Indeed, this work predates the discovery of aflatoxin B_1 by 30 years.

The present book represents the combined and coordinated efforts of several authors, each of whom is an expert of a particular area of mycotoxin research in Japan. The contents are arranged according to the characteristic features and properties of mycotoxins, offering general reviews from the mycological, chemical, toxicological and pathological viewpoints. Moreover, one chapter is devoted specifically to the carcinogenicity of mycotoxins, since this problem is one of the most serious in relation to human health.

As editors for this book, we wish to thank all authors for their contributions, and the staff of Kodansha for their advice and practical assistance.

March, 1978

Kenji URAGUCHI
Mikio YAMAZAKI

Introduction

Kenji URAGUCHI
Tokyo University of Agriculture, Sakuragaoka, Setagaya-ku, Tokyo 156, Japan

There are still diseases, some with distinct clinical features, that have unknown etiology. Most important in practice, however, is the group of undetermined diseases, the clinical characteristics of which are insufficiently distinct for them to be clearly differentiated from diseases of different etiology. Due to their unknown etiology, such diseases are often treated or analyzed by improper methods. Their variety must in fact be great, as history indicates.

Medical progress in the 19th century clarified the role of certain natural environmental factors in disease development. Poisons of insects, snakes, fish, shellfish and mushrooms, other natural products such as alkaloids and heavy metals were identified as causative agents of certain diseases. Since their occurrence and distribution is naturally restricted in most cases, the risks and casualities tend to be localized, and from the epidemiologic viewpoint, such cases of natural product poisoning are essentially incidental or sporadic, occasionally endemic, and rarely large in magnitude.

In the latter half of the 19th century, the role of microbes as an etiologic factor in the development of diseases was established, and later the modern concept of infectious or communicable diseases caused by specific pathogens was developed. Such pathogens are generally prolific microbes with a strong tendency to propagate in the host and environment, and show a generally wider distribution than natural poisons. Microbial agents are thus able to act more positively upon the host under favorable ecological conditions, so that there is greater risk of disease development among unspecified persons. Epidemiologically, there are various disease patterns, from endemicity to pandemicity, as illustrated by cholera, plague, influenza and tuberculosis, and mass outbreaks can occur under favorable circumstances, in striking contrast to usual natural product poisoning.

In epidemiologic studies of infectious diseases, initial studies emphasized host-parasite relationships. However, recently, it has been shown that the microbial agent tends to act upon the host rather as an environmental factor, and even with these animate pathogens, their action can be educed from the observed response in the host, just as with a pure chemical entity. The toxicologic concept of the dose-response relationship between agent and host should thus be accepted as a fundamental principle even in research on microbial diseases. By expressing the intensity of the active agent in a quantitative way, the relationship between this enviromnental factor and the host could be scrutinized rather as in the general field of toxicology, and through quantitative studies on the action of the environmental factors, reactions of hosts, and their interrelationships, it should be possible to analyze accurately the problem of pluralistic disease formation. The host's reaction to a particular environmental factor is conditioned by its physical and ecological situation, and the response to the host's reaction elicited in the environment is also modified ecologically by the total environmental conditions. All ecologic factors controlling such responses on either side are thus worthy of study in elucidating the detailed process of disease formation. Supposing a certain etiologic agent affecting a host experiences a variety of ecologic conditions advantageous to the action-reaction system producing the disease *in vivo*, disease formation must be accelerated and the host must become more liable to the disease. This indicates an increased susceptibility in the host, and statistics will show a higher morbidity. The magnitude of an epidemiologic occurrence of a disease is thus controlled by complex interactions, and the observed pattern of occurrence of the disease is a collective expression of a dynamic conflict between the active agent, host, and environment, including natural, physical and socioeconomic factors.

Mycotoxicosis may be a somewhat unfamiliar term. However, it represents nothing new to Europeans familiar with past cases of ergotism through rye and other cereals, or to Japanese concerned over "moldy rice", i.e. fungus-infected rice, and the ambiguous etiology of the beriberi which prevailed in the late 19th century. The early vague fears or views of this mycotoxicosis prompted the first research in Japan, and the complex idea of mycotoxins gradually evolved as a working hypothesis.

With the aim of establishing the role of "moldy rice" in the beriberi, Sakaki[1)] investigated the rice then available on the market, and revealed heavy fungal contamination of this staple food in Japan. Samples of a certain moldy rice ("Sawate-mai") were subjected to toxicity tests, and proved to be neurotoxic to rabbits, indicating the presence of a mycotoxin. However, Sakaki did not examine the taxonomy of the toxin-producing fungus or chemistry of the toxic product, and final conclusions were not reached on the precise relation of the mycotoxin to the beriberi or the equivalence of the experimentally induced symptoms and actual disease. Mortality still persisted at a high level after Sakaki's publication in 1893. However, 20 yr later, with more advanced methods for rice storage and transportation as well as official rice inspection in producing districts throughout Japan, the situation regarding fungal infection improved and the high incidence of beriberi in urban districts began to fall. In and around 1910, there was thus a marked drop in mortality from acute cardiac beriberi.[2,3]

This improvement in the epidemiologic situation preceded the discovery of "vitamine"[4,5] and there is no reason to ascribe the decrease in acute beriberi to the proposed effects of vitamin B. Rather, the magnitude and scope of the acute beriberi and those of the fungal pollution of rice ran parallel to each other,[2] extending over half a century from the Meiji Restoration, According to the author's opinion, acute cardiac beriberi should therefore be differentiated from other types of beriberi, not only clinically but also etiologically. Cardiac beriberi represents an independent disease having a particular pathogenesis with no direct relation to vitamins or vitamin deficiency.[6] Clinically, "acute beriberi of the cardiac type" differs from other types of beriberi in its abrupt onset and rapidly developing signs and symptoms of paralysis, and also in the quick passage (a few days) from the initial signs to death, usually terminating in agonized writhing of the body, aggravated cardiovascular damage and irreversibly developing respiratory failure. Such a clinical picture unaccompanied by any incubation period, is clearly suggestive of an intoxication, and is not experimentally reproducible in animals with thiamin deficiency.

In this setting, exhaustive phytopathologic research was begun. From 1918, I. Miyake investigated fungi growing on rice, and successivley discovered several species. His surveys were somewhat intermittent but thorough, and finally resulted in 1938 in the discovery of a toxic fungus growing on rice, i.e. *Penicillium* sp., the so-called "yellowed rice fungus".[7] The species was identified as *Penicillium citreo-viride* Biourge by his collaborators, although it was first published as *P. toxicarium* Miyake.[8] This species was detected originally in a foreign rice sample transported from Formosa, and then in 9 samples of domestic rice produced in 1934–9 in different parts of northern Honshu, mostly on the Japan Sea side[7] These districts are situated in the chief rice-producing area of Japan and are well known for their tasty "soft rice": the climate is characterized by lesser insolation and rather low temperatures compared with the remainder of

Honshu. From the ecological viewpoint, climatic consitions such as regional temperature and humidity are known to control fungal growth and activity, and moderate temperatures (*ca.* 20°C) have been shown to favor the metabolic production (yield) of mycotoxin by the fungus.[2] In these studies, fungal growth was found to continue even after the temperature had exceeded the optimum range for mycotoxin production, i.e. the actual yield does not necessarily increase in parallel with the observed fungal growth.

According to Miyake,[7] the fungus grows mostly on rice during storage and transportation after harvesting. A surface scratch on the unpolished grains, particularly around the embryo bud, permits entry of the fungus. Infection usually begins on stored grain, when the water content is 14.6%, and if the content increases by as little as 1%, other fungi usually commence growth and overwhelm the fungus. When rice is kept in the narrow but favorable moisture range, an earlier infection by the fungus is likely to occur. These ecologic conditions thus had a direct bearing on the particular distribution of the fungus in northern Honshu, and so indirectly on the epidemiologic pattern of the mycotoxicosis occurring there.

When discussing causative agents of disease above, the author considered toxic natural products and pathogenic microbes, discussed differences in their modes of occurrence and behavior as well as their respective natural restrictions and ecologic conditions, and related these to differences in the magnitude and scope of the diseases. In the case of mycotoxins, there are further differences in modes of occurrence, behavior, distribution and ecologic conditions. First, the mycotoxin-producing fungus is generally far less restricted in its regional distribution and occurrence than the toxic natural product and is able to circulate more freely in the environment than other microbial agents, without dependence on selective parasitism to the human body as in the case of the pathogenic microbe. This favors a wider occurrence or broader distribution. Second, the produced mycotoxins tend to occur wherever the fungus grows and may persist in affected materials even after fungal eradication. This situation favors widespread pollution of the environment by mycotoxins, so that mass outbreaks of disease exceeding those of other natural agents in epidemiologic magnitude and scope, tend to occur. Nevertheless, mycotoxicoses show a great variety of epidemiologic patterns from small-scale occurrences (accidental, isolated or sporadic) to large-scale ones (endemic or epidemic), together with seasonal and annual variations and regional ecologic differences.

Although most mycotoxicoses have a potential for widespread outbreak, large-scale occurrences are in practice not always realized so frequently in man and animals as smaller-scale ones. Historically, however, the long prevalence of ergotism in Europe illustrates the definite possibility for certain mycotoxicoses to develop into huge, mass diseases where the people lack any scientific understanding of fungi and toxins or exact knowledge of the agent, host and environmental aspects of the disease in question. The epidemiologic event with *Claviceps purpurea* is thus an apt illustration of the degree to which a certain mycotoxicosis can grow in the absence of suitable preventive measures. The case of acute beriberi in Japan and other rice-growing countries of Asia may be viewed similarly. Although there were exploratory tests for bacterial infections and so-called "blue fish" poisoning, the long search for a toxic "moldy rice" finally bore fruit, as mentioned above. However, this discovery came too late to be of practical use in controlling the fungus or its mycotoxin in Japan since improvements in methods had already minimized the dangers.

Following Miyake's discovery of the fungus, however, the mycotoxin citreoviridin was isolated chemically by Hirata[9] and its chemical properties, molecular formula and structure were clarified.[10] In 1940, concurrently with such chemical studies, Uraguchi began toxicological experiments with crude extracts of the mycotoxin, and later with the purified compound. A series of reports[11–5] were published in 1947–55 giving fundamental data on the toxicity, comparative toxicity to animals, symptoms in poisoned mammals and other vertebrates, experimental analyses of the pharmacokinetic behavior and pharmacodynamic action of the mycotoxin, etc. The LD_{50} of the crude extract varied with the preparation, ranging from 0.1 to 1 g/kg s.c. in mice. Similar values were seen in 5 other mammals and 2 birds, but values were low in fish, frogs or snakes. The i.p., s.c. and p.o. LD_{50} values of the extract were in the ratio 8:10:30, and the respective times to death were 1.5–3, 3–6 and 3–8 h. The LD_{50} values of citreoviridin in mice have recently been given as 2.0 mg/kg i.v. and 8.3 mg/kg s.c. (dds); i.p. 7.2, s.c. 11.0 and p.o. 29.0 mg/kg.

The symptoms of acute, experimentally induced mycotoxicosis were characterized by progressive ascending paralysis associated with a gradual lowering of the body temperature, blood pressure depression and other cardiovascular damage, and respiratory failure.[16] Nervous symptoms were particularly evident in monkeys, cats and dogs, but rather obscure in fish, amphibia and reptiles. The toxin was found to be distributed freely in the liver, kidneys, adrenal cortex, adipose tissue and central nervous system, and excreted through the milk, urine, bile and vomitus. Pharmacologic experiments on nerve-muscle preparations and isolated nerves ruled out any peripheral nerve affection. Oscillographs of action currents revealed progressive depression of spinal and medullary function. The diaphragm and thorax became paralyzed, concurrently with the onset of advanced respiratory and circulatory failure. The dilated heart was found by X-ray fluoroscopy to float high above the immobilized diaphragm. Polygraphs in cats immobilized by curare under artificial respiration, disclosed selective depression of the blood pressure, with or without associated electrocardiographic changes. Neurograms of the phrenic nerve indicated that inhibition of the respiratory center was the cause of death in acute mycotoxicosis. The brain stem was affected but there was no definite evidence concerning higher brain affection, although in some cases catalepsy-like symptoms were observed transiently. Blindness often occurred in the later stages. Spastic, hyperkinetic, or convulsive signs and symptoms, vomiting and decerebrate rigidity were noticed in the course of progressive paralysis. The animals became astasic as paralysis advanced, and onset of astasia was indicative of hopelessness for recovery: this prognostically significant symptom generally occurred prior to the appearance of diaphragm paralysis and cardiac insufficiency.

Based on the above findings and developmental sequence of the clinical picture, the author concluded that the mycotoxicosis observed in animals was an acute progressive paralysis of ascending type associated with circulatory and respiratory failure. There was a close resemblance between the experimental symptoms and clinical manifestations of acute cardiac beriberi reported earlier in man by pathologists and doctors in Japan and abroad. In 1969, at the lst International Congress of Plant Pathology in London, the author[2,3,16)] set froth his view that the acute cardiac beriberi, which could be differentiated as a particular disease ("shoshin-kakke") from so-called beriberi in the broad sense, must have been a mycotoxicosis characterized by acute progressive paralysis.

In order to conclude finally that a certain mass disease of unknown etiology is actually a human mycotoxicosis, it is necessary to establish definitely that the agent acting to cause the disease in man is in fact mycotic in nature, i.e. *Penicillium citreo-viride* or citroeviridin in the present case. However, with the disappearance of the beriberi epidemic, direct verification of the etiologic function of this fungus or mycotoxin became entirely impossible. There has been no recurrence of the disease in Japan and intentionally induced mycotoxicosis in man, for comparative purposes, is of course impermissible. The only course left is to try to reconstruct the circumstances surrounding former outbreaks, while further ecologic analysis of the agent, host and its environment may prove useful, not only from the etiologic and epidemiologic viewpoint but also from that of toxicology, preventive medicine and disease control.

High incidences of liver cancer, cirrhosis and related diseases of the liver are known to occur in Asia and Africa, and recently the geographic pathology of some such diseases has been extensively studied. Clearly, anticipation of potential outbreaks is important for disease control. From the standpoint of environmental toxicology, mycotoxins and fungi are worthy of investigation as possible etiologic agents of unknown diseases, particularly in rice-growing countries of Monsoon Asia, since such areas with their huge populations and well-cultivated land, enjoy climatic conditions favorable for rice cultivation and fungal growth. Before the end of World War II, there were presumably few people actually engaged in studies of mycotoxins or with any established view of mycotoxicoses and toxic moldy products. However, the whole rice-producing region of the world, particularly so-called "Wet Asia", represented an area where living and environmental conditions strongly favored the existence of undetermined mycotic agents. Moreover, even in "Dry Asia", i.e. the wheat-producing region, similar suspicions appeared valid. Thus, early anticipations of mycotoxicoses by us and by other groups materialized in succession, as follows:

(1) In 1948, *Penicillium islandicum* Sopp[17] and its mycotoxins, luteoskyrin and cyclochlorotine,[18–20] were detected in the hepatotoxic yellowed rice distributed widely over the rice-producing countries of the world.

(2) In 1952, *Penicillium citrinum* Thom, producing citrinin, was isolated from the nephrotoxic yellowed rice occurring in S. E. Asia.[21,22]

(3) In 1949, *Fusarium graminearum*[23,24] and in 1954, *Fusarium roseum*[25] were detected in imported flour and red mold wheat, respectively, both causing sporadic events with nausea, vomiting and other toxic symptoms in Japan. This work served as a forerunner for later investigations on *Fusarium nivale* and related species (including also *Gibberella* sp.) as well as on their mycotoxins.

(4) In this connection, Forgacs *et al.*[26] reported that, in addition to frequent poisoning by the overwintered moldy crop of Eastern Siberia, noticeable incidents of severe poisoning (finally diagnosed as alimentary toxic aleukia, ATA) had occurred between 1941 and 1945 in Western Siberia. Regarding the causative agent, *Fusarium sporotrichioides*, *F. poae* and related species[27] were reported to produce sporo- and poae-fusarinogenins, etc.

(5) In 1960, Turkey X disease occurred in Great Britain, victimizing about 100,000 turkeys in 3 months. *Aspergillus flavus* was isolated from toxic groundnut meal[28] and aflatoxins were extracted from the contaminated materials and cultured fungus.[29–31] Succeeding experiments established the acute toxicity of the toxins, their fate and distribution *in vivo*, pathological changes such as liver necrosis and bile duct proliferation,[32,33] and biochemical data on the inhibition of protein, DNA and RNA synthesis in biological materials[34–7] and on enzyme induction in the liver.[37–9]

It is easy to be wise after the event, but in many ways these events represented fears come true. Further comments will therefore be given from the viewpoint of preventive medicine and food pollution control.

Concerning yellowed rice, a series of diseased rices that turned yellow are known, of which the *Penicillium citreo-viride*-infected material mentioned above was the original and only known toxic moldy rice before World War II. The other kinds of yellowed rice were encountered later. After the War, Japan suffered for several years from a shortage of domestic rice, and was so forced to depend on imported rice from various countries. However, most of these countries were themselves in a distorted socioeconomic condition, and the rice crop, means of storage and transportation had by no means returned to normal. It was thus difficult to supply rice of sufficient quality and quantity. Japan, however, endeavored to purchase rice from every possible source by every possible means, and was so able to collect foreign rice samples from most producing countries of the world. They included an all-time record of moldy rice samples for study. Rice rationing was still in force and all imported lots were subject to mycological inspection by the authorities on entry. As a result, more than 70 kinds of diseased rice, known and unknown, were found. Among these was another yellowed rice with typical pigmentation, on which *Penicillium islandicum* Sopp was discovered. The first strain was isolated from rice exported by Egypt in 1948, but the fungus was soon found to be distributed widely over Africa, Europe, N. and S. America, and most densely in Asia, particularly the countries of S. E. Asia. These were the findings of official inspections during the period 1948–53. Lots contaminated with yellowed rice grains represented a total of almost 150,000 tons, and were retained in official warehouses. They were finally excluded from the governmental rationing scheme due to their undeniable toxicity and carcinogenicity to the liver.[40–3)]

Concerning disposal of the diseased rice accumulated in the warehouses, the so-called "yellowed rice dispute" developed in the summer of 1954. Conflicting opinions existed among officials, experts, consumers and other parties over policy measures for rice unqualified for rationing, and over the proposed permissible dose of toxic rice. Hot discussions took place and for a time it seemed impossible to formulate a practical solution. However, the authorities finally realigned their policy in recognition of experimental data on the toxicity and carcinogenicity of the diseased material. They proposed to improve their mycotic inspection method towards more reliable sampling, and the disqualified rice was almost entirely converted to alcohol for industrial use. The settlement took four full years to reach but in the end the suspected carcinogenic rice was essentially witheld from nationwide distribution.

For the author and his co-workers, whose animal experiments had indicated not only the possibility of cancer or cirrhosis but also of various chronic liver diseases, this result was of course satisfactory. However, the ultimate aim was to enforce measures giving protection against the whole spectrum of potential hazards due to mycotoxins. In order to evaluate the efficacy of preventive steps, it is of course necessary to set up appropriate control areas, and the author made a concerted effort to determine the actual state of occurrence of liver cancer of mycotoxic origin and of other allied liver diseases in areas where the yellowed rice was proved to have been distributed most widely. Since the main source of such rice was foreign, inspection of rice imported for rations was intensified, but this was not practical in the case of animal feed. In Japan, no fatal case of human mycotoxicosis due to the particular fungus or its mycotoxins was observed, but in Sept. 1963, 2891 of 13,610 chicks fed with commercial rice-cake at 7 poultry farms in Nagano Pref., died from an acute poisoning.[44) *Penicillium islandicum* was detected in the feed at a level of about 15,000 colonies/g of rice-cake. Mice fed with this material became depressed and died with jaundice after 3–6 days. Histopathologically, centrilobular necrosis and subsequent collapse of the stroma with interconnection of necrotic areas were observed, so indicating typical early changes of the yellowed rice intoxication. A "contraband" policy for the distribution of the rice, mixed with normal rice, was daringly adopted by the Ministry of Welfare in advance of the appearance of any case of human mycotoxicosis. This was not a mere corrective expedient to gloss over belated treatment, but was considered a safety policy in its basic idea since it was established in conformity with achievements in experimental toxicology. Thus, in the case of yellowed rice, scientific research can be said to have forerun the political decisions on preventive policy.

The aflatoxin case, i.e. the Turkey X disease in the United Kingdom, although affecting turkeys not man, was a shocking event. It inspired later experimental research for effective preventive measures. Furthermore, with *Fusaria* in Siberia, the occurrence of human poisoning prompted detailed experimental research, as was true in the earlier cases of beriberi in Japan and ergotism in Europe.

The above examples illustrate the variety of epidemiologic conditions that may surround mycotoxicoses, and also reveal certain differences in the scientific and administrative responses in different countries. In general, the author has the impression that mycotoxicosis represents an essentially expanding disease, as indicated; however, it is practically controllable so long as there is adequate information or awareness about the mycotic agent and its ecology. Research on experimentally induced mycotoxicosis for comparison against actual mycotoxicoses in man and animals, is clearly essential for clarifying the nature of and controlling such diseases. The biologic and toxicologic action of the agents has been found to be influenced strongly by the general ecologic situation or conditions, incorporating factors related not only to traditional fungal and crop ecology but also to human ecology, including personal and socioeconomic factors. The socioeconomic situation clearly affects the size and pattern of disease formation and epidemiologic characteristics, and the overall ecologic situation or conditions may serve to promote or inhibit the developing diseases as a whole, and so affect steps for disease prevention. The combined influence of all these factors must endow mycotoxicoses with their character of being both expanding and controllable diseases.

Finally, the author wishes to return to the question of the nature of the so-called causative "mycotic agent". Both a toxin produced by a fungus, and the toxin-producing fungus itself, may be implicated in the development of a certain disease. However, the toxin, though active, is inanimate, while the living fungus has the capacity not only to produce the active compound but also to undergo movement through migration, growth and parasitic transport. In this respect, the mycotic agent should be regarded as a "toxin with vehicle". Clearly, the active agents in past human and animal mycotoxicoses were of this type, as reflected in their varied epidemiology. To gain a better understanding of mycotoxicosis formation and epidemiologic patterns, and to develop effective preventive measures, it is thus better to consider the etiologic agent not simply as an inanimate mycotoxin but rather as a "mycotoxin with vehicle", that is, a product of an active, ever-changing fungus whose activity and success depend sensitively on the ecologic conditions of its environment.

REFERENCES

1. J. Sakaki, *Z. Tokio-Med. Ges.* (Japanese), **5**, 21 (1891); J. Sakaki, *Tokyo. Med. Wochschr.* (Japanese), **779**, 6 (1893).
2. K. Uraguchi, *J. Stored Prod. Res.*, **5**, 227 (1967).
3. K. Uraguchi, *International Encyclopedia of Pharmacology and Therapeutics*, section 71, part V, Pharmacology of mycotoxins, p. 153, Pergamon Press (1971).
4. U. Suzuki and T. Shimamura, *Tokyo Kagaku Kaishi* (Japanese), **32**, 4 (1911).
5. C. Funk, *J. Physiol.*, **43**, 395 (1911).
6. T. Ogata, S. Kawakita, S. Suzuki and S. Kagoshima, *Nisshin Igaku* (Japanese), **13**, 742 (1924).
7. I. Miyake, H. Naito and H. Tsunoda, *Beikoku-Riyo-Kenkyujo Hokoku* (Japanese), **1**, 1 (1940).
8. I. Miyake, *Nisshin Igaku* (Japanese), **34**, 161 (1947).
9. Y. Hirata, *J. Jap. Chem. Soc.*, (Japanese), **68**, 63, 74, 104 (1947).
10. N. Sakabe, T. Goto and Y. Hirata, *Tetr. Lett.* **27**, 1825 (1964).
11. K. Uraguchi, *Folia Pharmacol. japon.* (Japanese), **34**, 39 (1942).

12. K. Uraguchi, *Nisshin Igaku* (Japanese), **34**, 155, 224 (1947); **35**, 166 (1948); **36**, 13 (1949); **37**, 337 (1950).
13. K. Uraguchi and F. Sakai, *ibid.*, **42**, 512 (1955).
14. K. Uraguchi, F. Sakai and S. Mori, *ibid.*, **42**, 690 (1955).
15. F. Sakai and K. Uraguchi, *ibid.*, **42**, 609 (1955).
16. A. Ciegler, S. Kadis and S. J. Ajl(ed.), *Microbial Toxins*, vol. VI, p. 371, Academic Press (1971).
17. H. Tsunoda, *Jap. J. Nutr.* (Japanese), **8**, 186 (1951); **9**, 1 (1952).
18. T. Tatsuno, M. Tsukioka, Y. Sakai, Y. Suzuki and Y. Asami, *Pharm. Bull.* (*Tokyo*), **3**, 476 (1955).
19. K. Uraguchi, T. Tatsuno, M. Tsukioka, Y. Sakai, F. Sakai, Y. Kobayashi, M. Saito and M. Miyake, *Jap. J. Exptl. Med.*, **31**, 1 (1961).
20. K. Uraguchi, T. Tatsuno, F. Sakai, M. Tsukioka, Y. Sakai, O. Yonemitsu, H. Ito, M. Miyake, M. Saito, M. Enomoto, T. Shikata and T. Ishiko, *ibid.*, **31**, 19 (1961).
21. H. Tsunoda and O. Tsuruta, *Nippon Shokubutsu-Koho* (Japanese), **18**, 4 (1954).
22. F. Sakai, *Folia Pharmacol. Japon.*, **51**, 431 (1955).
23. S. Hirayama and M. Yamamoto, *Bull. Hyg. Lab.* (Japanese), **66**, 85 (1950).
24. Y. Nakamura, S. Takeda, K. Ogasawara, T. Karashimada and K. Ando, *Hokkaido Eisei Kenkyujoho* (Japanese), **2**, 35 (1951); Y. Nakamura, S. Takeda and K. Ogasawara, *ibid.*, **2**, 47 (1951).
25. H. Tsunoda, O. Tsuruta, S. Matsunami and S. Ishii, *Proc. Fd. Res. Inst.*, (Japanese), **12**, 27 (1957).
26. I. Forgacs and W. T. Carll, *Advan. Vet. Sci.*, **7**, 272 (1962).
27. A. Z. Joffe, *Bull. Res. Council Israel*, **9**, 101 (1960); *Mycopath. Mycol. Appl.*, **16**, 201 (1962); *Plant Soil*, **18**, 31 (1963).
28. K. Sergeant, J. O'Kelly, R.B.A. Carnaghan and R. Allcroft, *Vet. Record*, **73**, 1219 (1961).
29. G. N. Wogan, E. L. Wick, C. G. Dunn and N. S. Scrimshaw, *Fed. Proc.*, **22**, 611 (1963).
30. J. H. Broadbent, J. A. Cornelius and G. Shone, *Analyst*, **88**, 214 (1963).
31. J. J. Coomes and J. C. Sanders, *ibid.*, **88** ,209 (1963).
32. P. M. Newberne, W. W. Carlton and G. N. Wogan, *Path. Vet.*, **1**, 105 (1964).
33. W. H. Butler, *Brit. J. Cancer*, **18**, 756 (1964).
34. R. H. Smith, *Biochem. J.* **88**, 50 (1963).
35. R. C. Shank and G. N. Wogan, *Fed. Proc.*, **23**, 200 (1964); *Toxicol. Appl. Pharmacol.*, **9**, 468 (1966).
36. M. B. Sporn and C. W. Dingman, *Proc. Am. Assoc. Cancer Res.*, **7**, 67 (1966).
37. J. I. Clifford and K. R. Rees, *Nature*, **209**, 312 (1966).
38. G. N. Wogan and M. A. Friedman, *Fed. Proc.*, **24**, 627 (1965).
39. M. B. Sporn, C. W. Dingman, H. L. Phelps and G. N. Wogan, *Science*, **151**, 1939 (1966).
40. M. Miyake, M. Saito, M. Enomoto, T. Shikata, T. Ishiko, K. Uraguchi, F. Sakai, T. Tatsuno, M. Tsukioka, Y. Sakai and T. Sato, *Acta Path. Japon.*, **10**, 75 (1960).
41. K. Uraguchi, F. Sakai, M. Tsukioka, Y. Noguchi, T. Tatsuno, M. Saito, M. Enomoto, T. Ishiko and M. Miyake, *Jap. J. Exptl. Med.*, **31**, 435 (1961).
42. K. Uraguchi, *Excerpta Med. Int. Congr. Ser.*, **87**, 465 (1965); *Tokyo Igakkai Zasshi* (Japanese), **74**, 295 (1966).
43. K. Uraguchi, M. Saito, Y. Noguchi, K. Takahashi, M. Enomoto and T. Tatsuno, *Fd. Cosmet. Toxicol.*, **10**, 193 (1972).
44. H. Tsunoda and S. Ito, *Shokuryo Kenkyusho Kenkyu Hokoku* (Japanese), **19**, 15 (1965).

CHAPTER 1

Current Scope of Mycotoxin Research from the Viewpoint of Food Mycology

Hiroshi KURATA
National Institute of Hygienic Sciences, Kamiyoga, Setagaya-ku, Tokyo 156, Japan

1.1. INTRODUCTION

The word mycotoxin derives from the Greek *mykes* (fungus) and the Latin *toxicum* (poison),[1] and is at present used to designate secondary fungal metabolites which cause pathological changes or physiological abnormalities in man and warm-blooded animals.[2] Mycotoxicosis is poisoning of the host which may follow ingestion of mycotoxin-contaminated foods and feed.

Forgacs, who is the modern pioneer of mycotoxin research, once described mycotoxicosis as a neglected disease.[3] Indeed, in spite of reports of earlier mycotoxicoses in several countries, little interest was shown in such diseases until the aflatoxicosis outbreaks of 1960 in the United Kingdom.[4] Nevertheless, since the turn of the century, it has been known that certain toxic fungal metabolites elicit biologic effects in both man and animals, and in the past two decades many diseases of previously unknown etiology have been shown to result from ingestion of fungus-contaminated foods and feed.

As discussed by Uraguchi in the Introduction, the early examples included the disease associated with yellowed rice in Japan in the 19th century[5] and the alimentary toxic aleukia associated with overwintered millet, wheat and barley in Russia,[6] while the much earlier ergotism in Europe may be recognized in retrospect as the first of many mycotoxicoses.[7] Moreover, the effects of poisonous basidiomycetes such as the mushrooms *Amanita muscaria* and *A. phalloides*, have long been known to fall within the category of mycotoxicosis.

This chapter will give a brief review of some of the mycotoxicoses occurring before 1960, and then proceed to discuss the important mycotoxin-producing fungi and their occurrence in various foodstuffs.

1.2. Brief History of Mycotoxicoses before 1960

Ergotism: This mycotoxicosis had its origin in the ingestion of rye and other grains infested with the mold, *Claviceps purpurea*. During the feudal days in Europe, this disease was known as "St Anthony's Fire" or "the Holy Fire" and periodic outbreaks resulted in thousands of deaths. The causal fungus infects prematured seeds of cereal crops and forms a dark-brownish fruiting body (sclerotium) at the diseased site. This sclerotium is called an ergot and contains toxic substances, ergot alkaloids.

Two characteristic forms of the poisoning were described: gangrenous and convulsion ergotism.[8] One of the latest epidemics occurred in Manchester, England, in 1925. The patients complained of coldness and numbness of the extremities, a sensation resembling insects creeping over or under the skin, dizziness, abdominal pains, depression, headaches, and a staggering gait.[9] Ergot poisoning is likely to occur during wet harvesting seasons when the grain is improperly cleaned prior to milling. Epidemics of ergotism are now rare and the last major outbreak was in 1825. However, serious outbreaks did occurr in Russia in 1926–7 and in England in 1928; an outbreak occurred in France as recently as 1951, and occasional isolated episodes are still reported.[10]

Facial eczema in ruminants: Since 1822, facial eczema in sheep and cattle in New Zealand has been found to be associated with the ingestion of pasture containing dried grass infected with the saprophytic mold, *Pithomyces chartarum*, formerly identifed as *Sporidesmium bakeri*.[3] The spores of this fungus contain sporidesmin, and outbreaks of the disease follow contamination of pasture during periods of abundant sporulation. Essentially, the sporidesmin causes liver damage, loss of weight, icterus, and photosensitivity. "Facial eczema" is a descriptive epithet for the inflammation and scabbing of unprotected skin which results from the photosensitivity. Such photosensitivity has been shown to be a secondary symptom, caused by build-up of the porphyrin, phylloerythrin, in the peripheral blood system. This build-up is a consequence of failure by the damaged liver to excrete phylloerythrin, a product of chlorophyll digestion. In the case of ruminants, apparently, the disease has so far been described as a chronic disease. Affected sheep develop lachrymation, salvation, a nasal discharge, hyperirritability, itchiness, and seek shade for relief. Subsequently, affected animals may develop leucocytosis, bilirubinemia, icterus, cachexia, and may die.[11,12]

Since 1952, a strikingly similar disease has been reported in the United States in cattle grazing on moldy Bermuda grass and moldy legume pasture.[13] The predominant fungus on such toxic forage is a strain of *Periconia minutissima*. However, confirmatory etiological evidence from animal experiments has not yet been obtained.

Stachybotryotoxicosis: The first reports of a disease in domestic animals attributable to *Stachybotrys* fungi are those of Vertinskii and Drobotko in the 1930's from the Ukrainian S. S. R.[14] Stachybotryotoxicosis is a mycotoxicosis of horses resulting from the ingestion of fodder contaminated with the fungus, *Stachybotrys alternans* (syn. *S. atra*). In 1958, Forgacs *et al.*[15] found that sheep, calves and swine were also affected by the toxin, although the calf appears to be more resistant than the horse. In 1959, Fortuskny *et al.*[16] reported a field outbreak of this toxicosis in cattle, and more recently, other similar outbreaks in cattle have been reported by Soviet scientists. Humans exposed to aerosols of the toxic substrate are also affected by the toxicosis. Typical symptoms in farm animals are most commonly observed in the field and develop during prolonged ingestion of small amounts of the toxic feed. During the first stage the animal salivates excessively, the submaxillary lymph nodes become greatly enlarged, the mucous membranes of the eyes and oral cavity become injected, and cracks develop at the mucocutaneous junction of the lips, followed by the formation of deep fissures. As the inflammatory process progresses, edema of the underlying tissue causes the lips to swell so that the animal assumes a hippopotamus-like appearance. This stage appears within 2–3 days after ingestion of toxic feed and may persist for 8–30 days. Stachybotryotoxins produced by the fungus have long been studied by several investigators[17] but recent work by the FDA group in the United States has provided strong, albeit indirect evidence that the compounds responsible for this toxicosis belong to the series of sesquiterpenes known as 12, 13-epoxy-Δ^9-trichothecenes.[18] It is interesting that the symptoms of this disease resemble those of the toxicosis caused by toxic trichothecenes.

Dendrodochiotoxicosis: In 1953, in the U.S.S.R., a fungal intoxication of horses was reported. It was found that *Dendrodochium toxicum* contaminating wheat straw had caused poisoning of stall-fed horses. The mycotoxin produced by the fungus, dendrodochin, is highly toxic. It acts principally on the central nervous and cardiovascular systems, and ingestion of the toxic straw can be quickly followed by death.[19,20] Growth of *D. toxicum* on forage saved for feeding to stalled animals depends, as with *S. alternans*, on the moisture content of the straw. Man is also susceptible to this toxin.[21] Quite recently, further investigations have indicated that the causal fungus may be one synonymous with *Myrothecium roridum*, a producer of the macrocyclic trichothecenes, roridins.

The toxicosis is thus apparently one of those caused by toxic trichothecenes, similarly to stachybotryotoxicosis.[22)]

Fusarium-mediated toxicoses: These toxicoses represent the most typical, but complex mycotoxicoses so far known. Since toxic trichothecenes have been chiefly implicated in them, fusariotoxicosis or trichothecene toxicosis may be the best descriptive term. From the historical viewpoint, an important report was published in 1891 by Woronin,[23)] describing in detail the "Taumelgetreide (staggering grain)" intoxication that affects man and domestic animals, and implicating *F. roseum* (= *Gibberella saubinetii*), *Helminthosporium* spp. and *Cladosporium herbarum* as causal fungi. Similar poisoning due to moldy cereals and their products also occurred widely in the U.S.S.R., Central Europe, Finland, North America, China, and other districts where wheat, barley, millet or corn was consumed as a staple food. Serious mass poisoning often developed when long rains and rather cold weather continued during the harvesting season. In Japan, "red-mold disease" or "wheat scab" has long been recognized as a disease of wheat, barley, oats and rye, and this type of so-called "Akakabibyo" (red-mold disease) causes huge damage to crops at intervals of several years or decades, often affecting one third or more of the national annual production. When the infected or molded cereals are used as foodstuffs or feed, serious diseases may develop in man and farm animals. Vomiting, diarrhea, feed refusal, congestion or hemorrhage in the lung, adrenals, intestine, uterus, vagina and brain, and destruction of bone marrow are the characteristic symptoms.[24–8)]

The scabby grain mentioned above is a deteriorated grain due to several species of the genus *Fusarium* (perfect stage is *Gibberella*). In the extensive work of Nishikado,[24)] the main causative pathogenic fungus was identified as *Fusarium graminearum* (*Gibberella zeae*). Our cultural examinations[26)] for scabby wheat indicated that most of the isolates were predominatly *Fusarium graminearum* but some other species of *Fusarium* were also isolated. In Japan, scabby grain is called "Akakabi-mugi or-kome," and this condition represents one of the important grain diseases that develop with the protracted rains of June and July coincident with the harvest time of wheat. In severe cases, the grain damage may be 100%. The symptoms of the diseased seeds involve the development of a reddish brown coloration, shrivelling and a prolonged immature state.

Several outbreaks of such toxicosis in man occurred in Japan after 1945, as shown in Table 1.1. Fortunately no lethal case was encountered in our work.[27)] The symptoms of poisoning appeared 1–2 h after eating the causative foods indicated above; however, the poisoning appeared to be of acute form. *Fusarium* species have been isolated from some of the

Table 1.1 Cases of poisoning by *Fusarium*-infested foods after 1945 in Japan

Year	Location	Causative foods	Diagnosis	No. of patients
1946	Tokyo	cooked wheat, flour balls	nausea, vomiting, diarrhea	60
1949 (Jan.)	Obihiro†	cooked noodles	vomiting, chills, diarrhea	?
1950 (Jan.)	Obihiro	cooked flour balls	nausea, vomiting, chills diarrhea, headaches, convulsions	33
1950	Obihiro	noodles	nausea, vomiting, chills diarrhea, headaches, convulsions	12
1950 (May)	Obihiro	noodles	nausea, vomiting, chills diarrhea, headaches, convulsions	19
1950	Fukagawa†	flour balls	as above, plus stomachache	31
1951	Tokachi†	noodles	as above, plus stomachache	53
1956	Honbetsu†	noodles, flour balls	as above, plus giddiness	male 40, female 3
1956 (Nov.)	Honbetsu	noodles, flour balls	as above, plus giddiness	male 51, female 9
1957	Sagamihara-shi, Kanagawa Pref.	cooked rice	nausea, vomiting, chills, convulsions, diarrhea, headaches, stomachache	male 6

†Hokkaido.

grain samples by plant pathologists, and animal experiments to determine the toxigenicity of the isolates have been conducted with suspected moldy wheat and artificial fungus-inoculated feed at the Hokkaido Institute of Health, Institute of Food Research, and the National Institute of Hygienic Sciences, Tokyo. Similar work has been continued by several investigators. *F. nivale* and certain other *Fusarium* species from the scabby wheat were proved to be toxic to mice and rats, but the actual toxic substances were not extracted chemically from cultures of the isolates or scabby grain until 1966.[29–33] In 1967, Morooka and Tatsuno[34] obtained toxic substances from isolates of *F. nivale* and named them nivalenol and fusarenon. Both are scirpene compounds belonging to the toxic trichothecenes.[35–37] Later, Y. Ueno studied the bioproduction of *Fusarium*-toxins among Japanese strains of *Fusarium* isolated from scabby grains, and then succeeded in the discovery of fusarenon-X, establishment of a bioassy method using rabbit reticulocytes and chemical detection of *Fusarium*-toxin producing fungi.

[38–41] Quite recently, he has also identified another new trichothecene from *F. solani* and named it neosolaninol.[42] The fungus has been isolated from diseased bean hulls suspected of *Fusarium*-mediated toxicosis of horses near Obihiro, Hokkaido.

In contrast with the rather sporadic *Fusarium*-toxicosis occurring in Japan, mass outbreaks of this toxicosis occurred epidemically in Orenburg and other districts of Russia from 1942 to 1947.[6] According to Joffe's report of 1963,[43] more than 10% of the population was affected and many fatalities occurred in nine of the 50 counties of the affected districts. The symptoms of the disease have been fully described and in the final stages, external necrotic effects appear in the patients. Further typical symptoms include spots on the skin, leucopenia, agranulocytosis, necrotic angina, hemorrhagic diathesis, sepsis, and exhaustion of the bone marrow. It is worthy of note that the disease affected children and all other age groups indiscriminately, although undernourished persons were attacked more severely. The causal agent of the disease was found to be associated with grain that had overwintered in the fields, and initially it was thought that millet was the most dangerous source material since many people had fallen sick after eating it. Later, however, it was shown that wheat and barley could also cause the disease. The disease was termed alimentary toxic aleukia (ATA) based on the symptoms. Samples of the grain consumed by affected patients were examined for their fungal flora by Joffe in 1960,[44,45] who isolated *Fusarium poae*, *F. sporotrichioides*, *Cladosporium epiphyllum* and *C. fagi* as dominant species from the samples. The toxicity of the isolates was established by fungal culture and confirmed by feeding trials and rabbit skin tests. Four toxins, sporofusariogenin, poaefusariogenin, epicladosporic acid, and fagicladosporic acid, were obtained from mycelial mats of the isolates. It is of interest to note that the toxins were produced at a low temperature (1.5–4°C), usually at the sporulating stage, and non-toxic strains of the same species did not grow at these temperatures. At the end of his report, Joffe mentioned that the practice of allowing grain to remain in the field over winter is not usual agronomic practice, but was necessited during the war years due to the manpower shortage.

In recent studies, it has been indicated that the causal fungi implicated in the ATA epidemic produced trichothecenes such as T-2 toxin and neosolaninol.[46] However, although the disease may well have been due to the same etiologic agent as other subsequent *Fusarium*-mediated toxicoses occurring thoughout the world, this disease, like other older mycotoxicoses, did not have a well-defined etiology at the time of occurrence.

Penicillium- and Aspergillus-mediated toxicoses: Publications in North America up to about 1953 indicate that although toxicoses of unknown etiology had occurred among domestic animals including livestock, the possiblity that fungi played a role in at least some of these cases was undecided among scientists.[47–9)]

(1) Hemorrhagic sweet clover disease. In about 1920, outbreaks of hemorrhagic sweet clover disease occurred among cattle in North Dakota and Canada. The animals bled to death after eating "spoiled" sweet clover hay. Schoefield[47)] verified experimentally that *Aspergillus*-infested clover hay could cause this toxicosis in rabbits; they died with typical hemorrhagic symptoms after eating the clover. This toxicosis is probably the first known animal toxicosis of this category.

(2) Moldy corn toxicosis in swine. In 1952, Sippel *et al.*[50)] studied this toxicosis intensively, and in 1953 described a widespread outbreak in swine foraging moldy corn in the fields. From previous laboratory reports, Sippel[51)] concluded that this toxicosis had probably first occurred as early as 1949, although it was not diagnosed as such at that time. Burnside *et al.*[52)] found that two fungi, one *A. flavus* and one *Penicillium rubrum* strain, when grown in pure cultures on sterilized corn and fed to swine, produced similar toxicosis. Some cattle that had foraged the moldy corn in the fields had also developed the syndrome, and under laboratory conditions the toxicosis was reproducible with both *A. flavus* and *P. rubrum*. Using *P. rubrum* as the test fungus, toxicosis was also produced in the horse, goat and mouse.[3)] It is possible that the causal mycotoxins were penicillic acid or rubratoxins or both. Swine affected chronically are depressed, anorexic and cachexic; they stand with their heads lowered, backs arched and flanks tucked in, and walk with a stiff gait. They usually have a low erythrocyte count, prolonged prothrombin time, and appear yellow and icteric, with a high icteric index.

(3) *Aspergillus*-toxicosis. This is not an actual toxicosis but was reproduced experimetally in calves using one strain each of *A. chevalieri*,[48)] *A. clavatus*[53)] and *A. fumigatus*.[54)] It resembles some field cases of bovine hyperkeratosis.[55,56)] Both the *A. chevalieri* and *A. clavatus* were in fact isolated from feedstuffs responsible for field cases of hyperkeratosis. Forgacs thus tentatively identified such bovine hyperkeratosis as *Aspergillus* toxicosis, due to the clinical and pathologic similarities in calves affected by the three *Aspergillus* toxins, and also to differentiate this toxicosis generically from stachybotryotoxicosis which is in many ways similar.

Under natural conditions, the symptoms appear to be most prevalent from fall until late spring and primarily affect cattle, although horses, rabbits and mice are also affected. The acute form develops following ingestion of large amounts of the toxic substratum and is characterized initially by abnormal lachrymation, pulse and respiration. As toxicosis progresses, the plasma vitamin A level decreases and the animal develops intense malaise, characterized by congested conjunctival mucous membranes, profuse fetid diarrhea, dehydration, anorexia, Cheyne-Stokes respiration, and prostration. Death may follow within 5–20 days after the onset of symptoms. In chronic cases in calves, the skin along both sides of the neck and cheeks may become thickened, the plasma vitamin A level drops, and agranulocytosis occurs.[3]

(4) Moldy feed toxicosis in poultry. It has been shown[50] that various fungi, particularly those isolated from feed spilled in litter from major broiler areas in the United States since 1953, are toxic. In areas where the poultry hemorrhagic syndrome was enzootic, a toxicosis strikingly similar to that observed in field cases of the syndrome, occurred. This mycotoxicosis has been produced experimentally in an acute and chronic form in battery birds, and in those maintained under field conditions. From results of etiologic examinations, the following fungi have been identified as toxic to chickens: *A. chevalieri*, *A. clavatus*, *A. flavus*, *A. fumigatus*, *Paecilomyces varioti*, *Penicillium citrinum*, *P. purpurogenum*, *P. rubrum*, several unidentified species of *Penicillium*, and *Alternaria* sp. The acute form of the disease is characterized by depression, diarrhea and paleness of the combs and wattles, followed by death. The chief necropsy finding is severe hemorrhage in a large number of organs, especially the thymus.[57–61]

Moldy rice toxicosis[62–75] This human toxicosis has been described and discussed by Uraguchi in the Introduction. It should be emphasized that the early work in Japan represented the first systematic mycotoxicological research in the world.

Animal mycotoxicoses associated with moldy feed have also occurred sporadically throughout Japan since 1953. Hori *et al.*[76] and Yamamoto[77] investigated an outbreak of poisoning among dairy cows involving over 100 deaths, and demonstrated experimentally with calves that the causative agent was patulin produced by *Penicillium urticae* in the feed. In 1955, Iizuka *et al.*[78] identified maltoryzine, a toxic metabolite of *Aspergillus oryzae* var. *microsporus*, as the cause of poisoning among dairy cows in Chiba Pref. and also determined its chemical structure. Pathological studies on poisoned cases were made by Okubo *et al.*[79] It should be emphasized that *A. oryzae* (which closely resembles *A. flavus*) had been found to produce this toxin before the discovery of aflatoxins in 1960.

The mycotoxicoses described above do not cover all outbreaks reported in the literature. However, the valuable knowledge gained in their study served as a foundation for the recent progress of mycotoxin research in the world. Detailed records of mycotoxicoses and causal fungi reported after 1960 are given in other parts of this book.

1.3. IMPORTANT MYCOTOXIN-PRODUCING FUNGI

In relation to public health problems, aflatoxin is of course one of the most important known mycotoxins, since its potent hepatocarcinogenicity has been demonstrated in various experimental animals and its natural occurrence has been shown by chemical analysis to include various foodstuffs.[80–2] Another important mycotoxin is sterigmatocystin which has a rather lower carcinogenicity than the aflatoxins.[83] Luteoskyrin and cyclochlorotine (which may be identical to islanditoxin) and rugulosin represent a third group with known carcinogenicity to animals in long-term feeding experiments but whose natural occurrence in foodstuffs has apparently not yet been varified.[84] The carcinogenicity of patulin and penicillic acid has been demonstrated only as subcutaneous sarcomas induced in rats following parenteral administration;[85] no evidence of carcinogenicity to animals following ingestion has yet been observed. Griseofulvin, a potent antibiotic usually applied to man in prolonged oral doses for treating dermatophytoses, has known hepatotoxicity and hepatocarcinogenicity after oral as well as parenteral administration.[86,87] However, its natural occurrence is obscure and it can thus be regarded only as a minor mycotoxin of the third group from the viewpoint of food hygiene.

Many non-carcinogenic mycotoxins and their producing fungi have been reported on the basis of various toxicological experiments.[25,88–102] Some have been indicated as etiological agents of certain human and animal mycotoxicoses, and may thus be ranked with the third group of important mycotoxins. Toxic trichothecenes, which are widely produced by the the genus *Fusarium*, represent a good example.

In order to give a proper understanding of mycotoxins, their producing fungi will be described in this section, in order of availability. A tremendous amount of information concerning mycotoxin research problems has been published, including many extensive reviews.[1,8,28,33,80,81,103–27] These reviews mostly provide details of chemical, biochemical, toxicological and clinical aspects. In this chapter, therefore, the mycoflora of foodstuffs will be emphasized.

1.3.1. Aflatoxin- and Sterigmatocystin-producing Fungi

A rather limited number of species of the *Aspergillus flavus* group, e.g. *A. flavus* (Fig.1.1), *A. flavus* var. *columnaris* and *A. parasiticus*, were first

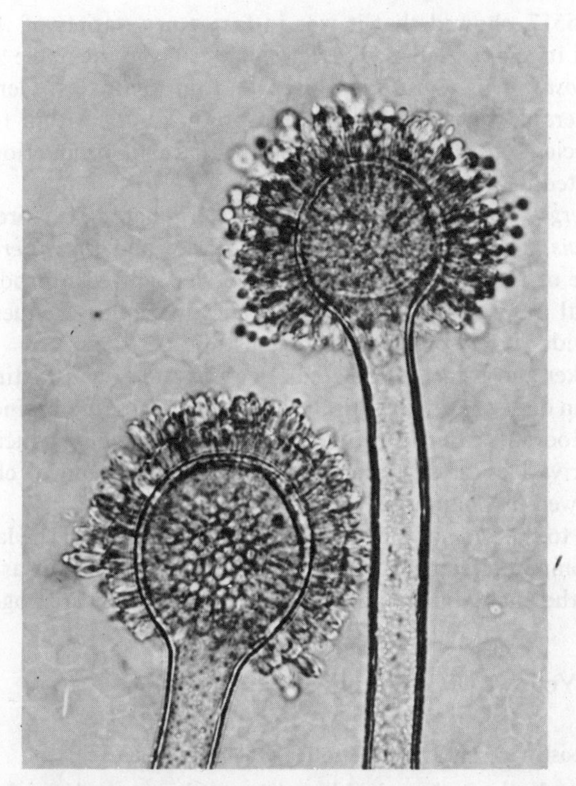

Fig. 1. 1. Aflatoxin-producing strain of *Aspergillus flavus* showing the conidial structure at maturity (× 400).

recognized as aflatoxin producers. Several investigators have studied the aflatoxin-producing ability of other fungal species and the following strains have been demonstrated to produce aflatoxins, although further confirmatory studies are required: *A. niger, A. oryzae, A. ruber, A. wentii, A. ostianus, Penicillium citrinum, P. frequentans, P. expansum, P. digitatum, P. variabile, P. puberulum, Rhizopus* sp., *Mucor mucedo*, and *Streptomyces* sp.[128–37)]

A. oryzae, the fungus used in miso and soy sauce production, has been reported to produce a trace of toxin in Japan.[138)] However, Hesseltine *et al.*[139)] and Miyaki and Aibara,[140)] who examined over 300 strains of the species obtained from pure starter cultures used for the manufacture of miso, soy sauce and sake, found no aflatoxins.

Detailed mycological studies by Murakami *et al*[141)] of one of the first known aflatoxin-producing cultures, originally identified as *A. flavus*

ATCC 15517, showed that it was in fact a new variety of *A. parasiticus*. Based on its globose-shaped vesicles, it was given the name *A. parasiticus* var. *globosus*. This culture produced all four aflatoxins. Generally there is considerable variation in aflatoxin producibility within the same species. Species-substrate relationships in aflatoxin production have been investigated in detail by several authors.[142–61)]

Aspergillus versicolor is the major sterigmatocystin producer while *A. nidulans, A. sydowi, A. rugulosus, A. flavus* and *Drechslera* sp. (imperfect stage of *Cochliobolus*) have also been recognized as producing strains by several bioproduction tests.[162–5)] They are quite frequently isolated from a wide variety of foods and feedustuffs.

Holker and Kagel[166)] isolated 5-methoxysterigmatocystin from a mutant strain of *A. versicolor*. Burkhardt and Forgacs[167)] obtained O-methylsterigmatocystin from a strain of *A. flavus* capable of producing aflatoxin. Such derivatives of sterigmatocystin and their chemical characteristics are reviewed in Chapter 2.

Due to the wide occurrence of producing fungi and the large amounts of the compound elaborated sterigmatocystin is perhaps as important a toxin as the aflatoxins despite its lower toxicity and carcinogenicity.

1.3.2. Yellowed Rice Fungi

A. Luteoskyrin- and Cyclochlorotine-producing Fungi

Luteoskyrin and cyclochlorotine are hepatotoxic and hepatocarcinogenic mycotoxins obtained from liquid or solid cultures of *Penicillium islandicum* (Fig. 1.2). These toxins, so-called yellowed rice toxins, have been well documented by several Japanese investigators.[168)] At present *P. islandicum* is the only known specises to produce these toxins.[169,170)] Saigo *et al.*[171)] have grouped the toxin-producing strains mycologically under the following three categories according to the suface appearance of the fungal colonies: (1) luteoskyrin-producing type strains appearing less floccose, green-yellow in color but brown in reverse, (2) skyrin- or flavoskyrin-producing type strains having a velvety and yellow-orange surface and reverse color, and (3) erythroskyrin-producing type strains having a velvety, dark orange surface color, with greenish orange or greenish grey shades, but appearing dark brownish grey in reverse.

P. islandicum was first isolated by Sopp in 1921 from soil collected on the island of Skyr, Norway. In 1930. Thom placed this species in the *P. funiculosum* series, but did not reidentify the strain as *P. funiculosum* since it differed from the original *P. funiculosum* culture in its more re-

Fig. 1. 2. Luteoskyrin-producing strain of *Penicillium islandicum* showing the typical biverticillately-symmetrical penicilli (× 400).

stricted growth, production of more orange-red aerial hyphae, and development of shorter hyphal ropes or funiculi.[172] This mold is commonly isolated from agricultural commodities, especially moist rice and corn.

B. Citrinin-producing Fungi

Citrinin is a representative mycotoxin among the yellowed rice toxins and was first detected as a pure compound from liquid cultures of *Penicillium citrinum* by Raistrick and Hetherington in 1931.[173] The following citrinin-producing fungi are known at present: *P. citrinum*, *P. viridicatum*, *P. implicatum*, *P. fellutanum*, *P. citreo-viride*, *P. velutinum*, *P. canescens*, *P. jenseni*, *P. steckii*, *P. notatum*, *P. palitans*. *P. expansum*, *P. claviforme*, *Aspergillus niveus*, *A. terreus*, and *A. flavipes*.[168,174]

It is interesting to note that the species *P. viridicatum* and *P. palitans* are producers of both citrinin and ochratoxin A, and that the two mycotoxins may even be isolated from a single strain.[175] The natural occurrence of citrinin has already been reported to include wheat, barley, rye and oats in Canada by Scott *et al.*[176]

C. Rugulosin-producing Fungi

Rugulosin, an anthraquinoid fungal metabolite, was first isolated from mycelia of *Penicillium rugulosum* by Breen *et al.*[177] This toxin is one of the yellowed rice toxins taken up for toxicological study in Japan.

So far, three rugulosin-producing fungi have been reported: *P. rugulosum*, *P. brunneum*, and *P. tardum*.[178,179] However, it is impossible to evaluate the true importance of rugulosin as a food contaminant at present since many problems still remain unsolved.

D. Citreoviridin-producing Fungi

Citreoviridin is a neurotoxic mycotoxin produced by a certain strain of *Penicillium citreo-viride*, formerly identified by Miyake[180] in Japan as *P. toxicarium*. This mold was isolated several years ago from rice collected in Taiwan, but is now less common in domestic rice in Japan.

Only four species of *Penicillium* are known to produce citreoviridin, i.e. *P. citreo-viride*, *P. ochrosalmoneum*, *P. fellutanum*, and *P. pulvillorum*.[65,181,182] The natural occurrence of the toxin has not yet been reported in food or feed.

1.3.3. Patulin- and Penicillic Acid-producing Fungi

Patulin was isolated from *Penicillium patulum* by Birkinshaw *et al.*[183] in 1943, and was expected to have some application as an antibiotic for treating various human infectious diseases.

In 1952, an outbreak of feed poisoning accidentally occurred in Japan among dairy cattle.[76,77] The mass death of 118 cows was attributed to consumption of moldy malt feed infected with a toxigenic strain of *P. urticae* (syn. of *P. patulum*). This fungus was shown to produce patulin, which was itself later proved to be a causal agent of the poisoning by animal experiments.

Several species of *Penicillium* and *Aspergillus* are capable of producing patulin in the laboratory. They are *P. patulum* (= *P. urticae*), *P. expansum*, *P. claviforme*, *P. lapidosum*, *P. melinii*, *P. equinum*, *P. novae-zeelandiae*, *P. divergens*, *P. griseofulvum*, *P. leucopus*, *P. cyclopium*, *A. clavatus*, *A. giganteus*, and *A. terreus*; as well as *Byssochlamys nivea*.[184] In nature, the major producer of patulin is probably *P. expansum*.

Penicillic acid was first isolated from *Penicillium puberulum* as a secondary metabolite by Alsberg and Black in 1913.[185] Prior to this, the Italian investigator, Gosio, had reported that one strain of *Penicillium* was toxic to various experimental animals and a compound resembling penicillic acid was obtained from cultures of it. Later, Alsberg and Black established that penicillic acid was toxic to various laboratory animals and conjectured that fungal toxins might be involved in the pellagra resulting from consumption of moldy corn. Further toxicity findings were given by Brinkinshaw and Raistrick in 1932.[186]

Producibility of this compound has been demonstrated chemically in the following fungal species: *P. puberulum, P. stoloniferum, P. cyclopium, P. martensii, P. thomii, P. suaveolens, P. palitans, P. baarnense, P. madriti, P. paraherquei, Aspergillus ochraceus, A. sulphureus, A. quercinus, A. melleus,* and *A. ostianus.* Such a wide distribution indicates that producibility of the toxin probably has little or no taxonomic significance.

P. martensii is recognized as a plant pathogenic agent of blue-eye disease of corn. It is significant that the natural occurrence of penicillic acid includes several samples of diseased corn studied by Kurtzman, Ciegler and Lillehoj.[188–90]

1.3.4. Griseofulvin-producing Fungi

Griseofulvin is presently employed as a systemic therapeutic agent for human cutaneous fungal infections. It was first discovered by Oxford *et al.*[191] in 1939 from mycelia of *Penicillium griseofulvum* as a colorless, neutral, crystalline material.

Other fungi reported to produce griseofulvin are: *Penicillium patulum, P. albidum. P. raistrickii, P. brefeldianum, P. viridi-cyclopium,* and *P. brunneo-stoloniferum.*[192]

No natural occurrence of this compound among foodstuffs has yet been reported, so that its significance for human health cannot be properly evaluated. However, the producing fungi mentioned above are frequently isolated from various agricultural commodities.

1.3.5. Ochratoxin A-producing Fungi

Ochratoxins are metabolites of *Aspergillus ochraceus* and were discovered as a result of laboratory screening of a large number of fungal cultures for toxicity by Scott *et al.*[88] in South Africa. Van der Merwe *et al.*[193] and Steyn and Holzapfel[194] isolated methyl and ethyl derivatives of ochratoxin A and of the non-toxic dechloro derivative, ochratoxin B. The known ochratoxin-producing fungi are: *A. ochraceus, A. ostianus, A. melleus, A.*

alliaceus, *A. petrakii*, *A. sclerotiorum*, *A. sulphureus*, *Penicillium viridicatum*, *P. cyclopium*, *P. commune*, *P. palitans*, *P. purpurescens*, and *P. variabile*.[194–6)]

P. viridicatum from ham was the first species outside the *A. ochraceus* group to be reported as an ochratoxin A producer.[197–9)] Ochratoxin A- and ochratoxin A/citrinin-producing strains of *P. viridicatum* and *P. palitans* were found to be present in moldy Canadian grain in which ochratoxin A and citrinin had been detected.[176)] In a survey of the mycotoxin-producing potential of *Penicillia* isolated from moldy fermented sausages, 17 isolates produced ochratoxin A.[200)]

The natural occurrence of ochratoxins has been reported to cover the following foodstuffs:[196,201,202)] corn, wheat, barley, heated grain, mixed feeds, dried white beans, and moldy peanuts. The highest ochratoxin A and citrinin concentrations in grain were 27 and 2 ppm, respectively.[203)] The kidney, liver and adipose tissue of swine consuming moldy barley and oats have been found to be naturally contaminated organs, the highest concentration of ochratoxin A being 67 ppb.[204)] Based on these findings, man may thus be exposed to this toxin via two routes: consumption of foodstuffs directly contaminated by mycotoxin-producing fungi, and consumption of residue-containing meat of slaughtered animals fed on mycotoxin-contaminated feed.

1.3.6. Rubratoxin-producing Fungi

Rubratoxins are toxic metabolites produced by isolates of *Penicillium rubrum* inhabiting soil, food- and feed-stuffs.

Rubratoxin B is produced by only two species of the genus *Penicillium*: *P. rubrum* and *P. purpurogenum*.[205–8)] Although both have frequently been isolated from feeds implicated in field outbreaks of mycotoxicoses, no direct information is yet available on the natural distribution of the rubratoxins in human foodstuffs or animal feeds.

1.3.7. Trichothecene Toxin-producing Fungi

Trichothecenes, the group of sesquiterpenoids previously known as scirpenes, include a number of fungal metabolites produced in cultures of various species of the genera *Fusarium*, *Cephalosporium*, *Myrothecium*, *Trichoderma* and *Stachybotrys*.[28,209,210)] Most trichothecene-producing fungi are *Fusarium* species (Table 1.2), which are commonly isolated from moldy cereals, especially low quality and moistened corn and wheat. According to the structural variation at the C-8 position and taxonomic situation of the toxin-producing fungi, Ueno *et al.*[211)] have divided the toxic trichothecenes into type A (T-2 toxin, neosolaniol, diacetoxyscirpenol,

TABLE 1.2 Major toxic trichothecenes and their producing fungi

Major toxic trichothecenes†	Producing species of *Fusarium*
1. Diacetoxyscirpenol (B-24 toxin)	*F. tricinctum* (= *F. poae*, *F. sporotrichioides*), *F. solani*, *F. lateritium*, *F. roseum* (*F. sambucinum*, *F. graminearum*, *F. equiseti*, *F. scirpi*, *Gibberella zeae*, *G. intricans*)
2. HT-2 toxin	*F. tricinctum*, *F. poae*
3. T-2 toxin	*F. tricinctum*, *F. roseum*, *F. lateritium*, *F. solani*, *Trichoderma viride*
4. Neosolaninol	*F. tricinctum*, *F. roseum*, *F. solani*, *F. rigidiusculum*
5. Fusarenon-X	*F. nivale*, *F. oxysporum*, *F. episphaeria* (= *F. aquaeductuum*), *F. merismoides*
6. Nivalenol	*F. nivale*
7. Diacetylnivalenol	*F. scirpi*, *F. nivale*, *F. equiseti*
8. Dihydronivalenol DHN	*F. nivale*
9. Deoxynivalenol (RD-toxin or vomitoxin)	*F. roseum*

†1. 4β,15-Diacetoxy-12,13-epoxy-trichothec-9-en-3α-ol.
2. 15-Acetoxy-3α, 4β-dihydroxi-8α-(3-methylbutyryloxy)-12, 13-epoxy-trichothec 9-one.
3. 4β, 15-Diacetoxy-8α-(3-methylbutyryloxy)-12, 13-epoxy-trichothec-9-en-3α-ol.
4. 4β15-Diacetoxy-3α, 8α-dihydroxy-12, 13-epoxy-trichothec-9-one.
5. 4β-Acetoxy-3α,7α,15-tetrahydroxy-12, 13-epoxy-trichothec-9-en-8-one.
6. 3α,4β,7α,15-Tetrahydroxy-12,13-epoxy-trichothec-9-en-8-one.
7. 4β,15-Diacetoxy-3α,7α-dihydroxy-12,13-epoxy-trichothec-9-en-8-one.
8. 3α,4β,7α,15-Tetrahydroxy-12,13-epoxy-trichothecan-8-one.
9. 3α,7α,15-Trihydroxy-12,13-epoxy-trichothec-9-en-8-one.

HT toxin) and type B toxins (fusarenon-X, nivalenol) (see also Chapter 2).

Some confusion still exists over the taxonomic criteria for *Fusarium* species. This common and widespread genus consists mostly of soil inhabitants and palnt pathogens, and has been intensively studied by plant pathologists throughout the world for over a century. Various classification systems and variations on them have been proposed (e.g. ref. 212–7), while Booth's recent book[218)] includes about 50 species. For ease of identification of this genus, Booth's system[218,219)] is more clearly understandable to the non-specialist of *Fusarium* taxonomy. Clearly, comprehensive studies on the toxic trichothecene production among *Fusarium* species associated with various foodstuffs are needed.

From the viewpoint of food hygiene, it is most likely that the *Fusarium roseum* group including *F. graminearum* (= *Gibberella zeae*) (Fig. 1.3) includes the most critical fungi as regards cereal grain contamination, while other species such as *F. tricinctum*, *F. nivale* and *F. oxysporum* are important for the general prevention of *Fusarium*-mediated toxicoses.

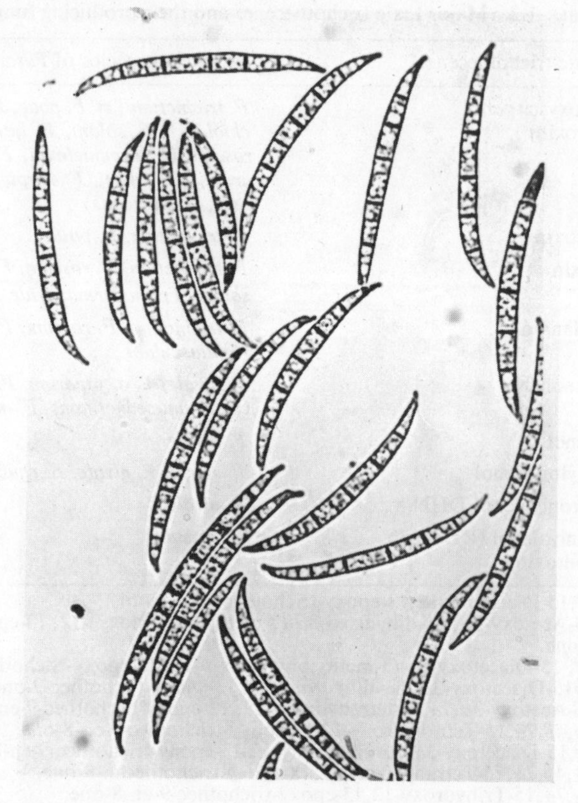

Fig. 1. 3. Conidia of *Fusarium graminearum* isolated from moldy corn (× 800).

1.3.8. Butenolide-producing Fungi

Butenolide has been isolated independently in three different laboratories.[220-2)] White[222)] was studying diacetoxyscirpenol but instead found that this toxin was produced by two isolates of *Fusarium equiseti* obtained from *Cortaderi* species. Brian *et al.*[223)] and Yates *et al.*[221)] characterized this toxin as 4-acetamido-4-hydroxy-2-butenoic acid δ-lactone.

Cattle grazing on pastures of tall fescue grass occasionally develop a syndrome called "fescue foot", a noninfectious disease characterized by signs including loss of weight, arching of the back, rough hair coat, lameness in the hind quarters, and dry gangrene of the tail and feet.[224,225)] Extracts of several species of *Fusarium* isolated from tall fescue grass caused death when injected intraperitoneally into mice. One strain of *F. tricinctum* (NRRL 3249) isolated from the hay produced the two toxins, butenolide and T-2 toxin, when grown on laboratory media.[226)] Grove *et al.*[227)] injected this butenolide intramuscularly into a heifer at 3.8 mg/kg for 90 days, and subsequently found dry gangrene at the end of the tail. However, the etiologic evidence is still circumstantial since other toxin-producing fungi are associated with hemorrhaging in farm animals and some of these produce other trichothecenes that could have similar toxicological action.[228] The butenolide has an i.p. LD_{50} in mice of 43.6 mg/kg and an oral toxicity of 275 mg/kg.[229] The following species of *Fusarium* have been reported as producers: *F. equiseti*, *F. tricinctum* (= *F. sporotrichioides*), *F. nivale*, *F. roseum*, *F. semitectum*, and *F. lateritium*.[41,211,221,226]

1.3.9. Zearalenone-producing Fungi

Zearalenone (F-2 toxin) is an estrogenic mycotoxin produced by several species of *Fusarium*. It was first isolated from cultures of *F. graminearum* (= *Gibberella zeae*) invading stored corn.[230,231] The molded corn appeared to cause estrogenic syndrome in swine.[232] *F. roseum* "Graminearum" is so far the major known zearalenone producer but *F. tricinctum* (= *F. sporotrichioides*), *F. nivale* and *F. roseum* (involving "Equiseti", "Gibbosum" and "Culmorum") are also known to produce this toxin. They are quite commonly isolated from food- and feed-stuffs, mostly corn, barley and wheat.[233–6] A survey for *Fusarium* toxins in the 1972 corn crop by the Food and Drug Administration of the midwest U.S.A. indicated a 17% incidence of zearalenone contamination at levels of 100–5000 ppb.[237]

1.3.10. Tremorgenic Toxin-producing Fungi

Several species of *Penicillium* and *Aspergillus* produce metabolites capable of inducing sustained tremors, convulsions and death in animals. Although the molecular formulas of these tremorgenic compounds show differences, some appear to have definite structural similarities. Recently identified tremorgenic mycotoxins and the producing molds are listed in Table 1.3.

1.4. Significance of the Mycoflora of Various Foodstuffs

It is well known that seeds of cereal grains, stored grains, and processed foods of both plant and animal origin harbor large numbers of fungi under certain natural conditions. These fungi include parasitic species that are host-specific, saprophytic species or both, and even the saprophytic fungi tend to be rather selective in their choice of living substrate and

TABLE 1.3 Tremorgenic mycotoxins and their producing fungi

Fungi	Toxin	Source	Ref.
Penicillium verruculosum	Verruculogen	Peanuts	238
P. palitans[†1]	Penitrem A[†2]	Foodstuffs	
P. cyclopium	Penitrem B	Foodstuffs	239–43
P. crustosum	Penitrem C	Foodstuffs	
P. paraherquei	Verruculogen	Barley, Corn, Peaunuts	244
P. citreo-viride	Citreoviridin	Rice grain	65, 252
P. paxili	Paxilline	Corn	245
P. roqueforti	Roquefortine	Foodstuffs	246
Aspergillus flavus	Unnamed		247
A. fumigatus	Fumitremorgin A Fumitremorgin B	Rice grain	248, 249
A. clavatus	Tryptoquivaline		250
	Tryptoquivalone		251

†1 *P. cyclopium, P. crustosum, P. granulatum, P. olivino-viride*, and *P. puberulum* are also producers of penitrem A (*cf.* A. Ciegler and J. I. Pitt, *Mycopath. Mycol. Appl.*, **42**. 119 (1970)).

†2 Formerly, tremortin A, B. and C.[241–243]

environment. Christensen[253] has stated that, in general, the fungi can be divided into rather distinct groups: (1) field fungi, (2) storage fungi, and (3) a possible third group tentatively designated as advanced decay fungi. The principal genera of field fungi are *Alternaria*, *Fusarium*, *Helminthosporium* and *Cladosporium*. All of the field fungi require a high seed moisture content for growth. The predominant fungi differ somewhat according to crop, region or geographic location, and climatic conditions. *Alternaria* is common in many grains and vegetable seeds, especially cereals, but is by no means limited to cereal seeds. If, in fact, wheat kernels harbor *Alternaria* in more than 90% isolation, and contain no storage fungi, this is considered good evidence that the seed is newly harvested and has been stored under conditions that prevent its deterioration. *Cladosporium* is also common in cereals that have been exposed to moist conditions during harvesting. It may cause darkening of the invaded hulls, but has no known effect on storability. *Helminthosporium* is common in many cereal seeds, such as rice, barley and oats, especially if moist weather has prevailed just before harvesting.

Fusarium is also very common in freshly harvested cereal seeds and causes "scab" in barley, wheat and corn. The scabby grain is sometimes implicated in human and animal intoxications, as mentioned above. The toxicity of grains invaded by this fungus is considered in Chapter 3.

The mycoflora of stored grains and other materials including various kinds of beans and nuts, mainly consists of the so-called storage fungi. These include *Penicillium*, *Aspergillus* and *Sporendonema*, plus possibly a few species of yeast. *Penicillium* and *Aspergillus* are ubiquitous fungi, and occur throughout most parts of the world. These storage fungi are the dominant type of molds associated with stored seeds. They are usually only superficially present on seeds at harvest, but there is some invasion of dead plant material such as blossom ends, leaf fragments, straw, etc.

The major factors influencing the development of storage fungi on stored grain are as follows: (1) the moisture content of the stored grain, (2) temperature, (3) storage period, (4) degree of earlier invasion of the grain by storage fungi before arrival at the given site, (5) amount of foreign material, and (6) the activity of insects and mites.[254–7)]

The microflora on and within foodstuffs is known to affect the quality, storage behavior and processing of various kinds of agricultural commodities. Research on the normal mycoflora of foodstuffs, based on an understanding of the microecology of growth and proliferation, should yield good estimates of the magnitude of mycotoxin contamination of foodstuffs. For this purpose, the mycotoxicological examinations should be performed in two parallel ways: fungal isolation using adequate media to determine the type and relative abundance of the mycofloral elements, and chemical determinations of the mycotoxin producibility of the fungal isolates. Several detailed studies on mycological examination methods for foodstuffs have been made.[256,258–69)] On the basis of the results, it should be possible to determine the distribution of mycotoxic fungi among the total mycoflora.

The mycoflora of various agricultural commodities, mainly cereal grains, has been thoroughly surveyed by plant pathologists in the U.S.A.[254,256,257,270)] The mycoflora of imported and domestic rice has been studied in Japan by various investigators.[114,262,263,267,268,271–5)] Similar aproaches are now being adopted with other foodstuffs including home items, in Japan,[98,100)] Thailand and Hong Kong,[276)] Malaya,[277)] Canada,[176)] and countries of southern Africa,[278,279)] especially with regard to the diet of the local people. Summaries of the mycoflora of various foodstuffs based on the results obtained are given in later sections.

The procedures for routine isolation of fungi from rice and other grains are as follows:

(1) Cereal grains. To obtain the internal mycoflora of cereal grains, the grain is plated following surface washing with sterile water, as described by Kurata *et al.*[267)] About 10 g of grain is washed aseptically with 10 successive volumes of sterile distilled water to remove surface contaminants. Occasionally, a surface sterilizer such as sodium hypochlorite is used for

surface sterilization. Five grains are placed into each of 10 petri dishes containing culture media. Czapek-Dox agar, and potato-dextrose agar (PDA) containing 100 mg/l of chloramphenicol to suppress bacterial growth, are recommended as suitable media. In addition, MY 20 agar and 7.5% NaCl-added malt agar[259] are employed to the estimate number of halophilic fungi such as members of the *Aspergillus glaucus* group. The dishes are incubated at 25°C, and inspected daily. The total number of tested grains in one sample is 100 to 300. The developed fungal colonies are transferred to adequate slant media for identification.

(2) Powders and semi-solid type foodstuffs. For the isolation, purification and identification of fungi from flour products, the quantitative dilution plate method was established by Hesseltine and Graves[266] in the U.S.A. and by Inagaki[261] in Japan. We use a slightly modified method as our routine procedure. After thoroughly mixing the sample to obtain an even distribution of mycoflora, 11 g of the material is aseptically weighed in a sterile aluminum weighing scoop (30 ml capacity) and transferred to a 250 ml dilution bottle containing 99 ml of sterile 0.1% agar water and approximately 10 g of minute glass balls, to help good dilution. Higher serial dilutions are then prepared from this primary mixture (1:10). Three media, PDA and peptone-glucose agar to which adequate bacteriostatics have been added, and MY 20 agar and malt salt agar for isolation of osmophylic fungi, are generally employed.

(3) Liquid type foods. The principal technique is almost the same as that used for milk, juice and drinking water.[269] The culture media employed are those described under the previous method.

1.4.1.. Rice

Rice is one of the most important agricultural commodities, especially as a staple food for the people of Asia. Freshly harvested rice, rough rice, carries chiefly field fungi, i.e. various species of *Helminthosporium*, *Curvularia*, *Piricularia*, *Alternaria*, *Cladosporium*, *Fusarium*, *Epicoccum*, *Phoma*, and other plant pathogenic fungi such as *Tilletia* and *Ustilaginoidea*. Of these, *Alternaria*, *Helminthosporium*, *Epicoccum* and *Fusarium* are predominant.[114,262,265,280–2]

Del Prado and Christensen[265] reported the isolation of *Fusarium* (32%), *Aspergillus niger* (31%) and *Cladosporium* (20%) from the hulls, and among 1950 crop samples from Louisiana, 40% of the isolates from washed hulls were *Curvularia*, 17% *Fusarium*, and 14% *Aspergillus glaucus*. The principal molds isolated from the caryopsis, in order of prevalence, were *A. glaucus*, *A. niger*, *Curvularia*, *Penicillium*, and *Fusarium*. After harvest, when the moisture content falls below 15% (water capacity,

$a_\omega = 0.75$), the mycoflora gradually changes to the storage fungus type, e.g. *Penicillium* and *Aspergillus*. Following a further decrease of the moisture content to below 14% ($a_\omega = 0.70$), the general species of the storage fungi decrease gradually, finally leaving osmophilic fungi such as *A. restrictus*, *A. chevalieri* and *A. amstelodami*.

A. repens, *A. ruber* and *Penicillium pusillum* remain in the internal mycoflora of stored rice. The dominant species constituting such mycofloras vary according to the storage conditions of the rice. However, *Penicillium* contamination tends to be more prevalent than *Aspergillus* contamination in the early period of storage, since the stored rice still retains a moisture content of about 18% ($a_\omega = 0.85$).[283–6]

The mycoflora of any given lot of rice may thus be composed of field fungi or storage fungi or combinations of them. The prevalence of groups or species at any given time is the cumulative result of the lot's entire history, but reflects most strongly the inoculum levels, available moisture, and temperature. Hence, there is ordinarily a continuous change in the specific make-up of the mycoflora which reflects the ability of various species to compete under a given set of environmental conditions. To clarify this phenomenon in detail, Schroeder and Sorenson[283] have studied the quality control problems of rice under aerated storage. They observed that the field fungi decreased as the storage fungi increased and that the rate of change was accelerated by passage through a drier, although it continued at a slow rate during tempering and/or storage.

The distribution of fungi in rice grains, especially polished rice, has been examined by seed pathologists in various parts of the world, as summarized in Table 1.4. It was apparent that the genus *Aspergillus* represents the major constituent of the mycoflora of rice, especially milled rice. However, the dominant genera and species vary according to the year surveyed. It is of interest to note that *A. flavus* and *A. candidus* appeared predominantly in rice samples examined in 1955–7.[114] Thailand rice also showed the same result.[263] These facts suggest that the rice examined was sufficiently moist during shipping for these two species to increase. Under similar circumstances at present, such samples should clearly be examined chemically for aflatoxin contamination. In contrast, the domestic rice in Japan harbors chiefly the *A. glaucus* group. In domestic rice stored in Japanese farms, Miyaki *et al.*[275] noted that *A. ochraceus*, an ochratoxin A-producing species, was recorded predominantly as the storage fungus group. Kurata *et al.*[114,272] reported results of a mycotoxicological survey of Japanese polished rice harvested in 1965 and 1966, indicating that the incidence of highly and mildly toxic fungi was 2.4 and 9.7% respectively. The domestic polished rice investigated under rationing was considered to have been conditioned and handled under satisfactory conditions.

TABLE 1.4. Summarized data of mycological examinations of imported and domestic rice grains in Japan

Year and origin	Fungi[†1]	Ref.
1955–7 Imported	*Aspergillus flavus* (27), *A. candidus*(27), *A. glaucus*(21), *Penicillium citrinum*(10), *P. cyclopium*(10), *P. funiculosum*(9) *Cladosporium*(5 >), *Alternaria*(1 >)	114
1958–9 Imported	*A. candidus*(39), *P. cyclopium*(25), *A. versicolor*(21), *P. chrysogenum*(19), *A. amstelodami*(18), *A. flavus*(17) *A. terreus*(16), *P. concavo-rugulosum*(14), *P. islandicum*(10), *A. niger*(10)	114
1964–7 Imported	*A. glaucus* group(36), *A. versicolor*(30), *P. viridicatum*(25), *P. chrysogenum*(11), *P. funiculosum*(11),[†2] *P. cyclopium*(7), *Alternaria*(97), *Fusarium*(2)	114
1965 Domestic	*A. repens*(8), *A. versicolor*(5), *A. candidus*(4), *A. ruber*(3), *P. citrinum*(3), *P. cyclopium*(3)	114
1966 Domestic	*A. repens*(22), *A. amstelodami*(17), *A. versicolor*(17), *A. candidus*(13), *P. cyclopium*(10), *P. expansum*(8)	114
1960 Imported	*A. flavus*(27), *A. candidus*(24), *A. chevalieri*(22), *P. citrinum*(19), *A. versicolor*(17), *P. islandicum*(16), *P. rugulosum*(10), *P. citreo-viride*(7), *P. cyclopium*(> 5), *P. urticae*(> 5), *A. clavatus*(> 5), *A. ochraceus*(> 5)	262
1968–9 Domestic	*A. ochraceus*, *A. versicolor*, *A. niger*, *A. candidus*, *A. flavus*	275
1952 U.S.A.	*A. glaucus*, *A. restrictus*, *A. niger*, *A. candidus*, *A. flavus*	265
1968 Valencia	*A. flavus-oryzae*(30), *P. italicum*(30), *P. expansum*(25), *A. niger*(14)	287

†1 Figures in parentheses are percentages for total number of fungi isolated.
†2 Involved *P. islandicum*.

Highly potent aflatoxin-producing fungi were first isolated by Boller and Schroeder[149] in the U.S.A. Shortly after, the development of aflatoxin in rough rice following prolonged storage in aerated bins was extensively studied by Calderwood and Schroeder.[288] The results obtained indicated that when rice with a moisture content of 24–6%, under aeration at 1 and 2 c.f.m. per barrel, was stored, the market grade of the rice fell and a high percentage of the kernels was infected with *A. flavus* before the 21st day. A relatively large amount of aflatoxin (about 500 ppb) developed when the high moisture content rice (24–6%) was stored during warm weather. They recommended that green rice be dried to a moisture content of 20% or less within a maximum of 48 h, particularly in southern rice-growing areas in August and September, when the ambient temperature is usually high. To prevent aflatoxin production in stored rice, Kurata *et al.*[289]

have suggested that, for long-term storage, the relative humidity in the storage house should always be kept below 70%. The critical time for contamination in rice can be given as the short period just before and after harvesting.

1.4.2. Wheat, Barley, Oats and Rye

Microbiological studies of stored wheat and barley have been intensively developed by American and Canadian seed pathologists over the past 40 yr. In 1954, Semeniuk[256] gave an excellent review of the microflora of cereal grains and cereal products, listing the kinds and abundances of fungi, bacteria and actinomycetes found in wheat, oats, barley, rye, corn (maize), flour, bran, and grain meals, together with the effects of environmental conditions on the growth of these microorganisms and the methods used to control their activities Later, two related and similar reviews covering the deterioration of stored grains by fungi, were given by Christensen *et al.*[254,290] Further investigations on the connection with mycotoxins have been made by several workers in Japan.[291–5] These various data on the mycoflora of cereal grains were obtained from results of isolation tests, mainly by means of the agar-plate test. The procedure for this test is almost the same as that used for the fungal examination of rice grains, as explained above. The blotter test, in which seeds are incubated on a moist blotter, may also be employed.[296] Consequently, it is only the mesophilic mycoflora of surface-disinfected cereal grains, and particularly the phytopathogenic composition of that mycoflora, which has been studied in any detail.

The chief species of fungi from wheat grains indentified by these authors may be summarized as follows:

(1) Parasitic flora: *Fusarium graminearum*, *F. culmorum*, *F. nivale*, *F. tricinctum*, *F. scirpi*, *F. semitectum*, *Helminthosporium sativum*, *Alternaria tenuis*, *Cephalosporium* spp., *Cladosporium herbarum*, *C. epiphyllum*, *C. fagi*, *Septoria nodorum*, and *Phoma* spp.

(2) Saprophytic flora: *Aspergillus candidus*, *A. flavus*, *A. oryzae*, *A. fumigatus*, *A. niger*, *A. versicolor*, *A. restrictus*, *A. glaucus* group, *Penicillium chrysogenum*, *P. frequentans*, *P. terrestre*, *P. rugulosum*, *P. palitans*, *P. flavi-dorsum*, *P. purpurogenum*, *P. spinulosum*, *P. steckii*, *P. brevi-compactum*, *Botrytis cinerea*, *Scopulariopsis brevicaulis*, *Scopulariopsis* spp., *Acrostalagmus cinnabarinus*, *Trichoderma lignorum*, *Torula* spp., and *Mucor* spp.[292,297,298]

The mycoflora of the kernels and stored grain of barley is described next. Principally, the constitution of the mycoflora of barley is not entirely different from that of wheat. According to surveys by Flannigan,[299] Mulinge and Apinis[300] and Mulinge and Chesters,[301] the following species of fungi are predominant in the dried kernels: *Alternaria tenuis*, *Penicillium cyclopium*, *P. roqueforti*, *Aspergillus glaucus* group, *A. candidus*, *A. fumigatus*, *A. terreus*, *Aureobasidium pullulans*, *Cladosporium* sp., *Epicocum nigrum*, *Dactylomyces crustaceus*, and *Absidia corymbifera*. They also indicated that a smaller percentage of kernels over a high population of samples were contaminated with *Botrytis cinerea*, *Cephalosporium* sp., *Fusarium avenaceum*, *Mucor pusillus*, and *Sporobolomyces roseus*. Thermophilic fungi have also been found associated with barley grain, as follows: *Mucor pusillus*, *A. fumigatus*, and *Dactylomyces crustaceus*. They can grow on media at 40–60°C.

The mycofloras of oat and rye grains show resemblances not only to the microfloras of other cereal grains but also to those of hay and straw.

1.4.3. Flour

Microbiological studies of flour are necessary for resolving various food hygiene problems due to its importance as a constituent of many refrigerated, frozen, and cooked convenience foods. Since such foods provide excellent nutritive media for the growth of bacteria and fungi, the flours used in them must have a low microbiological population to ensure shelf life, eliminate package swelling and breaking, and prevent slime formation, off-odor, and discoloration of the contents. Considerable efforts have been made by several investigators at the N.R.R.L. in the U.S.A.[302–4] to reduce the microbiological population of flours. However, detailed surveys on the microflora, especially fungal flora, of flour are still needed. In the case of wheat flour, up to 1938, there were several reports, mainly on the population and kinds of bacteria,[305–8] and the work in these papers was extensively reviewed by Clifford.[309] After 1945, detail investigations on the incidence of molds in wheat flour were initially made by several investigators,[259,310,311] although Barton-Wright[312] had briefly reported in Canada that over 90% of molds from wheat flour belonged to the genera *Penicillium*, *Aspergillus*, *Cladosporium* and *Botrytis*, and the chief *Penicillium* species were *P. brevi-compactum*, *P. patris-mei*, and *P. expansum*. In 1966, Hesseltine and Graves[270] completed an extensive review of the microbiology of flours, citing hundreds of references published after 1885.

Recently, comprehensive mycological surveys of wheat, buckwheat and other flours have been carried out by Japanese researchers.[261,268,271,313–5] In both flour and dough, field fungi were encountered infrequently. On the other hand, storage molds were quite common. It was found that *Aspergillus* species were far more common than *Penicillium* species.

Based on these reports, the general mycoflora of commercial wheat flours can be summarized as follows: the *Aspergillus* species are the *A. glaucus* group, *A. candius, A. flavus, A. versicolor, A. niger, and A. ochraceus*; the *Penicillium* species other than those reported by Barton-Wright are *P. cyclopium, P. citrinum, P. chrysogenum, P. funiculosum, P. frequentans, P. martensii, P. urticae, P. roqueforti*, and *P. viridicatum*; the other genera are *Alternaria, Fusarium, Cladosporium, Diplodia* and *Mucor*. Of these fungi, *A. fumigatus, A. candidus, A. flavus*, certain species of the *A. glaucus* group, *P. cyclopium, P. citrinum*, and *P. roqueforti* were recognized as the dominant fungi of commercial flours and fresh and spoiled refrigerated dough products.[316,317] Quantitative determinations of the flour molds have been made in detail,[259,266] using data arising from results of fungal evaluations based on the principal method for flour-type foodstuffs described above.

The types of fungi present in flour are not those found on wheat in the field but duplicate forms encountered during storage and milling. Hesseltine[266] has emphasized that if the moisture content is 12% ($a_\omega = 0.65$) or less, no growth of microorganisms occurs. Since flour with a low moisture content is kept in storage and during processing, the number of viable microorganisms decreases fairly rapidly. From the viewpoint of food sanitation, it is considered that the bacterial population represents a more important problem than the fungal population.

1.4.4. Buckwheat Grain, Flour and Noodles

Buckwheat noodles are a popular food among the Japanese, and form one element of the staple diet that is sometimes used in Japanese dried cakes. Recently, buckwheat grain has been imported mostly from Canada, South Africa and the Republic of China to Japan by ship. Hitokoto *et al.*[318] have examined the mycofloras of the grain, flour and noodles. Earlier and later investigations have been identified in only a few instances, and it was apparently shown that the fungal population of buckwheat flour considerably exceeds that of wheat flour. Most of the contaminant fungi appear to have derived from soil carried with the grain. Consequently, the following fungi were indicated as the dominant species, in order of isolation: *P. viridicatum, P. cyclopium, P. frequentans, P. casei, P. commune; A. candidus, A. flavus, A. glaucus* group, *A. ochraceus, A. melleus; Mucor, Cladosporium, Alternaria*, and *Phoma*.

1.4.5. Corn and Corn Products

Numerous studies have been made on the role of fungi in the deterioration of stored corn.[254,256,319,321] Little information, however, is available on the mycoflora of freshly harvested stored corns on the basis of general mycology. It is known that the ferquency of fungal infection for whole corn kernels usually ranges from 55 to 92%. The genera most responsible are *Aspergillus* and, when the grain is stored at a moisture level of 18% (w/w) or above, *Penicillium*. The major field fungi that invade corn are almost identical with those found in the external mycoflora of wheat and rice grains, i.e. species of *Alternaria*, *Cladosporium*, *Nigrospora*, *Helminthosporium*, *Trichoderma* and *Fusarium*. *Helminthosporium maydis* is frequently isolated from diseased kernels with southern corn blight.[322] The chief species of *Penicillium* associated with freshly harvested corn are *P. oxalicum* and *P. funiculosum*, while *P. cyclopium*, *P. brevi-compactum* and *P. viridicatum* are the chief species isolated from stored kernels. The following are frequently seen in either or both materials: *P. luteum*, *P. frequentans*, *P. implicatum*, *P. charlesii*, *P. purpurogenum*, *P. multicolor*, *P. variabile*, *P. citrinum*, *P. steckii*, *P. urticae*, *P. palitans*, *P. puberulum*, *P. chrysogenum*, *P. digitatum*, *P. janthinellum*, *P. expansum*, *P. granulatum*.[323,324] The dominant *Aspergilli* are *A. flavus*, *A. niger*, *A. fumigatus*, and the *A. glaucus* group (*A. chevalieri*, *A. amstelodami*, *A. repens*).[325–9]

Corn is harvested and often stored at an initial moisture content of 20–2% or above, which is favourable for both field and storage *Penicillia*. Such fungi are therefore important in stored corn due to the high moisture level at which the grain is usually stored.[330] *Fusaria* are common field fungi which tend to die during storage due to inadequate moisture, oxygen, or temperature levels. These same conditions which are inhibitory to *Fusaria* may, however, be conducive for the growth of *Penicillia*. As a result, *Penicillia* become the dominant storage fungi, while *F. moniliforme* is the predominant fungus on damaged seeds. Successive occurrences of *Fusaria* during harvesting, handling and storage should be studied mycologically to assess the process of possible toxic trichothecene contamination.

Over the past several years, various data on the natural occurrence of mycotoxins, especially aflatoxins, zearalenone and T-2 toxin, have been obtained with moldy corn in the U.S.A. and other countries.[203,331] Thus, corn is the most noticeable crop associated with human and animal mycotoxicoses. In fact, there have been many case reports of moldy corn toxicoses apparently induced in domestic animals following ingestion of feed containing *Aspergillus*- or *Fusarium*-infested corn.

Corn starch usually carries quite a small number of fungi (below 100 colonies/g). The mycoflora consists of *Penicillium* and *Aspergillus*, mostly carried over from the corn kernels.[261]

1.4.6. Sorghum

Sorghum (*Sorghum vulgare* Pers.) is a kind of cereal grain used for animal feed or human food in some countries. Although it is processed for starch and other materials, it is not a widely popular foodstuff. Sorghum seeds have a high natural infection rate of field fungi. Species of *Alternaria* often occur in 100% of the seeds, and *Fusarium, Chaetomium, Cladosporium, Curvularia*, and *Helminthosporium* are frequently encountered.[332)]

When sorghum is stored under moist conditions, *Fusarium* increases at a moisture content of 20–4%, *Penicillium* grows at 18–24%, *Trichoderma* grows at 22–4%, but *Alternaria* decreases with increasing moisture content and time. During storage, other principal fungi appear, such as *Verticillium, Scopulariopsis, Trichothecium*, and *Penicillium*. Of these, *A. flavus* shows the best growth at a moisture content of 29%, while *Fusarium moniliforme* appears to represent the best competitor among the mycoflora of moistened grains. The following fungi make up the internal mycoflora of sorghum grains (identified from surface-disinfected seeds): *Aspergillus candidus, A. flavus*, the *A. glaucus* group, *A. nidulans, Cephalosporium, Circinella, Epicoccum, Helminthosporium, Mucor, Phoma, Rhizopus, Thermomyces*, and *Trichoderma*. *Alternaria, Cladosporium* and *Fusarium* are the three predominant genera inhabiting surface-disinfected sorghum seeds immediately after harvesting.[333–5)] *Penicillium, Mucor* and *Rhizopus* are present as surface contaminants.

Based on this mycoflora, it is clearly possible that sorghum could be a causative food of mycotoxicoses. Particular caution is thus required with fungus-contaminated materials.

1.4.7. Cottonseed

The mycoflora of cottonseed has been determined only incompletely and imprecisely. In the U.S.A., however, aflatoxin comtamination of the seeds has been identified in some diseased samples infected with *Aspergillus flavus*.[336,337)] Before harvesting, cotton seeds including intact seeds and separated meats and hulls, mainly harbor pathogenic fungi such as *Fusarium moniliforme, F. roseum* and other *Fusaria, Colletotrichum gossypii, Alternaria, Cladosporium*, and *Mucor*. The saprophytic fungi, *A. flavus* and *A. niger*, are frequently associated with freshly harvested cottonseed.[338,339)] After harvesting, the fungi generally categorized as "storage fungi", i.e. the *Aspergillus glaucus* group, *A. versicolor*, *A. candidus* and other *Penicillia*, increase with storage period.[340]

1.4.8. Soybeans

Soybean is an important agricultural commodity as an oil and protein source for man and animals. It represents a valuable addition to usual foods in creating a balanced diet. Soybean curd ("tofu" in Japanese) is prepared from soy milk and is an essential fresh food in the meals of Japanese and many Asian people.

The microflora of soybeans is not greatly different from other seeds and cereals. *Alternaria* spp. are isolated most frequently from the pistil and stamens of unopened and opened flowers and from style and stigma remains adhering to pod tips. *Cercospora kikuchii* comprises the highest proportion of isolates from immature seeds within pods.[341] Other fungal genera isolated fresh seeds, in order of frequency, are as follows: *Fusarium*, *Aspergillus*, *Colletotrichum*, *Penicillium*, *Diaporthe*, and *Curvularia*.[342] In stored soybeans after harvesting, the *Aspergillus glaucus* group (such as *A. repens* and *A. amstelodami*), *A. restrictus*, and *A. halophilicus*-like xerophilic fungi at moisture contents below 13.0%, commonly occur.[343–7]

1.4.9. Peanuts and Peanut Products

As peanut fruits mature in the soil, they become increasingly susceptible to invasion by microbes from the surrounding soil. Numerous reports on the microorganisms associated with mature fruit, overmature fruit and damaged fruit have appeared in the literature.[348–56]

A. Endocarpic Mycoflora of Window-dried Peanuts

The dominant genera of shells of windrowed fruit have been reported to include *Chaetomium*, *Rhizoctonia*, *Fusarium*, *Sclerotium*, and *Alternaria*; the only dominant genus of the windrowed fruit was *Penicillium*.[354] Garren and Porter also examined the quiescent floral communities of cured mature peanuts, and the ecology of the important toxicogenic species, *A. flavus*, was discussed in their report.

(1) Isolation frequencies of 5% or more: *Aspergillus flavus*, *Alternaria tenuis*, *Chaetomium globosum*, *Cylindrocladium* sp., *Diplodia gossypina*, *Fusarium* spp., *Penicillium* spp., *Rhizoctonia bataticola* (= *Macrophomina phaseoli*), *R. solani*, and *Rhizopus stolonifer*.

(2) Isolation frequencies of up to 5%: *Aspergillus* spp., *A. niger*, *Nigrospora* sp., *Phoma* sp., *Thielavia* sp., and *Trichoderma viride*.

B. Endocarpic Microflora of Peanuts in Virginia[356)]

(1) Dominant fungi: *Penicillium* spp., *Trichoderma* spp., *Chaetomium* spp., and *Fusarium* spp.

(2) Sub-dominant fungi: *Aspergillus* spp., (*A. flavus* included), *Thielavia* spp., and *Rhizoctonia* spp.

(3) Other fungi: *Alternaria* spp., *Rhizoctonia* spp., *Epicoccum* spp., *Pythium* spp., and *Botrytis* spp.

C. Fresh Lifted Pods in Oklahoma[357)]

Several species of fungi never reported before as occurring in peanut pods, were found in the pods, as follows: *Actinomucor elegans*, *Aspergillus sclerotiorum*, *Sordaria fimicola*, *S. humana*, and *Sporormia australis*. The *Fusarium* species included *F. acuminatum*, *F. moniliforme*, *F. oxysporum*, *F. solani*, and *F.* spp.

D. Peanut Fruits

Penicillium (48.1%) and *Fusarium* (29.8%) are the most common genera in the shells; *Penicillium* (38.1%) and *Gliocladium* (2.3%) are the most common in seeds. The number of genera isolated from various samples was fairly constant during the season.[358–60)]

E. Peanut Fruits in Israel[350)]

Several investigators have reported briefly on the mycoflora of peanut fruits. It includes *Rhizopus oryzae*, *Aspergillus flavus*, *A. niger*, *Penicillium funiculosum*, *P. rubrum*, *Fusarium oxysporum*, and *F. solani*.

F. Peanut Fruits of Foreign Introduction[361)]

(1) Shell genera: *Botrytis* (33.0%) *Fusarium* (32.9%), *Phoma* (26.3%), *Penicillium* (17.4%), and *Mucor* (10.4%).

(2) Seed genera: *Penicillium* (26.6%), *Botrytis* (15%), *Fusarium* (11 %), *Mucor* (4.5%), and *Phoma* (21%).

G. Fresh and Stored Kernels in Israel[352)]

The species included *Aspergillus flavus*, *A. niger*, *Fusarium solani*, *Penicillium funiculosum*, *P. rubrum*, *Rhizoctonia* sp., *Rhizopus nigricans*, and *Sclerotium* sp. *A. flavus* was usually found in small quantities only. *A. niger* was the species almost invariably found in great numbers.

H. Stored Peanuts[348,352)]

The predominant fungi are several species of the *Aspergillus glaucus* group, *A. tamarii*, and *Penicillium citrinum*. The *A. glaucus* group involves

A. ruber, *A. repens*, *A. restrictus*, *A. chevalieri*, and *A. amstelodami*. The common species are *A. candidus*, *Cladosporium* sp., *Torula sacchari*, and *Penicillium funiculosum*. The *A. glaucus* group and *P. citrinum* are the principal molds of stored peanuts. *A. flavus* was not found, but *A. tamarii* occurred frequently and appeared to be important.[348)] On stored kernels, it was even more predominant than on fresh ones. *P. funiculosum* and *P. rubrum* were also prevalent.

I. Peanut Meal

The aflatoxicosis of turkeys caused by toxic peanut meal contaminated with *Aspergillus parasiticus* was first encountered in the United Kingdom in 1960.[362)] Peanut meal for feed commonly carries vast amounts of *Penicillium* and *Aspergillus*. The species include *A. chevalieri*, *A. amstelodami*, *A. fumigatus*, *A. flavus*, *A. tamarii*, *A. terreus*, and *P. notatum*, together with *Mucor* spp. and *Rhizopus* spp.

It is significant that *A. flavus* is the predominant fungus among the mycofloras of peanut meals.[128,363,364)] A relatively high proportion of the *A. flavus* isolates from all peanut cultivating areas are capable of producing aflatoxin. However, aflatoxins have been found only rarely in peanut fruits prior to or at the time of digging. Some peanuts from post-digging experiments in the U.S.A. have been found to be contaminated with aflatoxins. Prompt, steady drying to a seed moisture level of 7–8% within 3–5 days is highly effective for suppressing the development of *A. flavus* and contamination of the seeds by aflatoxins.[365–8)]

1.4.10. Spices

Spices ordinarily contain many kinds of plant materials including seeds, fruits and leaves. The chief materials are black and white peppers. So far as the author is aware, only two reports on mycobiological examinations of spices have so far been published, i.e. by Christensen *et al.*[369)] and by Horie *et al.*[370)] in Japan. The most common fungi are the *Aspergillus glaucus* group, *A. candidus*, *A. flavus*, *A. ochraceus*, *A. niger*, and *A. versicolor* as storage fungi; the other species include *Penicillium chrysogenum*, *P.* spp. and *Rhizopus* spp. Based on this knowledge of the mycoflora, the possibility of aflatoxin contamination exists, so that chemical surveys for mycotoxins should always be considered.

1.4.11. Fermented Foods

Representative fermented foods are the miso (soybean paste) shoyu (soy sauce) and vinegar consumed daily in quantity by most Japanese. The former two products are usually prepared by the use of a starter, "tane-koji", obtained commercially or individually prepared in the traditional way and preserved. In fact, "tane-koji" is a broth of dried fermented rice containing numerous fungal spores, usually of strains of *Aspergillus*.

Tempeh and ragi are known as interesting Indonesian fermented products in which Mucorales are inoculated with soybeans as substrate. Angkak in China, Indonesia and the Philippines is rice substrate fermented with *Monascus purpureus*.

The mycofloras of miso and shoyu have been examined mostly by Asian scientists.[98,100,372,373] Generally, miso has about 10% NaCl, so that a small number of fungi are associated with it. The normal mycoflora of miso is *Chaetomium*, *Eurotium*, *Alternaria*, *Arthrinium*, *Cladosporium*, *Curvularia*, *Epicoccum*, and *Scopulariopsis*, plus many species of yeasts. Fungal species of relatively high frequency are *Aspergillus candidus*, *A. versicolor*, *Cladosporium* spp., *Mucor* spp., *Rhizopus* spp., *Penicillium puberulum*, *P. steckii*, *P. cyclopium*, and *P. waksmanii*. Since shoyu contains about 18% NaCl, the fungal population remains at quite a low level. *Aspergillus repens*, *A. versicolor*, *Penicillium puberulum*, and *P. waksmanii* are the common contaminants. *Moniliella acetoabutens* is sometimes isolated from fermented vinegar. *A. oryzae*-utilizing starters in Japan have not yet been found to exhibit any aflatoxin producibility.[374–376] However, Lew[377] in Korea has reported that *A. flavus* and *P. cyclopium* from fermented foods including soybean paste can produce aflatoxins in liquid media. On mycotoxicological screening, Kinoshita *et al.*[378] showed that out of 37 fungal strains, 21 produced toxic culture filtrates and extracts, although none of the strains was detected as an aflatoxin producer.

1.4.12. Sugars

The microbial content of sugars varies with the kind of sugar. Highly refined granules and low moisture content sugars harbor few fungi (below 50 colonies per 10 g contaminated). The chief species fungi and yeasts are *Alternaria brassicae* (from beet sugar), *Aspergillus flavus*, *A. niger*, *A. sydowi*, *A. versicolor*, *A. wentii*, *A. nidulans*, *A. repens*, *Penicillium expansum*, *P. purpurogenum*, *P. luteum*, *Scopulariopsis brevicaulis*, *Geotrichum candidum*, *Neurospora sitophila*, and sugar-tolerant yeasts such as *Candida*, *Saccharomyces*, and *Torulopsis*.[174]

1.4.13. Honey

Yamazaki *et al.*[379] has reported that the mycoflora of honey includes *Chaetomium*, *Eurotium*, *Aspergillus*, *Cladosporium*, and *Penicillium* as dominant genera, based on a mycological survey of 23 samples.

1.4.14. Fruit Juice

The spoilage of fresh fruit juice with or without added sugar is limited to that caused by osmophilic microorganisms, i.e. those able to grow in the high concentrations. *Penicillia* are the predominant fungi, and strongly affect the quality of the juices. Some *Penicillium* species are susceptible to CO_2 gas. The sub-dominant species are *P. expansum* and *P. crustosum.* Stored juice is often contaminated with *Alternaria.*[380]

1.4.15. Meat and Meat Products

Meat represents ideal culture medium for many microorganisms due to its high moisture content and rich nutritive value. It can be preserved at low temperatures, e.g. below −10°C. Such cold meat sometimes allows the growth of psychrophilic fungi, such as *Chaetostylum fresenii, Chrysosporium pannorum, Thamnidium elegans, Mucor lusitanicus, M. mucedo, M. racemosus,* and *Rhizopus* spp.. In addition to these fungi, *Penicillium expansum, P. oxalicum, Eupenicillium shearii, Cladosporium herbarum,* and *Geotrichum candidum* are fairly common contaminants.

The mycoflora of chilled beef, as examined by Inagaki and Takahashi in Japan,[381,382] was found to include other species of fungi: *Alternaria alternata, Aspergillus sydowi, Penicillium commune, P. cyclopium, P. cyclopium* var. *echinulatum, P. frequentans, P. martensii, P. viridicatum, Phialophora* sp., *Trichoderma* sp., and *Verticillium* sp.[383] Extensive mycological surveys have been carried out on various kinds of minced meat collected from markets by Hitokoto *et al.*[384] The results indicated high frequencies of *Aureobasidium, Phoma, Fusarium, Cladosporium, Geotrichum, Aspergillus versicolor* and the *A. glaucus* group in samples of pork, chicken, beef, and mixed pork-chicken and pork-rabbit. Possible trichothecene toxin contamination of meat can thus be expected due to the presence of *Fusaria* as important fungi of the mycoflora of various meats and minced meats. Meat products, such as bacon, other smoked beef, fermented salami and sausages also deteriorate through the action of food-borne fungi. The mycoflora of these products consists mainly of the following saprophytic fungi: *Alternaria, Aspergillus, Botrytis, Fusarium, Monilia, Mucor, Penicillium,* and *Rhizoctonia.* In salami sausages, Takatori *et al.*[385] found that the major species were *P. cyclopium, P. miczynskii,* and *P. viridicatum.* The fungal contamination of meat products, especially salami, may originate from spices that are used as seasonings. Leistner and Tauchmann[386] have pointed out that some strains of *Aspergillus flavus* and *A. parasiticus* may produce large amounts of aflatoxins in fermented sausages.

1.4.16. Dried Fish Products

Dried fish products including "katsuobushi" (fermented dry bonito) are widely eaten as preserved foods in Japan. The dried fish are often used in soups or broiled. Under conditions of high moisture, they are sometimes spoiled by *Penicillia* and *Aspergilli.* The following fungi are frequently isolated from moldy dry fish: the *Aspergillus glaucus* group, *A. versicolor, A. ochraceus* group, *Penicillium expansum, P. olivino-viride, P. viridicatum,* and *Phoma* sp.. In "katsuobushi", special molds (*Aspergillus katsuobushi,* probably synonymous with *A. repens*) are artificially inoculated during processing to improve the flavour and enhance drying of the fish surface. *Aspergillus ochraceus, A. oryzae, A. ostianus, A. tamarii, Penicillium cyclopium* var. *echinulatum, P. putterillii,* and *Syncephalastrum racemosum* are known to occur as spoilage fungi.[98,100,373)]

1.4.17. Milk and Milk Products

Milk is also an excellent culture medium for many kinds of microorganisms. When milk "sours", it is often considered spoiled, especially if it curdles, although the lactic acid fermentation of milk is utilized in the manufacture of fermented milks and cheeses. Milk and milk products are commonly spoiled by molds, but essentially the spoilage effects of molds on milk are not cause for grave concern. Most of the molds of cheeses and butter grow in colored colonies on the surface or in crevices, without much penetration, but some actually produce rots. The dominant species of fungi affecting milk products are as follows.

A. Dry milk

A relatively limited number of fungal species have been isolated. They are *Aspergillus candidus, A. clavatus, A. flavus, A. fumigatus, A. niger, A. versicolor, Penicillium chermesinum, P. citrinum, P. cyclopium, P. frequentans, Geotrichum candidum, Cladosporium, Rhizopus, Paecilomyces,* and *Fusarium.*[277)]

B. Butter

Butter is always kept under low temperature conditions with certain added preservatives and NaCl. Growths of molds are therefore rarely seen, only on the surface of the butter. The general species of contaminant fungi in natural butter and margarine are *Geotrichum candidum, Monilia suaveolens* (= *Oospora* spp.), *Penicillium expansum,* and *Phialophora bubakii*.[380,387,388] The mycoflora of margarine is not greatly different from that of natural butter.[389]

C. Cheese

So-called blue cheese is prepared chiefly by inoculation of spores of two species of *Penicillium*, e.g. *P. roqueforti* and *P. camemberti*. In other cases, cheese is generally suitable as a substrate for the growth of saprophytic fungi involving *Penicillia* and *Aspergilli*. The following fungi frequently grow on the surface of cheese: *Aspergillus* spp., *Cladosporium herbarum*, *Geotrichum candidum*, *Geotrichum* spp., *Mucor* sp., *Neurospora sitophila*, *Penicillium aurantio-virens*, *P. casei*, *P. expansum*, *P. puberulum*, and *Scopulariopis* sp.[174,380]

1.4.18. Miscellaneous Foods

A. Green Tea Leaves

The predominant species of fungi are *Aspergillus niger*, *A. glaucus* group, *A. versicolor*, *Penicillium citrinum*, *P. expansum*, *Cladosporium* spp., and *Mucor* spp.[390]

B. Cacao Beans

The fungi present in raw cacao beans include *Aspergillus glaucus* and *Penicillium* spp. as the chief elements of the mycoflora.[391]

C. Pecans

Of the fungi isolated from pecans, the following species are the most prevalent: *Aspergillus chevalieri*, *A. flavus*, *A. niger*, *A. ochraceus*, *A. repens*, *Botryodiplodia theobromae*, *Epicoccum nigrum*, *Penicillium citrinum*, *P. implicatum*, *P. funiculosum*, and *Pestalotia* sp.[392]

D. Chestnuts

The most common fungi isolated from decayed chestnut tissues are *Phoma castaneae* and *Pestalotia* sp. Of minor importance are various species of *Phomopsis*, *Penicillium*, *Alternaria*, *Fusarium*, and *Rhizopus*.[393]

1.5. SURVEYS ON THE MYCOFLORA OF DIETARY FOODS

As regards mycotoxins that represent possible hazards to human health, comprehensive surveys of dietary foods consumed by particular populations have been conducted by mycological and chemical assay methods in the Philippines,[394)] tropical S.E. Asia, Thailand,[276,395–9)] S. Africa (Swaziland),[400,401)] Kenya,[402)] Uganda,[403)] and Japan.[98,100,373,404)] Most of the surveys were made at special localities with an epidemic background where high incidences of hepatomas had been recorded. However, a limited amount of work with a mycological emphasis has also been undertaken. The studies in S. Africa, tropical S.E. Asia and Japan demonstrated the distribution of fungi in foods in these countries and also distinguished the species capable of mycotoxin production. Such screening is of great value in determining the relative frequency of occurrence of particular toxin-producing fungi on particular agricultural products. The mycological survey of more than 3000 market foods collected in Thailand and Hong Kong [276)] showed *Aspergillus* to be the commonest contaminating fungus, with *A. flavus*, the predominant species. Other fungi included species of *Penicillium*, *Fusarium*, and *Rhizoctonia*. Bioassay of the acute and subacute toxicity to rats revealed infecting agents capable of producing mycotoxins other than aflatoxins. Among 162 isolates, 49 proved to be toxigenic, and these included species of all genera commonly identified in food samples.

Over 2000 wholesome foodstuffs sampled from the home in 1967–73 have been examined for their fungal population by Japanese mycologists, through the cooperative efforts of toxicological and eipidemiological research groups.[98,100,373,404)] The representative fungal strains isolated were tested for their acute toxicity to mice, and subjected to cytotoxic evaluation using HeLa cells. In the preliminary survey in 1967, 133 isolates were tested for their toxic response. Nine strains (6.8%) were toxic to both HeLa cells and mice, 15 (11.3%) to HeLa cells only, and 25 (18.5%) to mice only. In the subsequent studies in 1968 and 1969, a total of 257 strains were tested and the results showed that 36 and 55 strains were toxic to HeLa cells and mice, respectively. The highly toxigenic species found in these screening tests were *Aspergillus ochraceus*, *A. versicolor*, *A. amstelodami*, *A. chevalieri*, *A. niger*, *A. ostianus*, *Penicillium cyclopium*, *P. viridicatum*, *P. notatum*, *P. urticae*, *Chaetomium globosum*, and *Phoma* spp. Mycological data for household foodstuffs are summarized in Table 1.5.

Similar mycotoxicological surveys have also been carried out in the Eastern Transvaal and Swaziland, especially as regards the diet of the rural Bantu in 1966 and 1969.[279,400,401)] A total of 635 food samples including corn, peanuts, sorghum and assorted legumes were investigated mycologically.[279)] In the case of isolates of *Aspergillus flavus*, aflatoxin producibility was examined chemically and 531 strains of fungi isolated from the foodstuffs were tested for their toxigenicity by the duckling method. This mycological investigation indicated that *A. flavus*, the *A. glaucus* group, *A. ochraceus*, *Fusarium moniliforme*, *Penicillium citrimun*, *P. cyclopium*,

TABLE 1.5. Mycoflora of foodstuffs collected from four localities surveyed in Japan

1. Genera and species of fungi with high frequency of occurrence.

Species	Preferred foodstuffs
Aspergillus flavus group: *A. flavus*, *A. oryzae*	rice, "miso", mash, soy sauce
A. glaucus group: *A. amstelodami*, *A. chevalieri*, *A. repens*, *A. ruber*	rice, wheat, flour, dried fish products,† "miso", vegetable pickles
A. sydowi	mash
A. versicolor	"miso", rice, wheat, beans, dried fish products
Penicillium citrinum	wheat, flour
P. cyclopium	dried fish products, flour, "miso"
P. phoeniceum	wheat, rice
P. viridicatum	beans, flour
P. waksmani	"miso"
Candida spp.	vegetable pickles, rice
Chaetomium spp.	beans, dried fish products
Cladosporium spp.	cycas starch
Geotrichum spp.	vegetable pickles
Rhizopus spp.	"miso", mash

2. Genera and species of fungi with moderate frequency of occurrence.

Species	Preferred foodstuffs
Aspergillus candidus	rice, wheat, flour, beans, "miso"
A. mangini	dried fish products, flour
A. niger	"miso," starch
A. ochraceus	rice, beans, wheat, flour
A. ostianus	"miso", beans, dried fish products
A. restrictus	rice, wheat, flour
Penicillium meleagrinum	vegetable pickles
P. chrysogenum	wheat, flour, rice, "miso"
P. notatum	beans, flour
P. puberulum	rice, wheat
P. urticae	flour, mash
P. vinaceum	rice, wheat, flour
Alternaria spp.	wheat, beans, "miso," vegetable pickles
Diaporthe phaseolorum	beans
Epicoccum spp.	rice, wheat
Fusarium spp.	rice, wheat, beans, cycas starch
Paecilomyces varioti	"miso", flour
Pestalotia spp.	beans
Phoma spp.	rice, beans, dried fish products
Syncephalastrum racemosum	"miso"
Wallemia sebi	rice, dried fish products

3. Genera and species of fungi with low frequency of occurrence.

Species	Preferred foodstuffs
Aspergillus aculeatus	flour, beans
A. clavatus	beans
A. fumigatus	flour

TABLE 1.5—*Continued*

A. sclerotiorum	flour
A. sulphureus	rice, "miso"
A. tamarii	wheat, barley, "miso"
A. terreus	wheat
A. umbrosus	wheat
Penicillium cyaneo-fulvum	wheat, flour
P. expansum	dried sardines
P. islandicum	rice, flour
P. purpurogenum	wheat
P. palitans	rice, beans
P. paxilli	rice, flour
P. lividum	flour
Ascochyta spp.	beans
Actinomucor elegans	"miso"
Arthrinium spp.	flour, rice, dried tangle
Botrytis cinerea	barley, soy sauce
Curvularia spp.	wheat, rice
Epicoccum spp.	rice, wheat
Helminthosporium spp.	rice
Pitomyces chartarum	beans, dried sardines
Trichoderma spp.	vegetable pickles, beans
Nigrospora oryzae	"miso"
Zygosporium masonii	beans, dried sardines

†Dried small sardines and "katsuobushi".

P. meleagrinum, *P. purpurogenum*, and *P. crustosum* were the important fungi, i.e. species regularly present in all foodstuffs though in varying frequencies.

The above results from Japan, Thailand, other parts of tropical S.E. Asia, and Swaziland are in general agreement with those obtained in countries such as the U.S.A., Canada, Great Bretain and U.S.S.R. (mostly grain surveys for which mycological data were described above in section 1.4), showing similar genera and species of fungi with a high frquency of occurrence. However, some differences do exist among the countries, i.e. high frequencies of isolation of *Aspergillus flavus* and *A. candidus* were observed in the Thailand and Swaziland samples, and of *Fusarium sporotrichioides* in the U.S.S.R,[6,43–45] *F. moniliforme* in the U.S.A.[254,256,321] and *F. graminearum* in Japan[261,262,282] This may be due to geographic variability, differences in moisture content, quality and/or of kind of food samples.

A general view of the mycofloras of foodstuffs described in this chapter shows that the predominant species of molds making up the mycofloras invariably involve fungal groups with possible mycotoxigenicity, e.g. *Aspergillus flavus*, *A. ochraceus*, *A. versicolor*, *Penicillium citrinum*, *P. viridicatum*, *Fusarium* spp., etc. Up to the present, a vast number of mycotoxicological screening trials, mainly with food-borne fungi, have been carried out by several investigators in the world.[88,91–4,96–102,276,279,398,404–9)] The results have revealed several good examples of mycotoxin-producing fungi, including strains with both known and unknown mycotoxin producibility. It is conceivable that representative mycotoxins having such a history of origin include the ochratoxins, sterigmatocystin, penicillic acid, kojic acid, rubratoxins, etc.[98,100)] In principle, accurate information on the mycoflora of foodstuffs is a prerequisite or estimation of the mycotoxin contamination of all agricultural commodities. However, resolution of this problem is not simple. It is very difficult to determine which are the pure inhabitants of the substrate, and which the transients and incidentals. It must also be remembered that surface contaminants may be included in the mycofloral data, if inadequate procedures for mycological isolation are employed. Simple correlations between the substrates and mycofloras are difficult, since such comparisons become multidimensional. As described above, toxigenic molds are more or less ubiquitous and have been isolated from an extremely wide variety of materials. Many of these isolates have been described as having the capability to produce certain mycotoxins, but more often than not, the source from which they were isolated was not analyzed itself for the presence of toxin. In this respect, it should be emphasized that the presence of moldiness *per se* is not necessarily accompanied by any mycotoxin production. To assess the true contamination level of a mycotoxin in foodstuffs, chemical data on the natural occurrence of the mycotoxin are clearly the final requirement.

1.6. NATURAL OCCURRENCE OF MYCOTOXINS IN FOODSTUFFS

Recently, an enormous amount of information on natural contamination by mycotoxins has been collected, mostly in the U.S.A. Important naturally occurring mycotoxins and contaminated foodstuffs are listed briefly below.[203,323,331,410–20)]

Aflatoxins: peanuts and peanut products, rice, wheat, barley, corn (maize), oats, rye meal, sorghum, millet, cereal flours and flour products (spaghetti, noodles, grist, rice cake), bread, cottonseed, palm kernels, copra, soybeans, beans, peas, dried cassava and sweet potatos, molded rice cake, hazelnuts, walnuts, cashewnuts, almonds, pistachio nuts, sesami seeds, dried fish and shrimps, oranges, lemons, peaches, tomatoes, dried chili peppers and garlic, dried vegetables and fruit, spices (nutmeg), mixed commercial feeds.

Sterigmatocystin: grain samples, moldly stored rice.
Ochratoxin A: wheat, barley, oats, rye, corn, rice.
Citrinin: barley, wheat, oats, rye.
Patulin: apple juice.
Penicillic acid: corn.
T-2 toxin: corn.
Deoxynivalenol: barley, wheat.
Zearalenone: corn, feedstuffs.

It is an interesting fact that aflatoxins have frequently been found in prepared dietary foods and native beer consumed by certain people in Kenya[402] and Thailand.[395–9] In general, aflatoxins seem to be more of a problem in the tropics than in the temperate zone, but no region of the world can be considered aflatoxin-free, due to the free movement of various foodstuffs from one part of the globe to the other. Accordingly, in Japan, as in many other countries, considerable efforts have been made to determine the no-effect or tolerance levels of aflatoxins, and to investigate the possible adverse effects of various mycotoxins, particularly carcinogenic mycotoxins, on farm animals and the possible transmission of toxins to edible animal products and tissues.

When new or improved chemical assays for the important mycotoxins have been established, as can be expected in the near future, advanced data on the naturally occurring mycotoxins will become available, and accurate estimates of the significance of mycotoxin hazards to human and animal health will become possible. For the present, however, data on the mycoflora of foodstuffs otbained by fungal examinations and by the toxicological approach must serve for the prediction of the risks involved.

1.7. Control of Mycotoxins

To control the mycotoxin contamination of food- and feed-stuffs, the following four kinds of measures are basically needed:[410,421] (1) prevention of the initial growth of molds and subsequent contamination by mycotoxins, (2) detection of mycotoxins in food materials and selective removal of the contaminated portions, (3) inactivation or destruction of all toxin (s) that may be present, by chemical, physical and/or enzymatic means, and (4) advance identification of possible new mycotoxin threats. Item (3) is extremely difficult to effect in practice and ideal techniques for detoxication are not expected within a short period of time. Items (1) and

(2) therefore represent more useful and realizable procedures. First, agricultural technology must be improved so as to prevent infestation by fungi, especially mycotoxin-producing fungi, in the pre- and post-harvest crops and in all agricultural commodities during storage, handling, processing and transport. Next, special attention should be given to the detection of lots containing mycotoxins and to diverting them from the normal food and feed channels as early as possible in the marketing process. Clearly, the early detection and diversion of small consignments of contaminated materials may aid the prevention of contamination of much larger supplies. However, to achieve this, rapid detection methods are required. As regards selective removal, some procedures have been developed to ensure peanut and peanut product quality during inspection of samples as they come from the farm to the primary market and during grading on the basis of visibly damaged or discolored kernels and through *Aspergillus flavus* fruiting on them. Electric-eye and hand-sorting are used to eliminate damaged kernels from lots destined for food. Also, a check system for aflatoxins has been organized in developed countries, and WHO has given a recommended guideline for total aflatoxins (15 ppb) for peanuts, corn and other foodstuffs. Limits of tolerance for other carcinogenic mycotoxins must thus be established as soon as possible. Item (4) is not a direct method for mycotoxin control, but is very important for preventive food hygiene. In this respect, new mycotoxin screening tests with as many strains of food-borne fungi are being developed through the cooperative efforts of various plant pathologists, mycologists, chemists and toxicologists in Japan, the U.S.A., Canada, S. Africa, and other countries.

REFERENCES

1. C. W. Hesseltine, *Mycopath. Mycol. Appl.*, **39**, 371 (1969).
2. J. Forgacs and W. T. Carll, *Advan. Vet. Sci.*, **7**, 273 (1962).
3. J. Forgacs, Proc. U.S. Livestock Sanitary Ass., 66th Ann. Mtg. (1962).
4. K. Sergeant, A. Sheridan, J. O'Kelly and R.B.A. Carnaghan, *Nature*, **192**, 1096 (1961).
5. R. Kinoshita and T. Shikata, *Mycotoxins in Foodstuffs* (ed. G.N. Wogan), p. 111, MIT Press (1965).
6. A. Z. Joffe, *ibid*, p. 77, MIT Press (1965).
7. S. J. van Rensburg and B. Altenkirk, *Mycotoxins* (ed. I. F. H. Purchase), p. 69, Elsevier, (1974).
8. P. M. Newberne, *Environment Health Perspectives*, **9**, 1 (1974).
9. M. T. Morgan, *J. Hyg.*, **29**, 51 (1929).
10. L. S. Goodman and A. Gilman, *Pharmacological Basis of Therapeutics*, Macmillan (1955).
11. N. T. Clare, *Advan. Vet. Sci.*, **1**, 182 (1955).
12. D. C. Dodd, *Mycotoxins in Foodstuffs* (ed. G. N. Wogan), p. 105, MIT Press (1969).

13. W. J. Gibbons, *Auburn Vet.*, **II**, 177 (1953).
14. J. Forgacs, *Mycotoxins in Foodstuffs* (ed. G. N. Wogan), p. 87, MIT Press (1969).
15. J. Forgacs, W. T. Carll, A. S. Herring and W. R. Hinshaw, *Trans. N. Y. Acad. Sci.*, **20**, 787 (1958).
16. V. A. Fortuskny, A. M. Govrov, I. Z. Tebybenko, A. S. Biochenko and E. T. Kalitenko, *Veterinariya*, **36**, 67 (1959).
17. Eeva-Liisa Korpinen, *Dept. Microbiol. and Epizootol. Academic Dissertation*, College of Veterinary Medicine, Helsinki, Finland (1974).
18. R. M. Epply and W. J. Baily, *Science*, **181**, 758 (1973).
19. D. C. Gajdusek, *Med. Sci. Publ.*, **2**, 82, Walter Reed Army Med. Center (1953).
20. P. J. Brook, *Ann. Rev. Phytopath.*, **3**, 172 (1966).
21. H. F. Kraybill and R. E. Shapiro, *Aflatoxin* (ed. L. A. Goldblatt), Academic Press (1969).
22. E. B. Smalley and F. M. Strong, *Mycotoxins* (ed I.F.H. Purchase), p. 199, Elsevier (1974).
23. A. D. Dickson, *Phytopath.*, **20**, 132 (1930).
24. Y. Nishikado, *Agr. Inst. Okayama Univ.* (Japanese), **45**, 14, 59, 159 (1957); **46**, 1 (1958).
25. H. Tsunoda, *Toxic Microorganisms* (ed. M. Herzberg), p. 143, UJNR Joint Panels on Toxic Microorganisms and the U.S. Department of the Interior (1970).
26. H. Kurata, F. Sakabe and S. Udagawa, *Bull. Natl. Inst. Hyg. Sci.* (Japanese), **82**, 123 (1964).
27. H. Kurata, 6th Joint Conf. of UJNR, Tokyo, **27a** (1973).
28. Y. Ueno, *J. Fd. Hyg. Soc. Japan* (Japanese), **14**, 403; **14**, 501 (1973).
29. N. Morooka, *ibid.*, **12**, 459 (1971).
30. N. Morooka, *ibid.* **13**, 368 (1972).
31. S. Hirayama and M. Yamamoto, *Bull. Natl. Hyg. Lab.* (Japanese), **66**, 85 (1948).
32. K. Ogasawara, *J. Fd. Hyg. Soc. Japan* (Japanese), **6**, 81 (1965).
33. K. Okubo, *J. Japan. Vet. Med. Assoc.* (Japanese), **22**, 453 (1969).
34. N. Morooka and T. Tatsuno, *Toxic Microorganisms* (ed. M. Herzberg), p. 114, UJNR Joint Panels on Toxic Microorganisms and the U.S. Department of the Interior (1970).
35. T. Tatsuno, M. Saito, M. Enomoto and H. Tsunoda, *Chem. Pharm. Bull.*, **16**, 2519 (1968).
36. T. Tatsuno, Y. Morita, H. Tsunoda and M. Umeda, *ibid.* **18**, 1485 (1970).
37. N. Morooka, N. Nakano, S. Nakazawa, and H. Tsunoda, *J. Agr. Chem. Soc. Japan* (Japanese), **4**, 151 (1971).
38. Y. Ueno, K. Saito and H. Tsunoda, *Toxic Microorganisms* (ed. M. Herzberg), p. 120, UJNR Joint Panels on Toxic Microorganisms and the U.S. Department of the Interior (1970).
39. Y. Ueno, I. Ueno, T. Tatsuno, K. Ohkubo and H. Tsunoda, *Experientia*, **25**, 1062 (1969).
40. Y. Ueno, Y. Ishikawa, K. Saito-Amakai and H. Tsunoda, *Chem. Pharm. Bull.*, **18**, 304 (1968).
41. Y. Ueno, *Mycotoxins in Human Health* (ed. I. F. H. Purchase), p. 163, Macmillan (1971).
42. K. Ishii, K. Sakai, Y. Ueno, H. Tsunoda and M. Enomoto, *Appl. Microbiol.*, **22**, 718 (1971).
43. A. Z. Joffe, *Pl. Soil*, **18**, 31 (1963).
44. A. Z. Joffe, *Bull. Res. Counc. Israel*, **8D**, 81 (1960).
45. A. Z. Joffe, *ibid.*, **9D**, 101 (1960).
46. C. J. Mirocha and S. Pathre, *Appl. Microbiol.*, **26**, 719 (1973).
47. F. W. Schoefield, *J. Am. Vet. Med. Ass.*, **64**, 533 (1924).
48. W. T. Carll, J. Forgacs and A. S. Herring, *Am. J. Hyg.*, **60**, 8 (1954).
49. J. J. Christensen and H.C.H. Kernkamp, *Minn. Agr. Expt. Sta. Tech. Bull.*, **113**, 128 (1936).

50. W. L. Sippel, J. E. Burnside and M. B. Atwood, Proc. **90**th Ann. Mtg., Am. Vet. Med. Ass., Toronto, Canada, p. 174 (1953).
51. W. L. Sippel, *Iowa Vet.* **28**, 15, 42 (1957).
52. J. E. Burnside, W. L. Sippel, J. Forgacs, W. T. Carll, M. B. Atwood and E. R. Doll, *Am. J. Vet. Res.*, **18**, 817 (1957).
53. J. Forgacs, W. T. Carll, A. S. Herring and B. G. Mahlandt, *Am. J. Hyg.*, **60**, 8 (1954).
54. W. T. Carll, J. Forgacs, A. S. Herring and B. G. Mahlandt, *Vet. Med.* **50**, 210 (1955).
55. P. Olafson, *Cornell Vet.*, **37**, 279 (1947).
56. C. Olson, Jr. and R. H. Cook, *Am. J. Vet. Res.*, **12**, 261 (1951).
57. J. E. Gray, G. H. Snoeyenbos, and I. M. Rynolds, *J. Am. Vet. Med. Ass.*, **125**, 144 (1954).
58. J. Forgacs and W. T. Carll, *Vet. Med.* **50**, 172 (1955).
59. J. Forgacs, H. Koch and W. T. Carll, *Poultry Sci.*, **34**, 1194 (1955).
60. J. Forgacs, H. Koch, W. T. Carll and R. H. White-Stevens, *Avian Diseases*, **7**, 363 (1962).
61. J. Forgacs, H. Koch, W. T. Carll and R. H. White-Stevens, *Am. J. Vet. Res.*, **19**, 744(1958).
62. J. Sakaki, *J. Tokyo Med. Soc.* (Japanese), **5**, 1097 (1891).
63. I. Miyake, H. Naito and H. Tsunoda, *Beikoku Riyo Kenkyujo Hokoku* (Japanese), **1**, 1 (1946).
64. Y. Hirata, *J. Chem. Soc. Japan*, **68**, 105 (1947).
65. N. Sakabe, T. Goto and Y. Hirata, *Tetr. Lett.* **27**, 1825 (1964).
66. H. Tsunoda, *Japan. J. Nutrition* (Japanese) **9**, 1 (1951).
67. Y. Kobayashi, K. Uraguchi, F. Sakai, T. Tatsuno, M. Tsukioka, Y. Noguchi, H. Tsunoda, M .Miyake, M. Saito, M. Enomoto, T. Shikata and T. Ishiko, *Proc. Japan Acad.*, **35**, 501 (1959).
68. M. Miyake, M. Saito, M. Enomoto, T. Shikata, T. Ishiko, K. Uraguchi, F. Sakai, T. Tatsuno, M. Tsukioka and Y. Noguchi, *Gann* **50**, 117 (1959).
69. M. Saito, *Acta Path. Japon.*, **9**, 785 (1959).
70. M. Enomoto, *ibid.* **9**, 189 (1959).
71. T. Ishiko, *Trans. Soc. Path. Japan*, **48**, 867 (1959).
72. S. Shibata and I. Kitagawa, *Pharm. Bull.*, **4**, 309 (1956).
73. S. Shibata and I. Kitagawa, *ibid.*, **8**, 884 (1960).
74. S. Marumo, *Bull. Agr. Chem. Soc. Japan*, **23**, 428 (1959).
75. K. Uraguchi, F. Sakai, M. Tsukioka, Y. Noguchi, T. Tatsuno, M. Saito, M. Enomoto, T. Ishiko, T. Shikata and M. Miyake, *Japan. J. Exptl. Med.* **31**, 435 (1961).
76. M. Hori and Y. Yamamoto, *J. Pharm. Soc. Japan* (Japanese), **73**, 1097 (1953).
77. T. Yamamoto, *ibid.*, **74**, 810 (1954).
78. H. Iizuka and M. Iida, *Nature*, **196**, 681 (1962).
79. Y. Okubo, N. Uragami, T. Hayama, Y. Seto, T. Miura, Y. Kano, S. Motoyoshi, S. Yamamoto, K. Ishida, H. Iizuka and M. Iida, *Japan. J. Vet. Sci.* (Japanese), **17**, 144 (1955).
80. L. A. Glodblatt (ed.), *Aflatoxin*, Academic Press (1969).
81. R. W. Detroy, E. B. Lillehoj and A. Ciegler, *Microbial Toxins*(ed. A. Ciegler, S. Kadis and S. J. Ajl), vol. VI, p. 3 (1971).
82. W. H. Butler, *Mycotoxins* (ed. I. F. H. Pruchase), p. 1, Elsevier (1974).
83. I. F. H. Purchase and J. J. van Der Watt, *Fd. Cosmet. Toxicol.*, **8**, 289 (1970).
84. K. Uraguchi, M. Saito, Y. Noguchi, K. Takahashi, M. Enomoto and T. Tatsuno, *ibid.*, **10**, 193 (1972).
85. F. Dickens, H. E. H. Jones and H. B. Waynforth, *Brit. J. Cancer*, **20**, 134 (1966).
86. F. W. Hurst and G. E. Paget, *Brit. J. Derm.*, **75**, 105 (1963).
87. S. S. Epstein, J. Andrea, S. Joshi and N. Mantel, *Cancer Res.*, **27**, 1900 (1967).
88. De B. Scott, *Mycopath. Mycol. Appl.*, **25**, 213 (1965).
89. K. T. van Warmelo, *Onderstepoort J. Vet. Res.*, **34**, 439 (1967).
90. H. Kurata, S. Udagawa, M. Ichinoe, Y. Kawasaki, M. Tazawa, H. Tanabe and M. Okudaira, *J. Fd. Hyg. Soc. Japan*, **9**, 385 (1968).

91. H. Kurata, S. Udagawa, M. Ichinoe, Y. Kawasaki, M. Tazawa, J. Tanaka, M. Takada and H. Tanabe, *ibid.*, **9**, 379 (1968).
92. C. M. Christensen, G. H. Nelson, C. J. Mirocha and F. Bates, *Cancer Res.*, **28**, 2293 (1968).
93. P. B. Hamilton, G. B. Lucas and R. E. Welty, *Appl. Microbiol.*, **18**, 579 (1968).
94. J. L. Richard, L. H. Tiffany, and A. C. Pier, *Mycopath. Mycol. Appl.*, **38**, 313 (1969).
95. A. Ciegler and J. T. Pitt, *ibid.*, **42**, 119 (1970).
96. R. A. Meronuk, K. H. Garren, C. M. Christensen, G. H. Nelson and F. Bates, *Am. J. Vet. Res.*, **31**, 551 (1970).
97. G. Semeniuk, G. S. Harshfield, C. W. Carlson, C. W. Hesseltine and W. K. Kwolek, *Mycopath. Mycol. Appl.*, **43**, 135 (1971).
98. M. Saito, K. Otsubo, M. Umeda, M. Enomoto, H. Kurata, S. Udagawa, F. Sakabe and M. Ichinoe, *Japan. J. Exptl. Med.*, **41**, 1 (1971).
99. M. F. O. Marasas and E. B. Smalley, *Onderstepoort J. Vet Res.*, **39**, 1 (1972).
100. M. Saito, T. Ishiko, M. Enomoto, K. Otsubo, M. Umeda, H. Kurata, S. Udagawa, S. Taniguchi and S. Sekita, *Japan. J. Exptl. Med.*, **44**, 63 (1974).
101. M. Archer, *Mycopath. Mycol. Appl.*, **54**, 453 (1974).
102. K. Ohtsubo, M. Enomoto, T. Ishiko, M. Saito, F. Sakabe, S. Udagawa and H. Kurata, *Japan. J. Exptl. Med.*, **44**, 477 (1974).
103. H. F. Kraybill and M. B. Shimkin, *Advan. Cancer Res.*, **8**, 191 (1964).
104. G. N. Wogan (ed.), *Mycotoxins in Foodstuffs*, M. I. T. Press (1965).
105. R. D. Conveney, H. M. Peck and R. J. Townsend, *Microbiological Deterioration in the Tropics, S. C. I. Monograph*, no. 23, 31 (1966).
106. G. N. Wogan, *Bact. Rev.*, **30**, 460 (1966).
107. H. Bösenberg, *Naturwissenschaften*, **56**, 350 (1969).
108. E. Borker, N. F. Insalata, C. P. Levi and J. S. Witzeman, *Advan. Appl. Microbiol.*, **8**, 315 (1966).
109. T. Arai, *J. Fd. Hyg. Soc. Japan* (Japanese), **7**, 289 (1966).
110. I. Uritani, *J. Ass. Off. Anal. Chem.*, **50**, 105 (1967).
111. R. I. Mateles and G. N. Wogan (ed.), *Biochemistry of Some Food-borne Microbial Toxins*, M. I. T. Press (1967).
112. I. F. H. Purchase and J. J. Theron, *Bull. Int. Acad. Path.*, **8**, 3 (1967).
113. L. A. Goldblatt, *Economic Botany*, **22**, 51 (1968).
114. H. Kurata, *J. Fd. Hyg. Soc. Japan* (Japanese) **9**, 431 (1968).
115. D. E. Wright, *Ann. Rev. Microbiol.*, **22**, 269 (1968).
116. A. Ciegler and E. B. Lillehoj, *Adv. Appl. Microbiol.*, **10**, 155 (1968).
117. E. Ichijima, *Hakko Kyokaishi* (Japanese), **28**, 482 (1970).
118. A. Ciegler, S. Kadis and S. J. Ajl (ed.) *Microbial Toxins*, vol. VI, Academic Press (1971).
119. S. Kadis, A. Ciegler and S. J. Ajl (ed.) *ibid.*, vol. VII (1971).
120. I. F. H. Purchase (ed.), *Mycotoxins in Human Health*, Macmillan (1971).
121. G. N. Wogan and R. C. Shank, *Advan. Environmental Sci. Technol.*, **2**, 321 (1971).
122. M. Saito, *Trans. Soc. Path. Japan* (Japanese), **61**, 33 (1972).
123. M. Enomoto and M. Saito, *Ann. Rev. Microbiol.*, **26**, 279 (1972).
124. W. B. Buck and R. W. Oehme (ed.), *Clinical Toxicology*, **5**, 437 (1972).
125. P. Krogh (ed.), *Control of Mycotoxins*, Butterworths (1973).
126. I. F. H. Purchase, *Mycotoxins* (ed. I. F. H. Purchase), Elsevier (1974).
127. A. Ciegler, *Lloydia*, **38**, 21 (1975).
128. P. K. C. Austwick and G. Ayerst, *Chem. Ind.*, **55** (1) 12 (1963).
129. F. A. Hodges, J. R. Zust, H. R. Smith, A. A. Nelson, B. H. Armbrecht and A. D. Campbell, *Science*, **145**, 1439 (1964).
130. C. W. Hesseltine, O. L. Shotwell, M. L. Smith, J. J. Ellis, E. E. Vandegraft and M. L. Goulden, *Mycologia*, **60**, 304 (1968).
131. C. W. Hesseltine, O. L. Shotwell, M. L. Smith, J. J. Ellis, E. E. Vandegraft and G. Shannon, *Toxic Miroorganisms* (ed. M. Herzberg), p. 202, UJNR Joint Panels on Toxic Microorganisms and the U.S. Department of the Interior (1970).

132. C. W. Hesseltine, W. G. Sorenson and M. L. Smith, *Mycologia*, **62**, 123 (1970).
133. M. M. Kulik and C. E. Holaday, *Mycopath. Mycol. Appl.*, **30**, 137 (1966).
134. P. M. Scott, W. van Walbeek and J. Forgacs, *J. Appl. Microbiol.*, **15**, 945 (1967).
135. S. K. Mishra and H. S. R. Murthy, *Curr. Sci.* (*India*), **37**, 406 (1968).
136. H. W. Schroeder and M. J. Verrett, *Can. J. Microbiol.*, **15**, 895 (1969).
137. B. J. Wilson, T. C. Campbell, A. W. Hayes and R. T. Hanlin, *Appl. Microbiol.*, **16**, 819 (1968).
138. H. Kurata, F. Sakabe, S. Udagawa, M. Ichinoe, M. Tazawa, S. Natori and M. Saito, 20th Ann. Mtg. Fd. Hyg. Soc. Japan (1970).
139. C. W. Hesseltine, O. L. Shotwell, J. J. Ellis and R. D. Stubblefield, *Bact. Rev.*, **30**, 795 (1966).
140. K. Aibara and K. Miyaki, Ann. Mtg. Agr. Chem. Soc. Japan (1965).
141. H. Murakami, K. Owaki and S. Takase, *J. Gen. Appl. Microbiol.*, **12**, 195 (1966).
142. H. W. Schroeder and L. J. Ashworth, Jr., *Phytopath.*, **55**, 464 (1965).
143. H. W. Schroeder, *Appl. Microbiol.*, **14**, 381 (1966).
144. H. W. Schroeder and L. J. Ashworth, Jr., *J. Stored Prod. Res.*, **1**, 267 (1966).
145. O. L. Shotwell, C. M. Hesseltine, R. D. Stubblefield and W. G. Sorenson, *Appl. Microbiol.*, **14**, 425 (1966).
146. N. D. Davis, U. L. Diener and D. W. Eldridge, *ibid.*, **14**, 378 (1966).
147. N. D. Davis, U. L. Diener and V. P. Agnihotri, *Mycopath. Mycol. Appl.*, **31**, 251 (1966).
148. U. L. Diener and N. D. Davis, *Phytopath.*, **56**, 1390 (1966).
149. R. A. Boller and H. W. Schroeder, *Cereal Sci. Today*, **11**, 342 (1966).
150. N. D. Davis and U. L. Diener, *Appl. Microbiol.*, **15**, 1517 (1967).
151. J. D. Wildman, L. Stoloff and R. Jacobs, *Biotech. Bioeng.*, **9**, 429 (1967).
152. D. L. Calderwood and H. W. Schroeder, **ARS 52–26**, 9 Jan. (1968).
153. L. B. Bullerman, P. A. Hartman and J. C. Ayers, *Appl. Microbiol.*, **18**, 714 (1969).
154. L. B. Bullerman, P. A. Hartman and J. C. Ayers, *ibid.*, **18**, 718 (1969).
155. J. L. Richard and J. Cysewski, *Mycopath. Mycol. Appl.*, **44**, 221 (1971).
156. V. Nagarajan and R. V. Bhat, *J. Agr. Fd. Chem.*, **20**, 911 (1972).
157. V. Nagarajan, R. V. Bhat and P. G. Tulpule, *Experientia*, **29**, 1302 (1973).
158. C. N. Shih and E. H. Marth, *Appl. Microbiol.*, **27**, 452 (1974).
159. E. B. Lilliehoj, W. J. Garcia and M. Lambrow, *ibid.*, **28**, 763 (1974).
160. R. J. Bothast and C. W. Hesseltine, *ibid.*, **30**, 337 (1975).
161. J. R. Buchanan, N. F. Sommer and R. J. Fortlage, *ibid.*, **30**, 238 (1975).
162. Y. Hatsuda and S. Kuyama, *J. Agr. Chem. Soc. Japan* (Japanese), **28**, 989 (1954).
163. F. Bergel, A. L. Morrison, A. R. Moss and H. Rinderknecht, *J. Chem. Soc.*, 415, (1944).
164. E. Bullock, J. C. Roberts and J. G. Underwood, *ibid.*, 4179 (1962).
165. J. J. van der Watt, *Mycotoxins* (ed. I. F. H. Purchase), p. 370, Elsevier (1974).
166. J. S. E. Holker and S. A. Kagel, *Chem. Commun.*, 1574 (1968).
167. H. J. Burkhardt and J. Forgacs, *Tetrahedron*, **24**, 717 (1968).
168. M. Saito, M. Enomoto, T. Tatsuno and K. Uraguchi, *Microbial Toxins* (ed. A. Ciegler, S. Kadis and S. J. Ajl), vol. VI, p. 299, Academic Press (1971).
169. B. H. Howard and H. Raistrick, *Biochem. J.*, **44**, 227 (1949); **51** 56/75, 221(1954).
170. T. Tatsuno, M. Tsukioka, Y, Sakai, Y. Suzuki and Y. Asami, *Pharm. Bull.* **3**, 497 (1955).
171. M. Saigo, K. Ogura and H. Tsunoda, *J. Fd. Hyg. Soc. Japan* (Japanese), **4**, 383 (1957).
172. K. B. Raper, C. Thom and D. I. Fennel, *A Manual of the Penicillia*, p. 625, Williams & Wilkinson Co. (1949).
173. H. Raistrick and A. C. Hetherington, *Phil. Trans. Roy, Soc.*, **B220**, 269 (1931).
174. O. Tsuruta and S. Udagawa, *Food and Molds* (Japanese), p. 299, Ishiyaku Publishing Inc. (1975).
175. A. Ciegler, D. I. Fennel, G. A. Sansing, R. W. Detory and G. A. Bennett, *Appl. Microbiol.*, **26**, 271 (1973).

176. P. M. Scott, W. van Walbeek, B. Kennedy and D. Anyeti, *J. Agr. Food Chem.*, **20**, 1103 (1972).
177. J. Breen, J. C. Dacre, H. Raistrick and G. Smith, *Biochem. J.*, **37**, 726 (1943).
178. S. Shibata and S. Udagawa, *Chem. Pharm. Bull.* **11**, 402 (1963).
179. Y. Yamamoto, S. Hamaguchi, I. Yamamoto and S. Imai, *Yakugaku Zasshi* (Japanese), **76**, 1428 (1956).
180. I. Miyake, H. Naito and H. Tsunoda, *Beikokuriyo Kenkyusho Hokoku* (Japanese), **1**, 1 (1940).
181. S. Udagawa, *J. Agr. Sci., Tokyo Nogyo Daigaku*, **5**, 10 (1959).
182. Y. Ueno, *Mycotoxins* (ed. I. F. H. Purchase), p. 283, Elsevier (1974).
183. J. H. Birkinshaw, S. E. Michael, A. Bracken and H. Raistrick, *Lancet*, **245**, 625 (1943).
184. P. M. Scott, *Mycotoxins*, (ed. I. F. H. Purchase), p. 383, Elsevier (1974).
185. C. L. Alsberg and O. F. Black, *U. S. Dept. Agr. Bull. Bur. Pl. Ind.*, **270**, 7 (1913).
186. J. H. Birkinshaw and H. Raistrick, *Biochem. J.*, **26**, 441 (1932).
187. A. Ciegler, R. W. Detory and E. B. Lillehoj, *Microbial Toxins* (ed. A. Ciegler, S. Kadis and S. J. Ajl), vol VI, p. 417, Academic Press (1971).
188. C. P. Kurtzman and A. Ciegler, *Appl. Microbiol.*, **20**, 204 (1970).
189. A. Ciegler and C. P. Kurtzman, *ibid.*, **20**, 761 (1970).
190. E. B. Lillehoj, M. S. Milburn and A. Ciegler, *ibid.*, **24**, 198 (1972).
191. A. E. Oxford, H. Raistrick and P. Simonart, *Biochem. J.*, **33**, 240 (1939).
192. B. Wilson, *Microbial Toxins* (ed. A. Ciegler, S. Kadis and S. J. Ajl), vol. VI, p. 489, Academic Press (1971).
193. K. J. van der Merwe, P. S. Steyn and L. Fourie, *J. Chem. Soc.*, 7083 (1965).
194. P. S. Steyn and C. W. Holzapfel. *J. S. African Chem. Inst.*, **20**, 186 (1967).
195. J. Harwig, *Mycotoxins* (ed. I. F. H. Purchase), p. 348, Elsevier (1974).
196. C. W. Hesseltine, E. E. Vandegraft, D. I. Fennell, M. L. Smith and O. L. Shotwell, *Mycologia*, **64**, 539 (1972).
197. W. van Walbeek, P. M. Scott and F. S. Thatcher, *Can. J. Microbiol.*, **14**, 131 (1968).
198. W. van Walbeek, P. M. Scott, J. Harwig and J. W. Lawrence, *ibid.*, **15**, 1281 (1969).
199. F. E. Escher, P. E. Koehler and J. C. Ayres, *Appl. Microbiol.*, **26**, 27 (1973).
200. A. Ciegler, D. J. Fennell, H.-J. Mintzlaff and L. Leistner, *Naturwissenschaften*, **59**, 365 (1972).
201. O. L. Shotwell, C. W. Hesseltine and M. L. Goulden, *J. Ass. Off. Anal. Chem.*, **52**, 81, (1969).
202. O. L. Shotwell, C. W. Hesseltine, M. L. Goulden. *Appl. Microbiol.*, **17**, 765 (1969).
203. C. W. Hesseltine, *Mycopath. Mycol. Appl.*, **53**, 141 (1974).
204. R. J. Townsend, M. O. Moss and H. M. Peck, *J. Pharm. Pharmacol.*, **18**, 471 (1966).
205. A. W. Hayes and B. J. Wilson, *Appl. Microbiol.*, **16**, 1163 (1968).
206. M. O. Moss, A. B. Wood and F. V. Robinson, *Tetr. Lett.*, **367** (1969).
207. S. Natori, S. Sakaki, H. Kurata, S. Udagawa, M. Ichinoe, M. Saito, M. Umeda and K. Ohtsubo, *Appl. Microbiol.*, **19**, 613 (1970).
208. M. O. Moss and I. W. Hill., *Mycopath. Mycol. Appl.*, **40**, 81 (1970).
209. J. R. Bamburg and F. M. Strong, *Microbiol Toxins*, (ed. S. Kadis, A. Ciegler and S. J. Ajl), vol. VII, p. 207 Academic Press (1971).
210. E. B. Smalley and F. M. Strong, *Mycotoxins* (ed. I. F. H. Purchase), p. 199, Elsevier (1974).
211. Y. Ueno, N. Sato, K. Ishii, K. Sakai, H. Tsunoda and M. Enomoto, *Appl. Microbiol.*, **25**, 699 (1973).
212. H. W. Wollenweber and O. A. Reinking, *Die Fusarien—ihre Beschreibung, Schadwirkung, and Bekämpfung*, Verlagsbuchhandlung Paul Parey (1935).
213. W. C. Snyder and H. N. Hansen, *Am. J. Bot.* **27**, 64 (1940); **32**, 657 (1945).
214. W. L. Gordon, *Can. J. Res.* (C), **22**, 282 (1944).
215. T. A. Toussoun and P. E. Nelson, *A Pictorial Guide to the Identification of Fusarium Species*, Pennsylvania State Univ. Press (1968).

216. A. Z., Joffe, *Mycopath. Mycol. Appl.*, **53**, 201 (1974).
217. T. A. Toussoun and P. E. Nelson, *Ann. Rev. Phytopath.* 13, **71** (1975).
218. C. Booth, *The Genus Fusarium*, Commonwealth Mycological Inst., Kew (1971).
219. C. Booth, *Ann. Rev. Phytopath.*, **13**, 83 (1975).
220. S. G. Yates, H. L. Tookey and J. J. Ellis and H. J. Burkhardt, *Phytochem.* **139**, (1968).
221. S. G. Yates, H. L. Tookey, J. J. Ellis, W. H. Tallent and I. A. Wolff, *J. Agr. Fd. Chem.*, **17**, 437 (1969).
222. E. P. White, *J. Chem. Soc.* (C), 346, (1967).
223. P. W. Brian, A. W. Dawkins, J. F. Grove, H. G. Hemming, D. Lowe and G. L. F. Norris, *J. Exptl. Bot.*, **12**, 1 (1961).
224. D. R. Jacobson, W. M. Miller, D. M. Seath, S. G. Yates, H. L. Tookey and I. A. Wolff, *J. Dairy Sci.*, **46**, 416 (1963).
225. S. G. Yates, *Microbial Toxins* (ed. S. Kadis, A. Ciegler and S. J. Ajl) vol. VII, p. 191, Academic Press (1971).
226. S. G. Yates, H. L. Tookey and J. J. Ellis, *Appl. Microbiol.*, **19**, 103 (1970).
227. M. D. Grove, S. G. Yates, W. H. Tallent, J. J. Ellis, I. A. Wolff, N. R. Kosuri and R. E. Nichols, *J. Agr. Fd. Chem.*, **18**, 734 (1970).
228. J. R. Bamburg, F. M. Strong and E. B. Smalley, *ibid.*, **17**, 443 (1969).
229. A. C. Keyl, J. C. Lewis, J. J. Ellis, S. G. Yates and H. L. Tookey, *Mycopath. Mycol. Appl.*, **31**, 327 (1967).
230. C. M. Christensen, G. H. Nelson and C. J. Mirocha, *Appl. Microbiol.*, **13**, 653 (1965).
231. C. J. Mirocha, C. M. Christensen and G. H. Nelson, *ibid.*, **15**, 497 (1967).
232. M. Stob, R. S. Baldwin, J. Tuite, F. N. Andrews and K. G. Gillette, *Nature*, **196**, 1318 (1962).
233. C. J. Mirocha, C. M. Christensen and G. H. Nelson, *Cancer Res.*, **28**, 2319 (1968).
234. R. W. Caldwell and J. Tuite, *Appl. Microbiol.*, **20**, 31 (1970).
235. K. Ishii, M. Sawano, Y. Ueno and H. Tsunoda, *ibid.*, **27**, 625 (1974).
236. C. J. Mirocha and C. M. Christensen, *Mycotoxins* (ed. I. F. H. Purchase), p. 147, Elsevier (1974).
237. R. M. Eppley, L. Stoloff, M. Truckness and C. W. Chung, *J. Ass. Off. Anal. Chem.*, **57**, 632 (1974).
238. R. J. Cole, J. W. Kirksey, J. H. Moore, B. R. Blankership, U. L. Diener and N. D. Davis, *Appl. Microbiol.*, **24**, 248 (1972).
239. B. J. Wilson, C. H. Wilson and A. W. Hayes, *Nature*, **220**, 77 (1968).
240. A. Ciegler, *Appl. Microbiol.*, **18**, 128 (1969).
241. C. T. Hou, A. Ciegler and C. W. Hesseltine, *Annal. Biochem.*, **37**, 422 (1970).
242. C. T. Hou, A. Ciegler and C. W. Hesseltine *Appl. Microbiol.*, **21**, 1101 (1971).
243. C. T. Hou, A. Ciegler and C. W. Hesseltine, *Can. J. Microbiol.*, **17**, 599 (1971).
244. T. Yoshizawa, *personal communication.*
245. R. J. Cole and J. W. Kirksey, *Can. J. Microbiol.*, **20**, 1159 (1974).
246. P. M. Scott, M.-A. Merrien and J. Polonsky, *Experientia*, **32**, 140 (1976).
247. B. J. Wilson and C. H. Wilson, *Science*, **144**, 177 (1964).
248. M. Yamazaki, S. Suzuki and K. Miyaki, *Chem. Pharm. Bull.*, **19**, 1739 (1971).
249. M. Yamazaki, K. Sasago and K. Miyaki, *J. Chem. Soc. Chem. Comm.*, 408 (1974).
250. J. Clardy, J. P. Springer, G. Büchi, K. Matsuo and R. Wightman, *J. Am. Chem. Soc.*, **97**, 663 (1975).
251. A. L. Demain, N. A. Hunt, V. Malik, B. Kobbe, H. Hawkins, K. Matsuo and G. N. Wogan, *Appl. Microbiol.*, **31**, 138 (1976).
252. K. Uraguchi, *J. Stored Prod. Res.*, **5**, 227 (1969).
253. C. M. Christensen, *Bot. Rev.*, **23**, 108 (1957).
254. C. M. Christensen and H. H. Kaufman, *Grain Storage*, p. 17, Univ. of Minnesota Press (1969).
255. C. M. Christensen, *The Molds and Man*, p. 131, Univ. of Minnesota Press (1972).
256. G. Semeniuk and A. W. Alcock, *Storage of Cereal Grains and Their Products*, p. 77, Am. Ass. Cereal Chem., St. Paul, Minnesota (1954).

257. C. Golumbic and M. M. Kulik, *Aflatoxin* (ed. L. A. Goldblatt), p. 307, Academic Press (1969).
258. A. E. Muskett, *Trans. Brit. Mycol. Soc.*, **30**, 74 (1948).
259. C. M. Christensen, *Cereal Chem.*, **23**, 322 (1946).
260. Y. Tanaka, S. Hirayama, H. Kurata, F. Sakabe, N. Inagaki, T. Matsushima and S. Udagawa, *Bull. Nattl. Hyg. Lab.* (Japanese), **75**, 443 (1957).
261. N. Inagaki and M. Ikeda, *ibid.*, **77**, 341 (1959).
262. O. Tsuruta, *Shokuryo Hokan Sosho* (Japanese), **17**, 1 (1960).
263. H. Iizuka, *Gen. Appl. Microbiol.*, **3** (2), 146 (1957).
264. H. Kurata, K. Ogasawara and V. L. Frampton, *Cereal Chem.*, **34**, 47 (1957).
265. F. A. del Prado and C. M. Christensen, *ibid.*, **29**, 456 (1952).
266. C. W. Hesseltine and R. R. Graves, 2nd Natl. Conf. Wheat Utilization Res. Peorea, Illnois (1963).
267. H. Kurata, F. Sakabe, S. Udagawa, M. Ichinoe, M. Suzuki and N. Takahashi, *Bull. Natl. Inst. Hyg. Sci.* (Japanese), **86**, 183 (1968).
268. H. Kurata and M. Ichinoe, *J. Fd. Hyg. Sci. Japan* (Japanese), **8**, 237 (1967).
269. F. S. Thatcher and D. S. Clark (ed.), *Microorganisms in Foods*, Univ. of Toronto Press (1974).
270. C. M. Hesseltine and R. R. Graves, *Economic Bot.*, **20**, 156 (1966).
271. N. Inagaki, *Japan. J. Pub. Health.* (Japanese), **12**, 1123 (1960).
272. H. Kurata, S. Udagawa, M. Ichinoe, Y. Kawasaki, M. Takada, M. Tazawa, A. Koizumi and H. Tanabe, *J. Fd. Hyg. Sci. Japan*, **9**, 23 (1968).
273. S. Udagawa, M. Ichinoe and H. Kurata, *Toxic Microorganisms* (ed. M. Herzberg), p. 174, UJNR Joint Panels on Toxic Microorganisms and the U. S. Department of the Interior (1970).
274. K. Miyaki, M. Yamazaki, Y. Kawasaki and H. Kurata, *Ann. Rept. Inst. Fd. Microbiol.* (Japanese), **21**, 133 (1968).
275. K. Miyaki, M. Yamazaki, Y. Horie and S. Udagawa, *ibid.*, **22**, 41 (1969).
276. R. C. Shank, G. N. Wogan and J. B. Gibson, *Fd. Cosmet. Toxicol.*, **10**, 51 (1972).
277. M. Saito, *Reports on Study of Mycotoxins in Foods in Relation to Liver Disease in Malaysia and Thailand*, Inst. Med. Sci., Univ. of Tokyo (1976).
278. G. A. Gilman, *Mycotoxins in Human Health* (ed. I. F. H. Purchase) p. 133, Macmillan (1971).
279. P. M. D. Martin, G. A. Gilman and P. Keen, *ibid.*, p. 281, Macmillan (1971).
280. A. Yanai, *Rept. Fd. Res. Inst.* (Japanese) **21**, 6 (1966).
281. H. W. Schroeder, *Phytopath.*, **53**, 843 (1963).
282. H. Kurata, *Shokunokagaku* (Japanese), **6**, 99 (1972).
283. H. W. Schroeder and J. W. Sorenson, Jr., *Rice, J.* July, 6 (1961).
284. H. A. Fanse and C. M. Christensen, *Phytopath.*, **60**, 228 (1970).
285. C. M. Christensen and L. C. Løpez, *ibid.*, **55**, 953 (1965).
286. C. M. Christensen, *ibid.*, **59**, 145 (1969).
287. E. Hernández, R. Vila and M. Hervás, *Rev. Agr. Tech. Alimentos*, **8**, 240, 510 (1968)
288. D. L. Calderwood and G. W. Schroeder, *ARS. 52–26*, 3, USDA (January 1968).
289. H. Kurata, F. Sakabe, S. Tanaka and K. Aibara, *Rept. Tottori Mycol. Inst.* (Japanese), **10**, 647 (1973).
290. C. M. Christensen and H. H. Kaufman, *Ann. Rev. Phyotopath.*, **3**, 69 (1965).
291. H. Tsunoda and O. Tsuruta, *Rept. Fd. Res. Inst.* (Japanese), **14**, 38 (1959).
292. M. Ichinoe, K. Takatori and H. Kurata, *Rept. Tottori Mycol. Inst.* (Japanese), **10**, 627 (1973).
293. O. Tsuruta, *Rept. Fd. Res. Inst.*, **29**, 16 (1974).
294. M. Ichinoe, K. Takatori, S. Tanaka, H. Kumata, T. Suzuki and H. Kurata, *J. Fd. Hyg. Soc. Japan* (Japanese), **16**, 381 (1975).
295. H. A. H. Wallace and R. N. Sinha, *Mycopath. Mycol. Appl.*, **57**, 171 (1975).
296. de Tempe, *Proc. Int. Seed Test. Ass.*, **27**, 819 (1962).
297. N. James, J. Wilson and E. Stark, *Can. J. Res.*, **24**, 224 (1946).
298. A. Z. Joffe, *Bull. Res. Counc. Israel*, **9D**, 101 (1959).
299. B. Flannigan, *Trans. Brit. Mycol. Soc.*, **53**, 371 (1969).

300. S. K. Mulinge and A. E. Apinis, *ibid.*, 361 (1969).
301. S. K. Mulinge and C. G. C. Chesters, *Ann. Appl. Biol.*, **65**, 285 (1970).
302. C. Vojnovich, V. F. Pfeifer and E. L. Griffin, Jr., *Cereal Sci. Today*, **11**, 16 (1966).
303. C. Vojnovich and V. F. Pfeifer, *The Northwesten Miller*, **273** (7), 12 (1966).
304. C. Vojnovich, V. F. Pfeifer and E. L. Griffin, Jr., *Cereal Sci. Today*, **12**, 54 (1967).
305. H. Turley, *Baking Technol.*, **1**, 327 (1922).
306. C. B. Gustafson and E. H. Parfitt, *Cereal Chem.*, **10**, 233 (1933).
307. D. B. Holtman, *J. Bact.*, **30**, 359 (1935).
308. C. S. Boruff, R. I. Claassen and A. L. Sotier, *Cereal Chem.*, **15**, 451 (1938).
309. H. W. Clifford, *Food*, **8**, 233 (1939).
310. C. M. Christensen, *Baker's Dig.*, **21**, 21 (1947).
311. C. M. Christensen and M. Cohen, *Cereal Chem.*, **27** 178 (1957).
312. E. C. Barton-Wright, *ibid.*, **15**, 521 (1938).
313. N. Inagaki and M. Ikeda, *Bull. Natl. Hyg. Lab.* (Japanese), **77**, 347 (1959).
314. F. Sakabe and H. Kurata, *Rept. Tottori Mycol. Inst.* (Japanese), **10**, 655 (1973).
315. K. Ogasawara, I. Sekijo, H. Sunagawa, M. Umemura and K. Mori, *Rept. Hokkaido Inst. Pub. Health* (Japanese), **25**, 67 (1975).
316. R. R. Graves and C. W. Hesseltine, *Mycopath. Mycol. Appl.*, **29**, 277 (1966).
317. C. W. Hesseltine, R. R. Graves, R. Rogers and H. R. Burmeister, *Appl. Microbiol.*, **18**, 848 (1969).
318. H. Hitokoto, S. Morozumi, T. Wauke, H. Zen-Yoji, H. Kurata and M. Ichinoe, *J. Fd. Hyg. Soc.* (Japanese), **14**, 364 (1973).
319. R. W. Lichtwardt and L. H. Tiffany, *Iowa State Coll. J. Sci.*, **33**, 1 (1958).
320. R. W. Lichtwardt and G. L. Barron, *ibid.*, **34**, 139 (1959).
321. J. Tuite, *Pl. Dis. Reptr.*, **45**, 212 (1961).
322. R. J. Bothast, G. H. Adams, E. E. Hatfield and E. B. Lancaster, *J. Dairy Sci.*,**58**, 386 (1975).
323. B. Koehler, *J. Agr. Res.*, **56**, 291 (1938).
324. P. B. Mislivec and J. Tuite, *Mycologia*, **62**, 67 (1970).
325. W. F. O. Marasas and E. B. Smalley, *Onderstepoort J. Vet. Res.*, **39**, 1 (1972).
326. B. L. Doupnik, *Phytopath*, **62**, 1367 (1972).
327. O. Tsuruta, T. Sugimoto, M. Minamisawa and M. Manabe, *Trans. Mycol. Soc. Japan* (Japanese), **15**, 258 (1974).
328. R. J. Bothast, R. F. Rogers and C. W. Hesseltine, *Cereal Sci. Today*, **18**, 22 (1973).
329. L. Stoloff, P. Mislivec and M. M. Kulik, *Appl. Microbiol.*, **29**, 123 (1975).
330. P. B. Mislivec and J. Tuite, *Mycologia*, **62**, 75 (1970).
331. O. L. Shotwell, C. W. Hesseltine and M. L. Goulden, *Cereal Sci. Today*, **18**, 192 (1973).
332. G. Swarup, E. D. Hausing and C. T. Rogerson, *Trans. Kansas Acad. Sci.*, **65**, 120 (1962).
333. C. M. Christensen, *Phytopath.*, **60**, 280 (1970).
334. C. M. Christensen, *Mycopath. Mycol. Appl.*, **44**, 277 (1971).
335. R. Burroughs and D. B. Sauer, *Phytopath.*, **61**, 767 (1971).
336. P. B. Marsh, M. E. Simpson, R. J. Ferretti, T. C. Campbell and J. Donoso, *J. Agr. Fd. Chem.*, **17**, 462 (1968).
337. P. B. March, M. E. Simpson, G. O. Craig, J. Donoso, and H. H. Ramey, Jr., *J. Environ. Qual.*, **2**, 276 (1973).
338. M. E. Simpson, P. B. Marsh, G. V. Merola, R. J. Ferretti and E. C. Filsinger, *Appl. Microbiol.*, **26**, 608 (1973).
339. D. E. Gardner, J. L. McMeans, C. M. Brown, R. M. Bilbrey and L. L. Parker, *Phytopath.*, **64**, 452 (1974).
340. C. M. Christensen, J. H. Olafson and W. F. Geddes, *Cereal Chem.*, **26**, 119 (1949).
341. R. A. Kilpatrick, *Phytopath.*, **47**, 131 (1957).
342. H. Kurata, *Bull. Natl. Inst. Agr. Sci.* (C) (Japanese), **12**, 1 (1960).
343. M. Milner and W. F. Geddes, *Cereal Chem.*, **23**, 225 (1946).

344. B. W. Kennedy, *Phytopath.*, **54**, 771 (1964).
345. C. M. Christensen and C. E. Dorworth, *ibid.*, **56**, 412 (1966).
346. C. M. Christensen, *ibid.*, **57**, 622 (1967).
347. C. E. Dorworth and C. M. Christensen, *ibid.*, **58**, 1457 (1968).
348. U. L. Diener, *ibid.*, **50**, 220 (1960).
349. K. H. Garren, *Phytopath. Z.*, **55**, 359 (1966).
350. A. Z. Joffe and S. Y. Borut, *Mycologia*, **58**, 629 (1966).
351. A. Z. Joffe, *Mycopath. Mycol. Appl.*, **37**, 150 (1969).
352. A. Z. Joffe, *ibid.*, **38**, 255 (1970).
353. C. McDonald, *Samaru Res. Bull.* (*Nigeria*), **114**, 465 (1970).
354. K. H. Garren and D. M. Porter, *Phytopath.*, **60**, 1635 (1970).
355. K. H. Porter and K. H. Garren, *App. Microbiol.*, 20, 133 (1970).
356. K. H. Porter and K. H. Garren, *Trop. Sci.*, **10**, 100 (1968).
357. G. L. Barnes, *Mycopath. Mycol. Appl.*, **45**, 85 (1971).
358. D. M. Porter and F. Scott-Wright, *Phytopath.*, **61**, 1194 (1971).
359. R. T. Hanlin, *Mycopath. Mycol. Appl.*, **38**, 93 (1969).
360. R. T. Hanlin, *ibid.*, **40**, 341 (1970).
361. R. T. Hanlin and W. L. Corley, *Trop. Sci.*, **13**, 147 (1971).
362. R. Allcroft and R. B. A. Carnaghan, *Chem. Ind.*, January 12, 50 (1963).
363. T. C. Tung and K. H. Ling, *J. Vitamin.*, **14**, 48 (1968).
364. M. Manabe, O. Tsuruta, T. Sugimoto, M. Minamisawa and S. Matsuura, *J. Fd. Hyg. Soc. Japan* (Japanese), **12**, 364 (1971).
365. H. W. Schroeder and L. J. Ashworth, Jr., *Phytopath.*, **55**, 464 (1965).
366. C. R. Jackson (1967) *ibid.*, **57**, 1270 (1967).
367. G. J. Griffin, *ibid.*, **62**, 1387 (1972).
368. G. J. Griffin and K. H. Garren, *ibid.*, **64**, 323 (1974).
369. C. M. Christensen, H. A. Fanse, G. H. Nelson, Fern Bates and C. J. Mirocha, *Appl. Microbiol.*, **15**, 622 (1967).
370. Y. Horie, M. Yamazaki, K. Miyaki and S. Udagawa, *J. Fd. Hyg. Soc. Japan* (Japanese), **12**, 516 (1971).
371. C. W. Hesseltine, *Mycologia*, **57**, 149 (1965).
372. M. Yamazaki, Y. Horie, H. Fujimoto, S. Suzuki, Y. Sakakibara and K, Miyaki, *J. Fd. Hyg. Soc. Japan* (Japanese), **11** (1970).
373. H. Kurata, S. Udagawa, M. Ichinoe, S. Natori and S. Sakaki, *Mycotoxins in Human Health* (ed. I. F. H. Purchase), p. 101, Macmillan Press (1971).
374. T. Yokozuka, Y. Asao, M. Sasaki and K. Oshita, *Toxic Microorganisms* (ed. M. Herzberg) p. 133, UJNR Joint Panels on Toxic Microorganisms and the U.S. Department of the Interior (1970).
375. M. Sasaki, K. Oshita and T. Yokotsuka, *Nippon Nogeikagaku Kaishi* (Japanese), **49**, 57 (1975).
376. H. Murakami and M. Suzuki, *Toxic Microorganisms* (ed. M. Herzberg) p. 198, UJNR Joint Panels on Toxic Microorganisms and the U.S. Department of the Interior (1970).
377. J. Lew, S.-P. Kwon, C.-W. Koh and Y. Chung, *Bull. Yonsei Univ.*, **7**, 191 (1969).
378. R. Kinoshita, T. Ishiko, S. Sugiyama, T. Seto, S. Igarashi and I. E. Goetz, *Cancer Res.*, **28**, 2296 (1968).
379. M. Yamazaki, Y. Horie, S. Udagawa, T. Echigo and M. Kimi, *J. Fd. Hyg. Soc. Japan* (Japanese), **16**, 1 (1975).
380. W. C. Frazier, *Food Microbiology*, p. 248, McGraw-Hill (1958).
381. N. Inagaki and Y. Takahashi, *Bull. Natl. Hyg. Lab.* (Japanese), **79**, 293 (1961).
382. N. Inagaki and Y. Takahashi, *ibid.*, **79**, 297 (1961).
383. J. C. Ayres, D. A. Lillard and L. Leistner, *Reciprocal Meat Conf. Proc.*, **20**, 156 (1967).
384. H. Hitokoto, S. Morozumi, T. Wauke and H. Zen-Yoji, *Ann. Rept. Tokyo Metr. Res. Lab. Pub. Health* (Japanese), **24**, 41 (1972).
385. K. Takatori, K. Takahashi, T. Suzuki, S. Udagawa and H. Kurata, *J. Fd. Hyg. Soc. Japan*, **16**, 307 (1975).

386. L. Leistner and F. Tauchmann, *Fleischwirtschaft*, **50**, 965 (1970).
387. Y. Sasaki, *J. Ferment. Technol.* (Japanese), **20**, 383, 505, 694 (1942); **21**, 96 (1943).
388. Y. Sasaki, *J. Appl. Microbiol.* (Japanese), **1**, 53 (1946).
389. I. Niiya, Y. Suzuki and M. Imamura, *J. Fd. Hyg. Soc. Japan* (Japanese), **10**, 381 (1969).
390. H. Hitokoto, S. Morozumi, T. Wauke, S. Sakai and H. Zen-Yoji, *Ann. Rept. Tokyo Metr. Res. Lab.* (Japanese), **25**, 17 (1974).
391. A. P. Hansen and R. E. Welty, *Mycopath. Mycol. Appl.*, **44**, 309 (1970).
392. B. Doupnik, Jr. and D. K. Bell. *Appl. Microbiol.*, **21**, 1104 (1971).
393. J. M. Wells and J. A. Payne, *ibid.*, **30**, 536 (1975).
394. T. C. Campbell, J. P. Caedo, Jr., J. Bulatao-Jayme, L. Salamat and R. W. Engel, *Nature*, **227**, 403 (1970).
395. R. C. Shank, G. N. Wogan, J. B. Gibson and A. Nondasuta, *Fd. Cosmet. Toxicol.*, **10**, 61 (1972).
396. R. C. Shank, J. E. Gordon, G. N. Wogan, A. Nondasuta and B. Subhamani, *ibid.*, **10**, 71 (1972).
397. R. C. Shank, N. Bhamarappavati, J. E. Gordon and G. N. Wogan, *ibid.*, **10**, 171 (1972).
398. R. C. Shank, S. Siddhichai, B. Subhamani, N. Bhamarappavati, J. E. Gordon and G. N. Wogan, *ibid.*, **10**, 181 (1972).
399. R. C. Shank, C. H. Bourgeois, N. Keschamras and P. Chandavimol, *ibid.*, **9**, 501 (1971).
400. I. F. H. Purchase and T. Gonçalves, *Mycotoxins in Human Health* (ed. I. F. H. Purchase), p. 263, Macmillan (1971).
401. P. Keen and P. Martin, *Trop. Geog. Med.*, **23**, 44 (1971).
402. F. G. Peers and C. A. Linsell, *Brit. J. Cancer*, **27**, 473 (1973).
403. M. E. Alpert, M. S. R. Hutt, G. N. Wogan and C. S. Davidson, *Cancer*, **28** (7), 253 (1971).
404. M. Saito, M. Enomoto, M. Umeda, K. Ohtsubo, T. Ishiko, S. Yamamoto and H. Toyokawa, *Mycotoxins in Human Health* (ed. I. F. H. Purchase), p. 179, Macmillan (1971).
405. M. Okudaira, *Japan. J. Med. Mycol.* (Japanese), **13**, 187 (1972).
406. Ben Doupnik, Jr. and D. K. Bell., *Appl. Microbiol.*, **21**, 1104 (1971).
407. C. E. Main and P. B. Hamilton, *ibid.*, **23**, 193 (1972).
408. H. Itakura, T. Ishiko, T. Mizunuma, Y. Kawasaki, J. Fujimoto and R. Kinoshita, *Trop. Med.*, **16**, 45 (1974).
409. J. W. Kirksey and R. J. Cole, *Mycopath. Mycol. Appl.*, **54**, 291 (1974).
410. E. Hassen and M. Jung, IUPAC Symp. Control of Mycotoxins, Göteborg, Sweden, p. 239 (1972).
411. O. L. Shotwell, C. W. Hesseltine, E. E. Vandegraft and M. L. Goulden, *Cereal Sci. Today*, **16**, 266 (1971).
412. O. L. Shotwell, M. L. Goulden and C. M. Hesseltine, *J. Ass. Off. Anal. Chem.*, **51**, 492 (1974).
413. M. Uchiyama, E. Isohata and Y. Takeda, *J. Fd. Hyg. Soc. Japan*, **17**, 103 (1976).
414. T. Yoshizawa and M. Morooka, *Agr. Biol. Chem.*, **37**, 2933 (1973).
415. I.-C. Hsu, E. B. Smalley, F. M. Strong and W. E. Ribelin, *Appl. Microbiol.*, **24**, 684 (1972).
416. P. W. Brian, C. W. Elson and D. Lowe, *Nature*, **178**, 263 (1956).
417. P. M. Scott, W. F. Miles, P. Toft and F. G. Dube, *J. Agr. Fd. Chem.*, **20**, 450 (1972).
418. J. E. Schade, K. McGreevy, A. D. King, Jr., B. Mackey, and G. Fuller, *Appl. Microbiol.*, **29**, 48 (1975).
419. C. J. Mirocha, J. Harrison, A. A. Nichols and M. McClintock, *ibid.*, **16**, 797 (1968).
420. O. L. Brekke, A. J. Pelinski, G. E. N. Nelson and E. L. Griffin, *Cereal Chem.*, **52**, 205 (1975).
421. C. W. Hesseltine, IUPAC Symp. Control of Mycotoxins, Göteborg, Sweden, p. 251 (1972).

CHAPTER 2

Chemistry of Mycotoxins

Mikio YAMAZAKI
Research Institute for Chemobiodynamics, Narashino-shi,
Chiba-ken 275, Japan

2.1. CHEMICAL CHARACTERISTICS OF MYCOTOXINS

Since fungi cannot fix CO_2 as green plants do, most fungi must absorb carbon compounds together with other nutritive substances from the environment to maintain growth.[1] They are heterotrophic, like most microorganisms and animals. Most carbon compounds utilized by fungi (carbohydrates are the most effective) are in principle metabolized ultimately to CO_2 and H_2O through various metabolic pathways; some of the most important are the Embden-Meyerhof-Parnas pathway for glycolysis, the pentose phosphate cycle for pentose metabolism and the Krebs cycle for tricarboxylic acid metabolism. Many kinds of metabolites are formed during the synthesis of essential chemical components and during the processes involved in the release of chemical energy to drive the synthetic reactions. Such reactions are essential in living organisms and are collectively called "primary metabolism" for convenience. The primary metabolism in fungi is very similar in principle to that in plants, animals and other organisms. Part of the carbon compounds taken into such organisms is not completely converted to CO_2 and H_2O, but converted to various intermediary metabolites during the metabolic processes. Certain sugars, organic acids, amino acids, aromatic compounds, etc. are regarded as common intermediary metabolites, and from these the so-called "secondary metabolites" are formed. Such compounds are often referred to as "natural products".[2] Unlike the primary metabolites, which are common to most organisms, the secondary metabolites are characteristic of particular organisms. For instance, the characteristic color and order of living organisms are often due to the properties of secondary metabolites. Although the significance of the secondary metabolites in organisms is obscure, the secondary metabolites of fungi and also of plants often exhibit characteristic physiological activity or toxicity toward other organisms.

Mycotoxins, fungal metabolites which exhibit toxic effects toward domestic animals, fowls and sometimes also human beings, are products of fungal secondary metabolism. Human beings seldom ingest large amounts of mycotoxins, since contaminated food becomes colored by the fungi and smells musty. Therefore, there are relatively few cases of acute poisoning in human beings, and chronic injury due to the ingestion of small amounts of toxins over a long period is usually of more significance than acute intoxication. On the other hand, there has been a comparatively large number of cases of acute intoxication in domestic animals or fowls due to the ingestion of feed infested by toxin-producing fungi. In some cases, causative mycotoxins have been detected and isolated but in other cases the causative agents of mycotoxicosis are still unknown.[3]

Ergotism occurred very widely in Europe in the Middle Ages. This was regarded as an epidemic in human beings, and was caused by the ingestion of rye bread contaminated with ergot toxins. Ergotism was greatly feared at that time, being known as "Holy Fire" burning the limbs of the people. The limbs of patients often became blackened and were lost in the late phase of heavy gangrenous ergotism. Although the date of the first occurrence of ergotism is uncertain, this may be the longest known mycotoxicosis of human beings. Ergot is a kind of plant disease usually caused by parasitism of the fungus *Claviceps purpurea* on rye. The active principle of ergot toxicity is now known to be alkaloids and they are known to cause heavy gangrenous and convulsive disease in animals after administration of a single large dose. Many kinds of ergot alkaloids have been detected and isolated and their physiological activities investigated. Accordingly, it is now known that the outbreak of ergotism is due to the effects of the ergot alkaloids.

Upon alkaline hydrolysis of ergot alkaloids, lysergic acid is obtained. Ergot alkaloids such as ergotamine, ergocristine, ergosine, ergokryptine or ergometrine (ergobasine) are peptides of lysergic acid (Table 2.1). The amino acids commonly found in ergot alkaloids are L-valine, L-leucine or L-phenylalanine. On the other hand, there are some biologically inactive peptides linking to isolysergic acid, a stereoisomer of lysergic acid in the configuration of the carboxyl group at C-8. These are ergotaminine, ergocristinine, ergosinine, ergokryptinine or ergobasinine, corresponding to ergotamine, ergocristine, ergosine, ergokryptine or ergobasine, respectively. The use of ergot in medicine by the Ancients in obstetrics is known. The ergot alkaloids are still used as ecbolic oxytocics at present.

Some important mycotoxicoses other than ergotism are listed in

TABLE 2.1. Structure of ergot alkaloids

Basic Structures X =	H COX N–CH_3 N H	XOC H N–CH_3 N H
OH	Lysergic acid	Isolysergic acid
R^1 R^2 –NH–CH H O C N N O O H R^3	Ergotamine $R^1 = R^2 = H$,	Ergotaminine $R^3 = -CH_2-C_6H_5$
	Ergocristine $R^1 = R^2 = CH_3$,	Ergocristinine $R^3 = -CH_2-C_6H_5$
	Ergosine $R^1 = R^2 = H$,	Ergosinine $R^3 = -CH_2-CH(CH_3)_2$
	α-Ergokryptine $R^1 = R^2 = CH_3$,	α-Ergokryptinine $R^3 = -CH_2-CH(CH_3)_2$
	Ergostine $R^1 = H$, $R^2 = CH_3$,	Ergostinine $R^3 = -CH_2-C_6H_5$
$-NH-CH(CH_3)(CH_2OH)$	Ergometrine (ergobasine)	Ergobasinine

Table 2.2. Principal mycotoxins recognized as causative for these diseases and the toxigenic or toxin-producing fungi involved are also shown.[3)]

As mentioned previously, most records of mycotoxicosis relate to outbreaks in domestic animals or fowls. In the case of alimentary toxic aleukia or stachybotryotoxicosis, however, outbreaks of heavy intoxication in human beings have been reported, although neither causative toxigenic fungi nor mycotoxins have been identified.

Most of the toxins shown in Table 2.2 are products of *Penicillium*, *Aspergillus* and *Fusarium*. *Penicillium* and *Aspergillus* fungi are generally known as "storage fungi" and tend to grow in stored cereals, legumes and other agricultural products. *Fusarium* fungi on the other hand are parasitic to higher plants and usually grow on living plants in the fields; they are designated as "field fungi". The possibility of contamination of foods by "storage fungi" is undoubtedly high, and increases with storage time. However, as in the case of ergotism, invasion of agricultural products by parasitic fungi in the fields before harvest is also possible. In this case, the food products may be intrinsically contaminated with toxins if the fungus is toxigenic. In any case, the mycotoxins causing various diseases are

TABLE 2.2. Major mycotoxicoses, toxins and fungi implicated in the diseases

Disease	Toxin	Fungi
Alimentary toxic aleukia	unknown	*Fusarium sporotrichioides*, *F. poae*
Stachybotryotoxicosis	unknown	*Stachybotrys atra*
Photosensitivity disease	sporidesmin	*Pithomyces chartarum*
Aflatoxicosis (Turkey X disease)	aflatoxins	*Aspergillus flavus* *A. parasiticus*
Fusarium toxicosis	nivalenol fusarenon-X T-2 toxin HT-2 toxin	*Fusarium nivale* *F.* spp.
Vulvovaginitis	zearalenone	*Fusarium graminearum* *F. roseum*
Patulin toxicosis	patulin	*Penicillium urticae*
Moldy malt sprout toxicosis	maltoryzine	*Aspergillus oryzae* var. *microsporus*
Hyperkeratosis (Cow X disease)	unknown	*Aspergillus chevalieri* *A. clavatus*
Ptyalism	slaframine	*Rhizoctonia leguminicola*
Yellowed rice toxicosis	cictreoviridin luteoskyrin cyclochlorotine	*Penicillium citreo-viride* *P. islandicum*
	citrinin	*P. citrinum*

characteristically different in their chemical properties and physiological activities.

2.2. Mycotoxins Derived from Amino Acids

Some important mycotoxins containing nitrogen are known to be produced from amino acids. In the pathways of amino acid metabolism in higher plants, decarboxylation of amino acids commonly occurs to form corresponding amines. These amines are regarded as "proto-alkaloids" and they act as precursors for alkaloid biosynthesis. However, decarboxylation is less important in fungi and various amides or peptides are produced in fungi in place of the amines or alkaloids formed in higher plants.

Sporidesmin from *Pithomyces chartarum* is one such nitrogenous metabolite which also contains sulfur in its structure. This unfamiliar polythia-2,5-dioxopiperazine compound was first isolated as a causative principle of facial eczema on sheep in New Zealand.[4] It causes jaundice in sheep on administration of a dose of 1 mg/kg and exhibits similar toxicity in the guinea-pig, rat and rabbit. From the same fungus, sporidesmins B to H have been isolated to date; these sporidesmins are structurally closely related, as shown in Table 2.3.[5] The yields of sporidesmins are very low in laboratory cultures. In general, an isolate which is highly sporulate is more likely to produce sporidesmins than one which does not produce spores. Obviously all the sporidesmins are basically derived from two amino acids, tryptophan and alanine, forming a cyclodipeptide. It has been shown that tryptophan and the methyl group of ^{14}C-methionine are incorporated into sporidesmin in cultures of *P. chartarum.*[6]

Dithia-2,5-dioxopiperazine, which is similar to sporidesmin, is comparatively common as a fungal metabolite. Gliotoxin, an antibiotic isolated from *Myrothecium verrucaria* (*Gliocladium fimbriatum*), also contains a thiadioxopiperazine moiety.[7] It was shown that phenylalanine and serine and also a C_1 unit from methionine provide the carbon skeleton of gliotoxin (Fig. 2.1).[8] The unusual amino-alcohol system in gliotoxin could arise from the interaction of an epoxide with the phenylalanine-derived nitrogen atom, but not by the direct hydroxylation of phenylalanine. In fact, phenylalanine was incorporated into gliotoxin with very high efficiency but *o*- or *m*-tyrosine or 2,3-DOPA was not.[9]

It can be speculated that the biosynthesis of sporidesmin occurs in a similar way to gliotoxin biosynthesis. The formation of the pyrroloindoline ring system in sporidesmin may be due to the interaction of an epoxy moiety at the 2,3-position of the indole ring of tryptophan with the

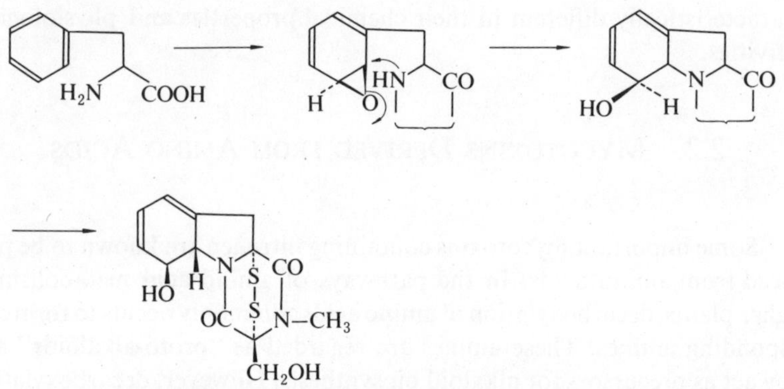

H_2N COOH → HN CO, H, O → N CO, HO, H → N S CO, HO H, S, OC S N–CH_3, CH_2OH

Gliotoxin

Fig. 2.1. Formation of gliotoxin.

amino nitrogen in the same molecule. It was recently shown[10] that 3α-hydroxypyrroloindoline can be obtained from dimethyltryptamine by photooxidation with pyridine-*N*-oxide in the bioorganic model reaction of monooxygenase (Fig. 2.2), and this suggests that tryptophan can be converted to hydroxypyrroloindoline *in vivo* by oxidation to the 2,3-epoxide catalyzed by monooxygenase.

A gliotoxin-producing strain of *Aspergillus fumigatus* does not produce gliotoxin when grown on media deficient in sulfur. The incorporation of $^{35}SO_3^{2-}$ into sporidesmin was demonstrated.[11] However, it is still not clear at which stage sulfur is introduced into sporidesmin and how the disulfide bridge is formed. The fact that the formation of a disulfide bridge in desthiobiotin accomplishes the biosynthesis of biotin suggests that the disulfide bridges in sporidesmin and gliotoxin might arise by the addition

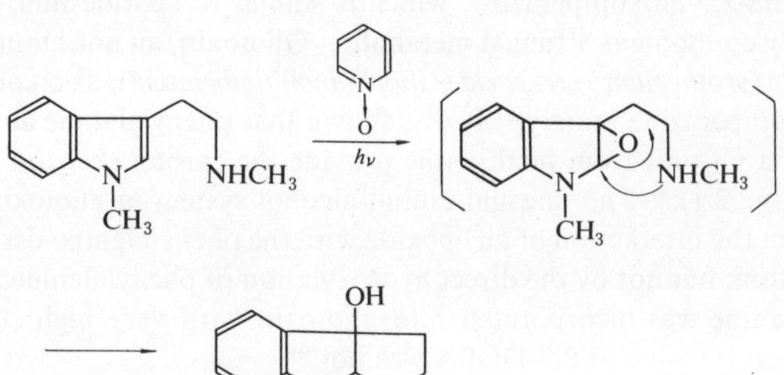

N, CH_3, $NHCH_3$ —(N–O, hν)→ [N, CH_3, O, $NHCH_3$] → OH, N, CH_3 CH_3

Fig. 2.2. Formation of 3α-hydroxypyrroloindoline from dimethyltryptamine.

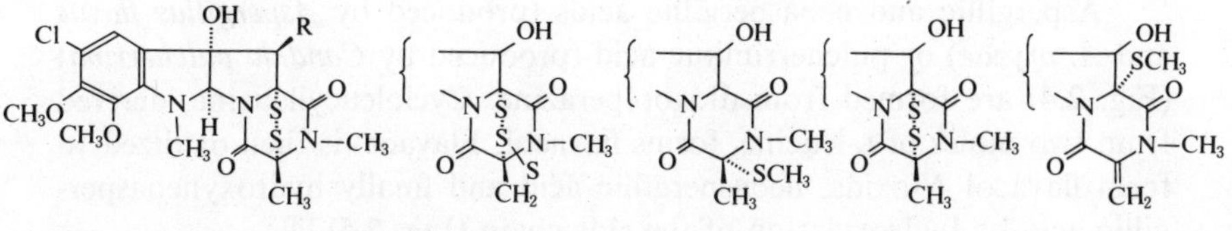

Sporidesmin (R=OH)
Sporidesmin B (R=H)

C D E F

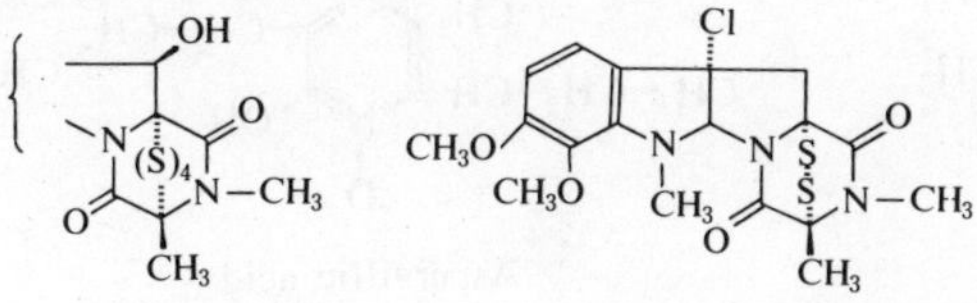

G H

TABLE 2.3. Structure and physical properties of sporidesmins

	mp(°C)	$[\alpha]_D$	UV $\lambda_{max}(\varepsilon)$	Molecular formula
Sporidesmin	179	−185°	218.5(4.60), 254(4.12), 302(3.45)	$C_{18}H_{20}N_3O_6S_2Cl$
Sporidesmin B	183	−78°	218 (4.50), 256(4.08), 307(3.41)	$C_{18}H_{20}N_3O_5S_2Cl$
Sporidesmin C	230–40 (diacetate)	−215° (diacetate)	222(4.35), 255(3.99), 309(3.46)	$C_{22}H_{24}N_3O_8S_3Cl$ (diacetate)
Sporidesmin D	110–20 (etherate)	+58° (ethanolate)	216 (4.45), 252(4.00), 300(3.28)	$C_{20}H_{26}N_3O_6SCl$–$C_4H_{10}O$(etherate)
Sporidesmin E	180–5 (etherate)	−132° (etherate)	217 (4.52), 252(4.22), 295(3.50)	$C_{18}H_{20}N_3O_6S_3Cl$–$C_4H_{10}O$ (etherate)
Sporidesmin F	65–75	—	216 (4.46), 250(4.14), 298(3.30)	$C_{19}H_{22}N_3O_6SCl$
Sporidesmin G	148–53	−217°	216 (4.65), 250(4.06), 298(3.47)	$C_{18}H_{20}N_3O_6S_4Cl$
Sporidesmin H	140–50	—	216 (4.37), 252(4.05), 290(3.81)	$C_{17}H_{20}N_3O_4S_2Cl$

of sulfur perhaps from methionine, methionine sulfoxide or *S*-adenosylmethionine to a diene compound as shown in Fig. 2.3.[12)]

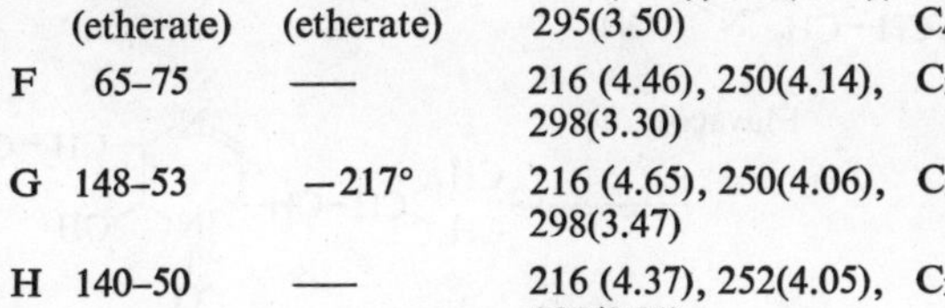

Fig. 2.3. Disulfide bridge formation in sporidesmin (R = H) and gliotoxin (R = OH).

Aspergillic and neoaspergillic acids (produced by *Aspergillus flavus* and *A. oryzae*) or pulcherriminic acid (produced by *Candida pulcherrina*) (Fig. 2.4) are formed from dioxopiperazine. Cycloleucylleucine, derived from two moles of L-leucine, forms flavacol. Flavacol is then oxidized to form flavacol *N*-oxide, neoaspergillic acid and finally hydroxyneoaspergillic acid by hydroxylation of the side chain (Fig. 2.5).[13]

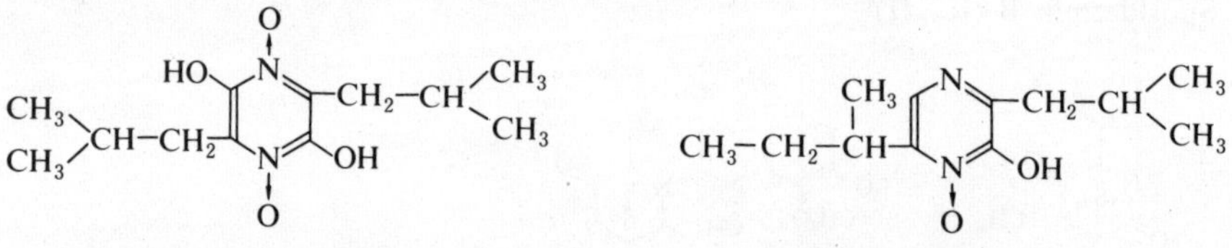

Pulcherriminic acid

Aspergillic acid

Fig. 2.4. Structure of pulcherriminic acid and aspergillic acid.

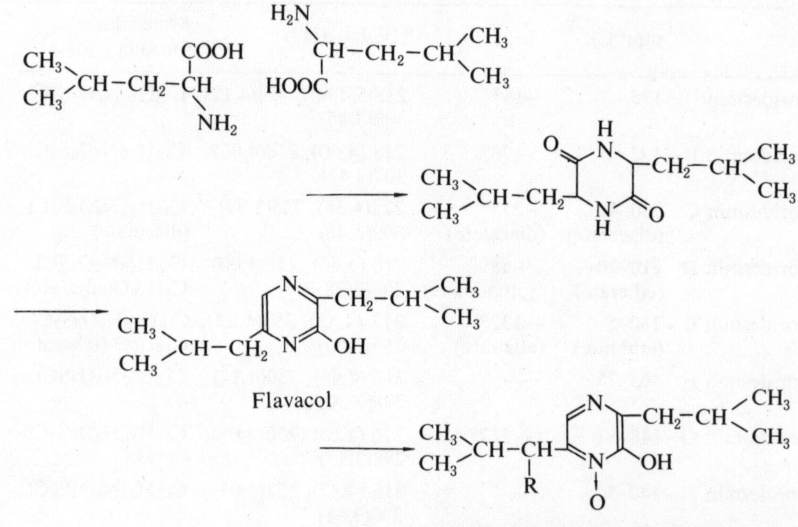

Flavacol

Neoaspergillic acid (R=H)
Hydroxyneoaspergillic acid (R=OH)

Fig. 2.5. Formation of flavacol, neoaspergillic acid and hydroxyneoaspergillic acid.

Viridicatin, a quinoline derivative obtained from *Penicillium viridicatum*, is perhaps an artifact produced from cyclopenin which is a cyclodipeptide of anthranilic acid and *N*-methyl phenylalanine. Cyclopenin was first isolated from *Penicillium cyclopium* and readily afforded viridicatin due to the action of an enzyme, "cyclopeptine dehydrogenase" (Fig. 2.6).[14]

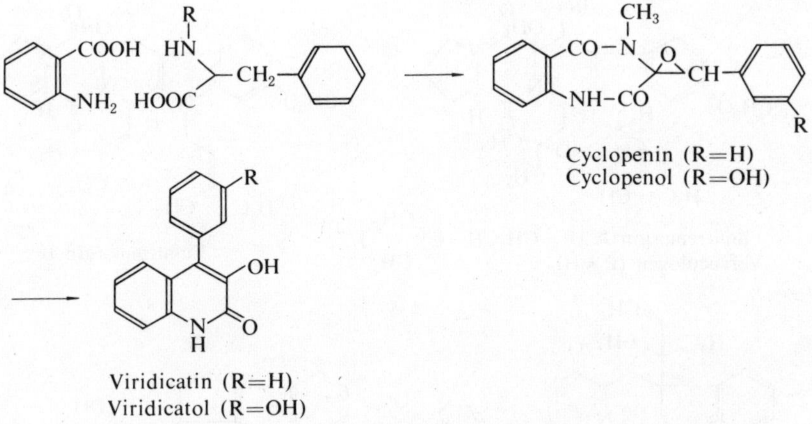

Cyclopenin (R=H)
Cyclopenol (R=OH)

Viridicatin (R=H)
Viridicatol (R=OH)

Fig. 2.6. Formation of cyclopenin and its conversion to viridicatin.

Several mycotoxins capable of causing sustained tremors in animals have recently been isolated from various strains of *Aspergillus* and *Penicillium*. The structures of verruculogen from *Penicillium verruculosum*[15] and fumitremorgin A and B from *Aspergillus fumigatus*[16] have been elucidated independently by American and Japanese investigators and it has become obvious that they all have a 2,5-dioxopiperazine skeleton formed from 6-methoxytryptophan and proline (Fig. 2.7).

Some evidence that fumitremorgin could arise by the condensation of tryptophan and proline has been obtained by feeding L-[3-^{14}C]-tryptophan and L-[U-^{14}C]-proline to cultures of *A. fumigatus*. [2-^{14}C]-Mevalonic acid is also incorporated into fumitremorgin A in similar experiments (M. Yamazaki, *unpublished data*).

Roquefortine (Fig. 2.7), a 2,5-dioxopiperazine compound formed from tryptophan and histidine has recently been isolated from *Penicillium roqueforti*. It also exhibits strong tremorgenic activity.[17]

Slaframine is an uncommon fungal nitrogenous metabolite. It is an octahydroindolizine alkaloid and gives a positive reaction to Dragendorff's reagent. This toxic metabolite of *Rhizoctonia leguminicola* was found as a result of a study of "slobber factor", which causes excessive salivation in cattle. A single dose of this toxin of 0.3 mg/kg to an anesthetized cat stimulates salivation for several hours. It is known that slaframine has no physiological activity itself, but is activated by enzymes in the liver. A deaminated ketone is perhaps formed as an active metabolite in the liver but no active metabolites have actually been isolated yet.[18] The biosynthesis of slaframine is thought to proceed through the condensation of lysine with acetyl-CoA followed by acetylation of the hydroxyl group and the incor-

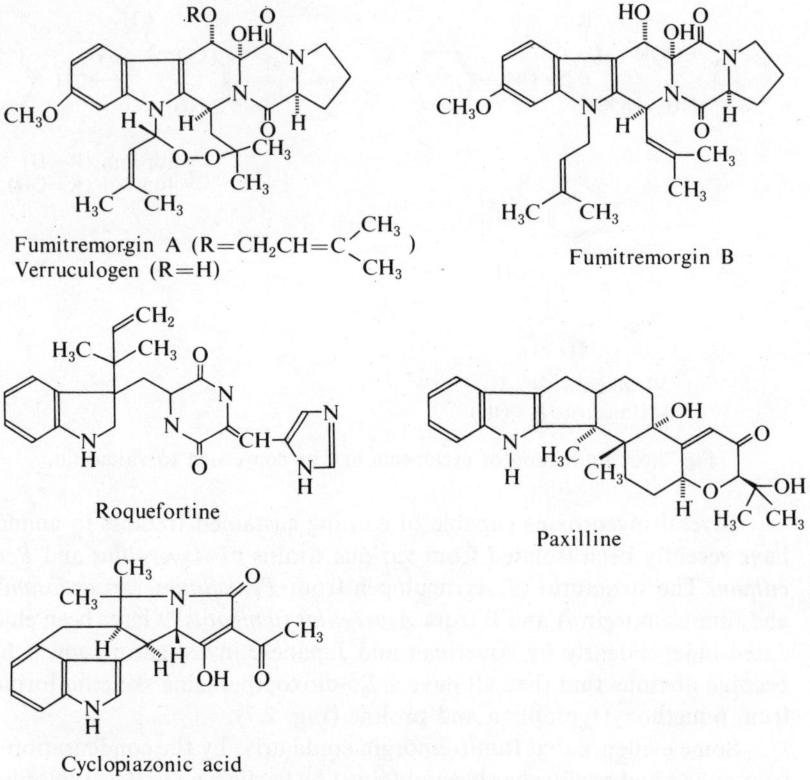

Fumitremorgin A ($R=CH_2CH=C(CH_3)_2$)
Verruculogen ($R=H$)

Fumitremorgin B

Roquefortine

Paxilline

Cyclopiazonic acid

Fig. 2.7. Structure of various tremorgenic mycotoxins.

poration of an amino group in the ring (Fig. 2.8), although the details of the mechanism of biosynthesis are obscure.[19]

As already mentioned, ergot alkaloids consist of peptides linked to the carboxyl group of lysergic acid. These peptides are, of course, derived from amino acids. Lysergic and isolysergic acids are now known to be derived from tryptophan by condensation with one "isoprene unit" formed from mevalonic acid (Fig. 2.9).

Oligopeptides such as cyclochlorotine[20] (islanditoxin[21]) or malfor-

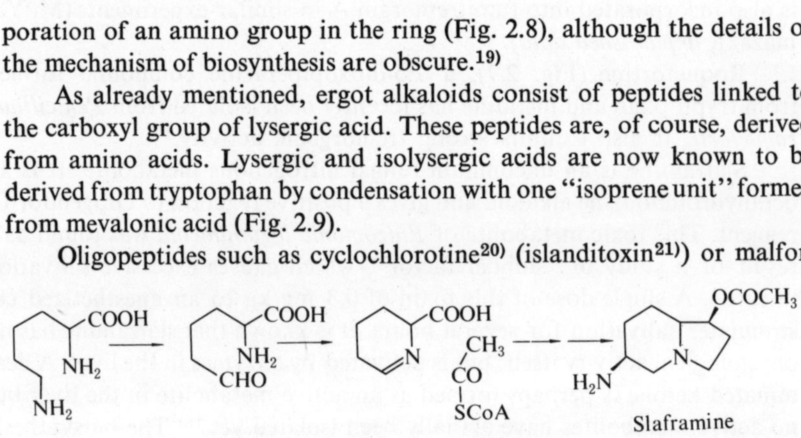

Slaframine

Fig. 2.8. Biosynthesis of slaframine.

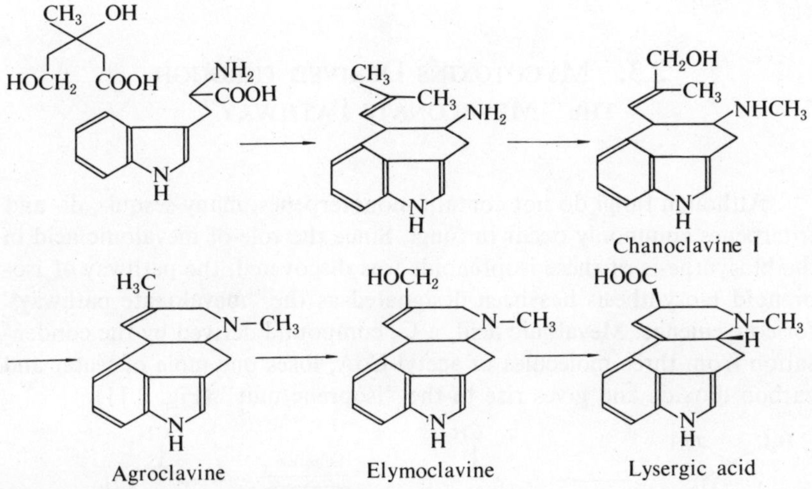

Chanoclavine I

Agroclavine Elymoclavine Lysergic acid

Fig. 2.9. Derivation of lysergic acid.

min A[22] (Fig. 2.10) are produced in some *Penicillia* and *Aspergilli*. Cyclochlorotine is responsible for the toxicity of foodstuffs contaminated by *Penicillium islandicum* and is a periportal cirrhogenic agent. Malformin A curiously contains D- together with L-amino acids. However, L-leucine and L-cysteine are well incorporated into this peptide, suggesting that the isomerization of L- to D-amino acid may occur at an intermediary stage of the biosynthesis.[23]

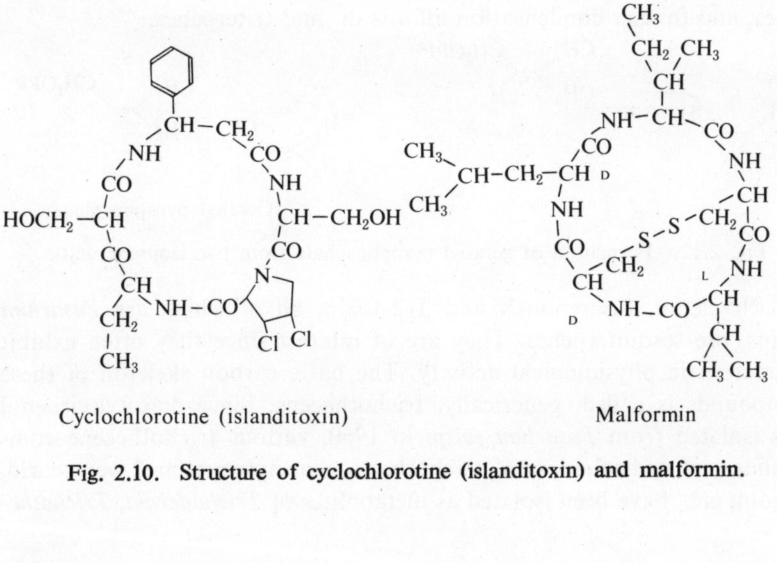

Cyclochlorotine (islanditoxin) Malformin

Fig. 2.10. Structure of cyclochlorotine (islanditoxin) and malformin.

2.3. MYCOTOXINS DERIVED THROUGH THE "MEVALONATE PATHWAY"

Although fungi do not contain monoterpenes, many sesqui-, di- and triterpenes commonly occur in fungi. Since the role of mevalonic acid in the biosynthesis of these isoprenoids was discovered, the pathway of isoprenoid biosynthesis has been designated as the "mevalonate pathway" for convenience. Mevalonic acid, a C_6 compound derived by the condensation from three molecules of acetyl-CoA, loses one mole of water and carbon dioxide and gives rise to the "isoprene unit" (Fig. 2.11).

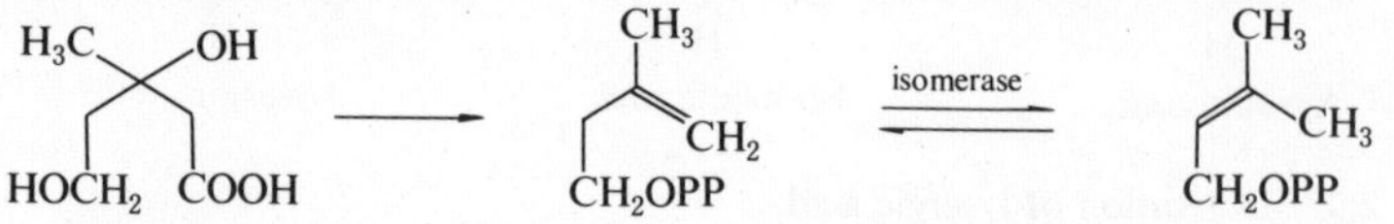

H_3C OH HOCH$_2$ COOH → CH_3 CH_2 CH_2OPP ⇌ (isomerase) CH_3 CH_3 CH_2OPP

Mevalonic acid Isopentenyl pyrohosphate ("isoprene unit") Dimethylallyl pyrophosphate

Fig. 2.11. Formation of "isoprene unit rom melavonic acid (PP = pyrophosphate).

Geranyl pyrophosphate is a product of the condensation of two isoprene units (Fig. 2.12). Extension of the geranyl unit by the condensation of another unit in the same way gives rise to C_{15} compounds, sesquiterpenes, and further condensation affords di- and triterpenes.

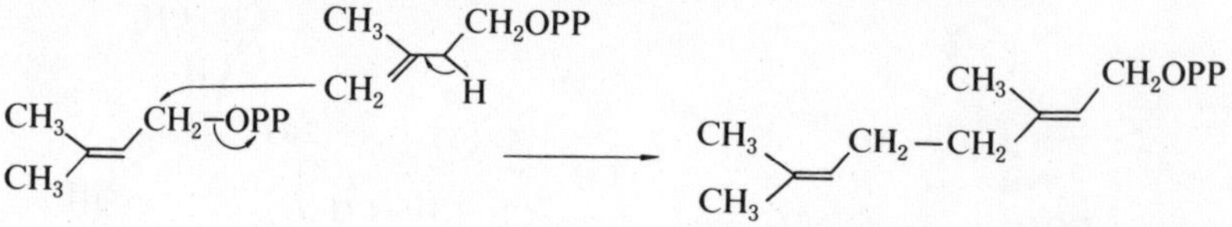

CH_3 CH_2OPP CH_2 H CH_3 CH_2–OPP CH_3 → CH_3 CH_2OPP CH_3 CH_2-CH_2 CH_3

Geranyl pyrophosphate

Fig. 2.12. Formation of geranyl pyrophosphate from two isoprene units.

Nivalenol, fusarenon-X and T-2 toxin, all of which are *Fusarium* toxins, are sesquiterpenes. They are of interest since they often exhibit characteristic physiological activity. The basic carbon skeleton of these compounds is called, generically, trichothecene. Since diacetylscirpenol was isolated from *Fusarium scirpi* in 1960, various trichothecene compounds such as trichodermol, trichothecin, diacetylverrucarol, verrucarin, roridin, etc., have been isolated as metabolites of *Trichoderma*, *Trichothe-*

cium, *Myrothecium* and *Cephalosporium* in addition to *Fusarium*. Most of these compounds are highly oxygenated and characteristically contain an epoxy group, as shown in Table 2.4.

TABLE 2.4. Structure of major fungal trichothecenes

	R^1	R^2	R^3	R^4	R^5	R^6
Scirpene	OH	OH	H	H	H	H
Trichodermol	H	OH	H	H	H	H
Trichodermin	H	OAc	H	H	H	H
Diacetylverrucarol	H	OAc	OAc	H	H	H
Diacetoxyscirpenol	OH	OAc	OAc	H	H	H
Neosolaniol	OH	OAc	OAc	H	OH	H
Trichothecolone	H	OH	H	H	=CO	
Trichothecin	H	isoCrot	H	H	=CO	
Nivalenol	OH	OH	OH	OH	=CO	
Fusarenon-X	OH	OAc	OH	OH	=CO	
T-2 toxin	OH	OAc	OAc	H	isoVal	H
HT-2 toxin	OH	OH	OAc	H	isoVal	H
Verrucarin A	$-OCOCH(OH)-CH(CH_3)-CH_2CH_2OCOCH{=}CH{=}CHCOO-$				from position 15 to 4, no other substituents	
Roridin A	$-OCOCH(OH)-CH(CH_3)-CH_2OCHCH{=}CHCH{=}CHCOO-$				from position 15 to 4, no other substituents	

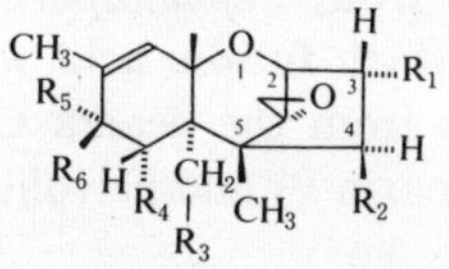

isoCrot = $-COCH{=}CH\langle^{CH_3}_{CH_3}$

isoVal = $-COCH{=}CHCH_3$

The formation of the trichothecene skeleton from farnesyl pyrophosphate is now well established. Since a hydrocarbon compound, trichodiene, has been isolated from cultures of *Trichothecium roseum* as a metabolite, biosynthesis involving this compound as a probable intermediate has been proposed.[26] Subsequently, another closely related epoxy alcohol, trichodiol, was obtained[27] and the incorporation of tritiated trichodiene into trichodiol as well as into scirpene and trichothecolone was confirmed by feeding experiments[28] (see Fig. 2. 13). The incorporation of 4(R)-[4-^{3}H, 2-^{14}C]-mevalonate into both trichodiol and trichothecolone has also been demonstrated.[29] Very recently, the formation of trichodiene from farnesyl pyrophosphate in a cell-free system from *T. roseum* has been reported.[30]

The toxicities of these trichothecenes are similar and the toxins cause radiomimetic injury to the intestinal tract on administration of lethal

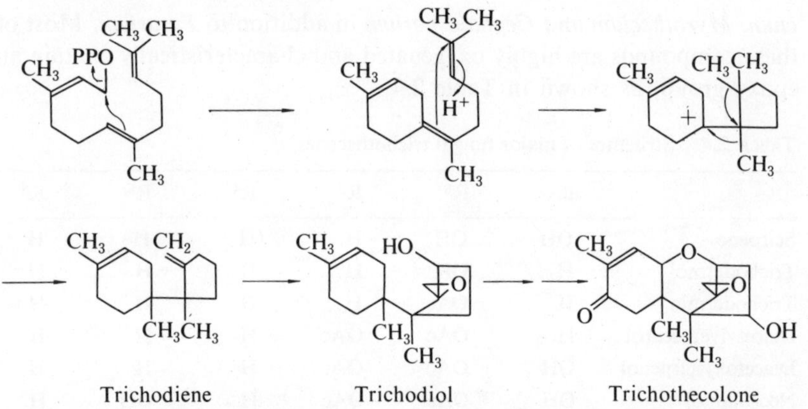

Trichodiene Trichodiol Trichothecolone

Fig. 2.13. Formation of the trichothecene skeleton from farnesyl pyrophosphate.

doses in mice. Mice injected with fusarenon-X at doses several times the LD_{50} died of shock in the early phase. The effects induced by the toxin may be compared with those seen in death after acute whole-body irradiation.[31]

Tritium-labeled squalene-2,3-oxide is incorporated into fusidic acid in *Fusidium coccineum* without randomization of radioactivity.[32] Helvolic acid was first isolated from *Aspergillus fumigatus* mut. *helvola*[33] and later from *Cephalosporium*, *Emericellopsis* and *Aspergillus oryzae*. Helvolic acid is closely related to fusidic acid and another C_{29} triterpene of the fusidane group, cephalosporin P_1 from *Cephalosporium* sp.[34] (see Fig. 2.14). *Fusidium coccineum*, from which fusidic acid was first isolated,[35] is taxonomically related to molds from the genera *Cephalosporium* and *Penicillium*. Fusidic acid, cephalosporin P_1 and helvolic acid all have antibiotic activity.

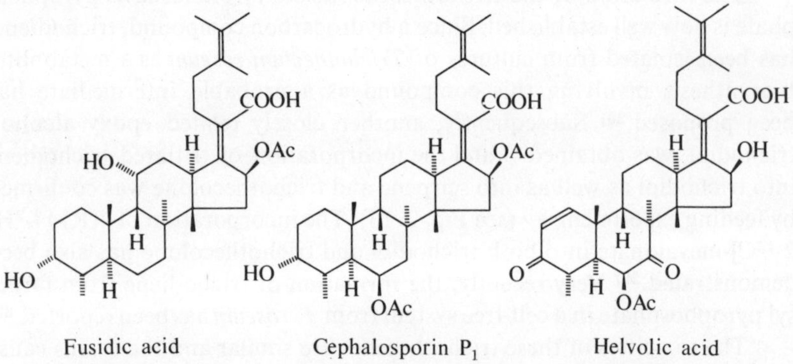

Fusidic acid Cephalosporin P_1 Helvolic acid

Fig. 2.14. Structure of fusidic acid, cephalosporin P_1 and helvolic aicd.

2.4. MYCOTOXINS DERIVED THROUGH THE "ACETATE-MALONATE PATHWAY"

Aflatoxin contains a coumarin skeleton and has the strong fluorescence characteristic of coumarin compounds. However, it is known that aflatoxin is derived through the "acetate-malonate" biosynthetic pathway. The acetate-malonate pathway for the biosynthesis of various compounds, including phenolics, was first proposed by Collie in 1907[36] based on the experimental demonstration that orsellinic acid was formed from dehydroacetic acid on treatment with alkali (see Fig. 2.15). However, the acetate hypothesis remained unconfirmed until quite recently.[37] At that time, the formation of acetate or acetyl-CoA in living organisms and the role of "active acetate" in metabolic processes were completely unknown. Birch in 1953[38] and Robinson in 1955[39] almost simultaneously revived the acetate hypothesis based on the finding that the structures of most naturally occurring phenolic compounds accord with their derivation by the cyclization of a polyoxomethylene chain, conveniently called a polyketide. Subsequently, support for the hypothesis was provided by the experimental finding that radioactively labeled acetate was incorporated into 6-methylsalicylic acid when the labeled precursor was fed to *Penicillium griseofulvum* (Fig. 2.16).[40]

Citrinin is a yellow crystalline antibiotic compound isolated from *Penicillium citrinum* by Raistrick in 1931,[41] and its proposed *p*-quinone methide structure has recently been confirmed by X-ray diffraction analysis.[42] The nephrotoxicity of citrinin was discovered from the experimental observation that the weight of kidneys of test rats increased to 150% of

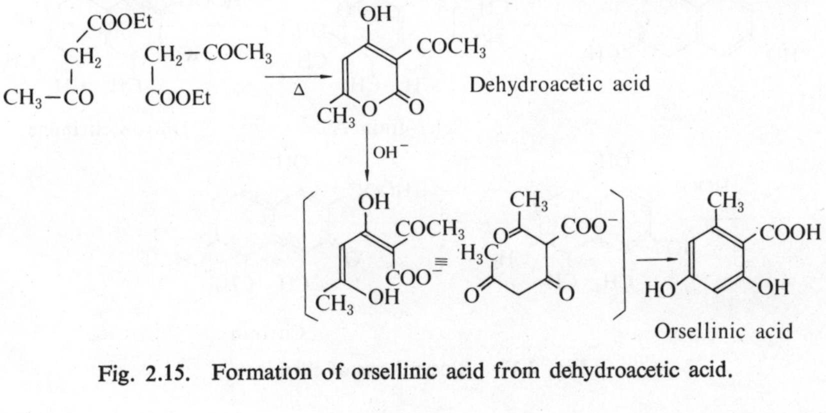

Fig. 2.15. Formation of orsellinic acid from dehydroacetic acid.

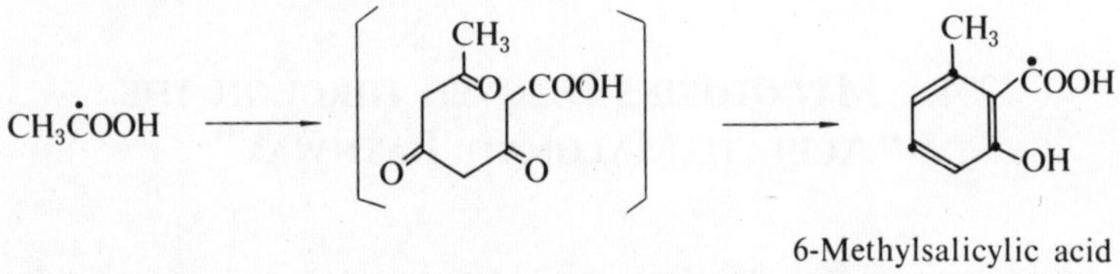

$CH_3\dot{C}OOH$ → [CH_3, O, COOH, O, O] → CH_3, $\dot{C}OOH$, OH

6-Methylsalicylic acid

Fig. 2.16. Incorporation of radioactively labeled acetate into 6-methylsalicylic acid.

control values when rice carrying *P. citrinum* was fed continuously to the rats for 3 weeks. Citrinin caused slight but definite renal damage.

The biosynthesis of citrinin can be rationalized from a pentaketide based on the acetate-malonate pathway with the addition of three extra carbons (two methyl groups and one carboxyl group). It was shown that all of these extra carbons are derived from C_1 units by experiments using $^{14}CH_3$-methionine and ^{14}C-formate.[43] Some citrinin-related compounds have been isolated from mutants of *P. citrinum*; these include sclerotinin A and B, dihydrocitrinone together with decarboxycitrinin, decarboxydihydrocitrinone and certain phenolic compounds.[44] The biosynthetic route shown in Fig. 2.17 has been proposed.[45]

Ochratoxin A was discovered in 1965 as a toxic metabolite of *Aspergillus ochraceus* which was found growing on foods and animal feed by a S. African research group during a survey of toxigenic fungi.[46] Later, a

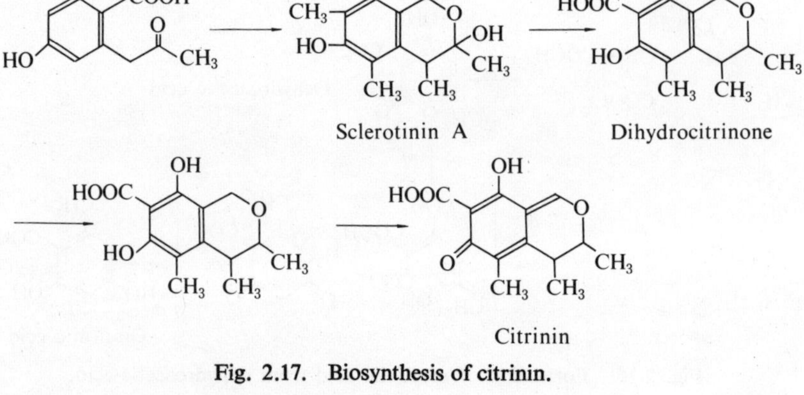

Sclerotinin A

Dihydrocitrinone

Citrinin

Fig. 2.17. Biosynthesis of citrinin.

disease in pigs involving kidney degeneration which had long been recognized in slaughterhouses in Denmark, was shown to be induced by feeding them moldy barley. Among the fungi isolated from the barley, *Penicillium viridicatum* was found to induce kidney disease in experimental animals resembling that found in naturally occurring mold nephrosis in pigs. Ochratoxin A has been isolated from this fungus together with oxalic acid and citrinin. Further, ochratoxin A has been detected in Danish barley and oats fed to swine. It seems likely that ochratoxin A is a nephrotoxic mycotoxin.

Ochratoxin A is a dihydroisocoumarin carboxylic compound linked through its carboxyl group to the amino nitrogen of phenylalanine. The biosynthesis of the dihydroisocoumarin skeleton is expected to be similar to that of citrinin and other fungal metabolites containing isocoumarin, such as oosponol or oospolactone (Fig. 2.18) from *Oospora* sp.[47] Some

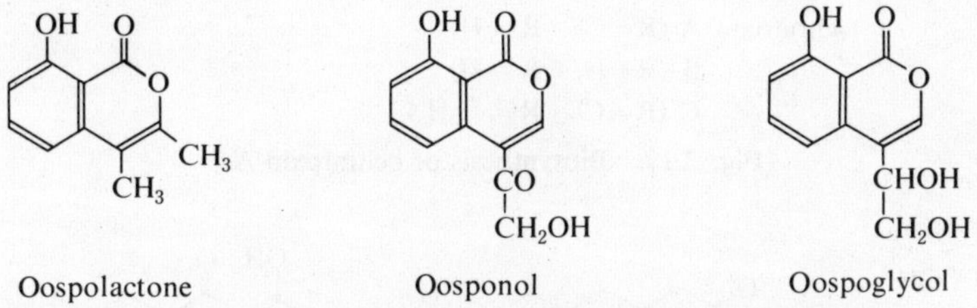

Oospolactone Oosponol Oospoglycol

Fig. 2.18. Structure of oospolactone, oosponol and oospoglycol.

evidence supporting this view has been obtained by isotopic tracer experiments in which ^{14}C-acetate, [2-^{14}C]-malonate, ^{14}C-formate and [1-^{14}C]-phenylalanine were incorporated into ochratoxin A in resting cultures of *A. ochraceus.*[48] The radioactivity of phenylalanine was incorporated solely into the amino acid moiety of ochratoxin A.[49] Recently, ^{13}C-nmr spectrometry was applied to the investigation of ochratoxin biosynthesis and the formation of the carboxyl group from a C_1 unit during the biosynthesis was proved[50] (Fig. 2.19). The stage at which the chlorine atom is introduced into ochratoxin A is not yet known. The highest incorporation of Na^{36}Cl into the toxin in cultures of *A. ochraceus* occurred when the salt was added on the 2nd or 3rd day of incubation.[51]

Maltoryzine, a phenol-like toxic principle of *Aspergillus* sp. IAM2950, was isolated in 1954 from the feed responsible for the poisoning of dairy cows in Japan.[52] This compound has an unusual oxygenation pattern in the phenol ring and also in the side chain. The carbon skeleton of this compound may well be derived from hexaketide (Fig. 2.20). However, no evidence to support the derivation of this toxin through the acetate-malonate pathway has yet been obtained.

Among fungal heptaketides, an antifungal antibiotic, griseofulvin,

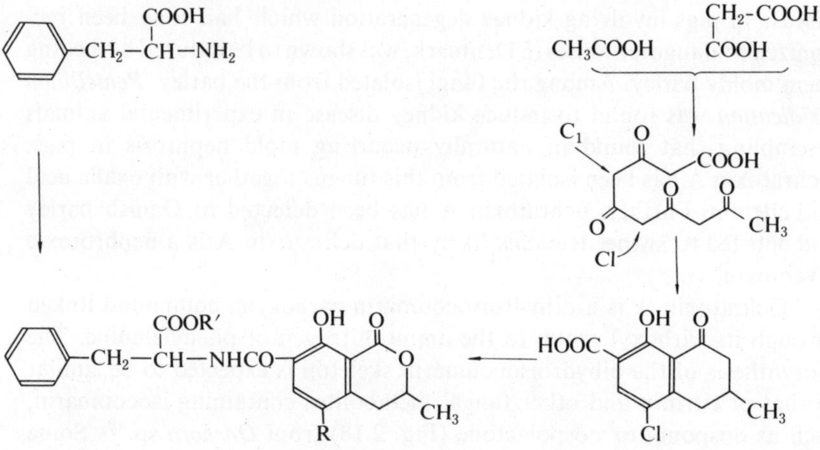

Ochratoxin A (R=Cl, R′=H)
B (R=H, R′=H)
C (R=Cl, R′=C_2H_5)

Fig. 2.19. Biosynthesis of ochratoxin A.

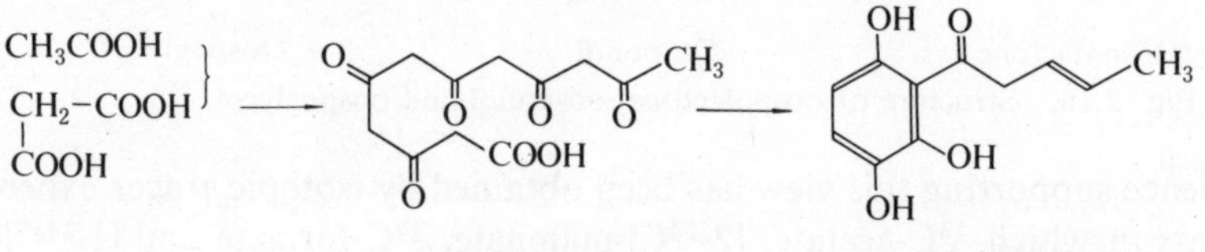

Maltoryzine

Fig. 2.20. Possible route of maltoryzine formation.

was first isolated by Raistrick[53] from a strain of *Penicillium griseofulvum* and was subsequently isolated from a large number of *Penicillium* species. Griseofulvin was actually used as an antifungal agent for disease of the skin because of its remarkable fungistatic properties. However, the medical use of this compound has now been prohibited since it was found to be carcinogenic.[54] The biosynthesis of griseofulvin from acetate was demonstrated at an early stage by Birch and his colleagues by tracer experiments using [1-^{14}C]-acetate.[55] The overall route of griseofulvin biosynthesis is shown in Fig. 2.21.[56] The route passes through griseophenone A, B and C and dehydrogriseofulvin, and all of these compounds have actually been isolated from *Penicillium patulum*.[57] When a strain of *P. patulum* was cultured in a medium deficient in chlorine, griseophenone C was produced predominantly instead of griseofulvin. It is of interest to note that the chemical synthesis of griseofulvin can be accomplished by the partial hydrogenation of dehydrogriseofulvin, which is obtained by oxidative coupling of griseophenone A.[58] This suggests the plausibility of the biosynthetic pathway shown in Fig. 2.21.

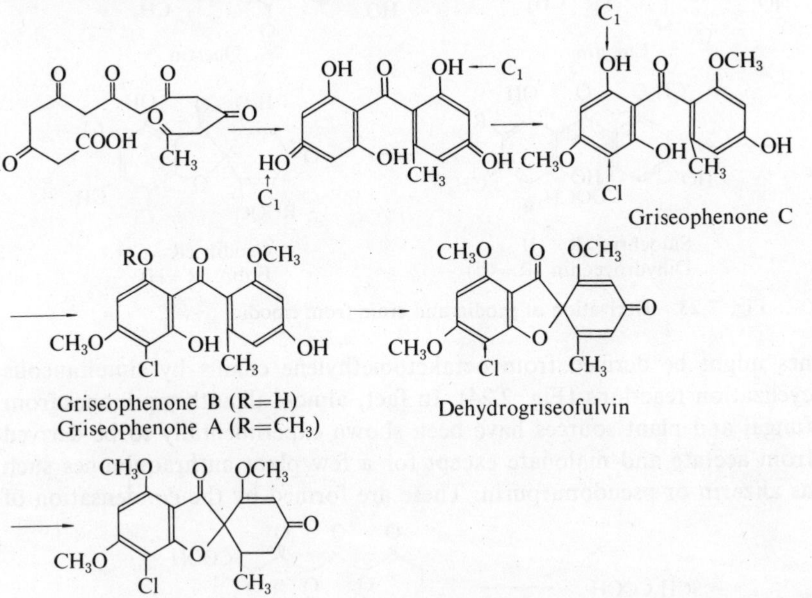

Griseophenone C

Griseophenone B (R=H)
Griseophenone A (R=CH_3)

Dehydrogriseofulvin

Griseofulvin

Fig. 2.21. Biosynthesis of griseofulvin.

Other fungal metabolites containing a "grisan" skeleton such as geodin and erdin from *Aspergillus terreus*[59] or trypacidin from *Aspergillus fumigatus*[60] (Fig. 2.22) may be biosynthesized in a similar manner. However, the biological conversion of emodin to geodin, dihydrogeodin and

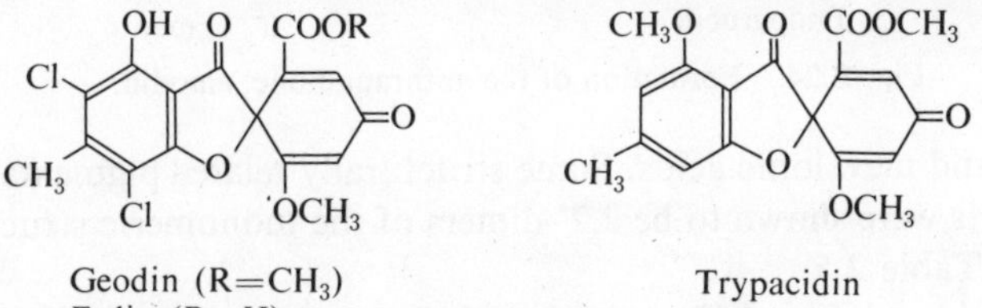

Geodin (R=CH_3)
Erdin (R=H)

Trypacidin

Fig. 2.22. Structure of geodin, erdin and trypacidin.

questin in *A. terreus* has recently been demonstrated by tracer experiments,[61] suggesting that geodin and erdin are derived from anthraquinones such as emodin by oxidative cleavage of the ring (Fig. 2.23).

Robinson in 1955[39] suggested that naturally occurring anthraquino-

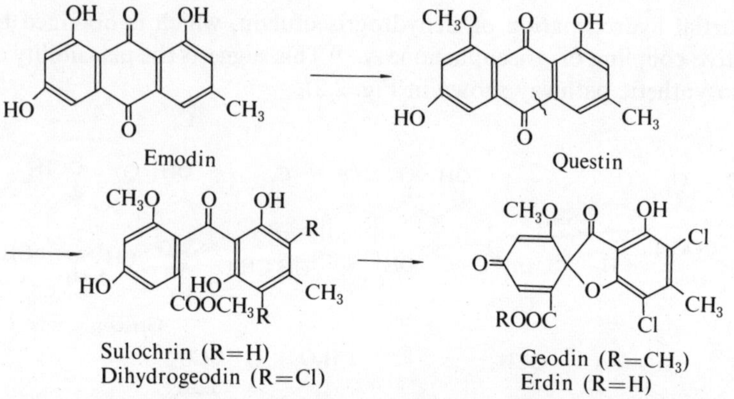

Emodin → Questin → Sulochrin (R=H), Dihydrogeodin (R=Cl) → Geodin (R=CH_3), Erdin (R=H)

Fig. 2.23. Derivation of geodin and erdin from emodin.

nes might be derived from octaketomethylene chains by simultaneous cyclization reactions (Fig. 2.24). In fact, almost all anthraquinones from fungal and plant sources have been shown experimentally to be derived from acetate and malonate except for a few plant anthraquinones such as alizarin or pseudopurpurin. These are formed by the condensation of

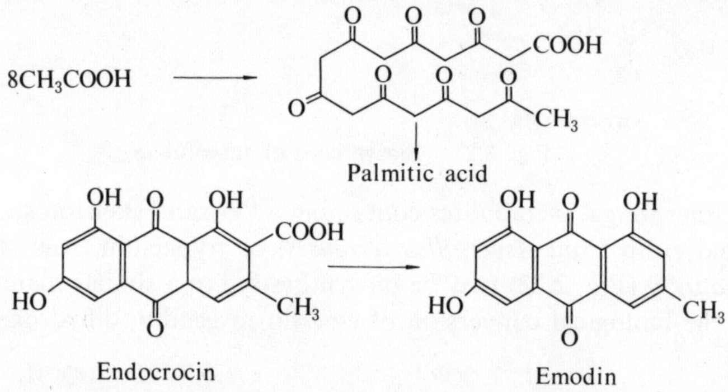

$8CH_3COOH$ → Palmitic acid → Endocrocin → Emodin

Fig. 2.24. Formation of the anthraquinone, emodin.

shikimic and mevalonic acids. Some structurally related pigments isolated from ergots were shown to be 2,2'-dimers of the monomeric structures, as shown in Table 2.5.

Franck and Flasch[62)] proposed the use of the term "ergochrome" to describe the pigments of Table 2.5. Among them, ergochrome AA (secalonic acid A) has recently been isolated from the lichen *Parmelia entotheiochroa*,[63)] and ergochrome EE (secalonic acid D) from the fungi *Penicillium oxalicum*[64)] and *Aspergillus ochraceus*[65)] (so-called secalonic acid A from *A. ochraceus* was recently shown to be D) as well as from ergots. Intraperitoneal injection of ergochrome EE causes severe peritonitis in mice

TABLE 2.5. Structure of ergochromes

Ergochrome AA (secalonic acid A)	Ergochrome AD
Ergochrome BB (secalonic acid B)	Ergochrome BD
Ergochrome CC (ergoflavin)	Ergochrome CD
Ergochrome AB (secalonic acid C)	Ergochrome DD
Ergochrome AC (ergochrysin A)	Ergochrome EE (secalonic acid D)
Ergochrome BC (ergochrysin B)	

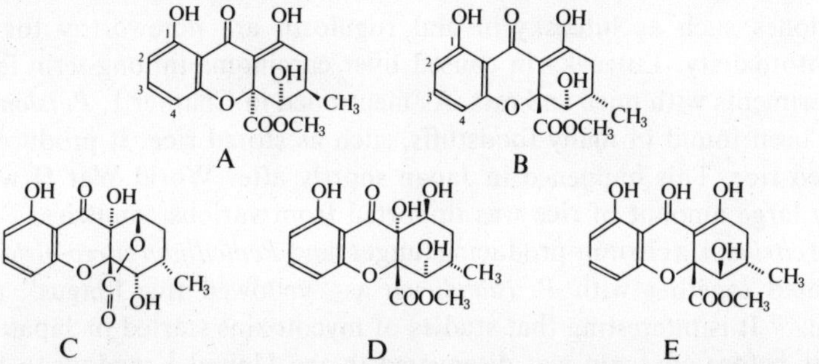

and rats and strong phlogistic properties have been demonstrated in this compound.[66] The LD_{50} of ergochrome EE was reported to be 42 mg/kg (intraperitoneal) for mice.[64]

The co-occurrence of ergochromes and anthraquinones such as endocrocin and clavorubin (Fig. 2.25) in ergots suggested that the ergot pigments might arise from anthraquinones by oxidative cleavage of the ring.[67]

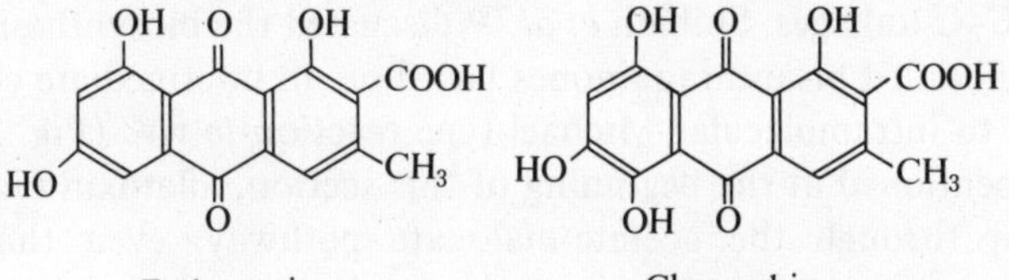

Fig. 2.25. Structure of endocrocin and clavorubin.

Actually, [U-^{14}C]-emodin biosynthetically prepared by feeding ^{14}C-acetate to *Penicillium islandicum* was shown to be converted to ergochrome BB (secalonic acid B) and BC (ergochrysin B) when emodin was added to the culture medium of *Claviceps purpurea* (Fig. 2.26). Thus, the above view was substantiated.[68]

Shibata and co-workers have been investigating anthraquinonoid pigments produced by *P. islandicum* and related fungi. Skyrin, for instance, has been established to be a dimer of emodin. The biosynthesis of skyrin from acetate has been demonstrated experimentally by Shibata's group.[69] Among many colored anthraquinonoids, the modified bisanthra-

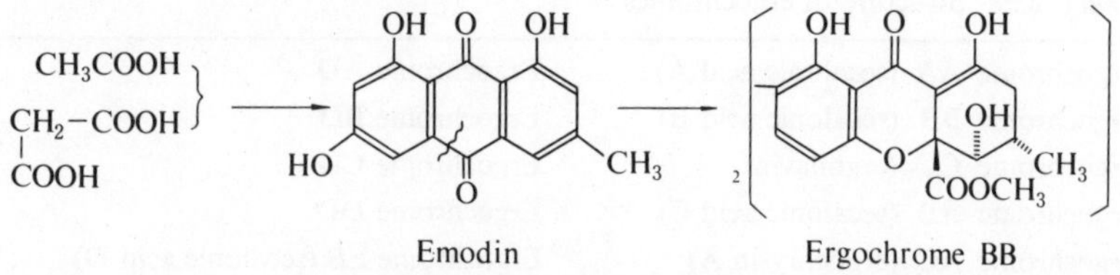

Emodin Ergochrome BB

Fig. 2.26. Formation of ergochrome BB from emodin.

quinones such as luteoskyrin and rugulosin are noteworthy for their hepatotoxicity. Luteoskyrin caused liver carcinoma in long-term feeding experiments with mice and rats. As mentioned in Chapter 1, *P. islandicum* has been found in many foodstuffs, such as stored rice. It produces yellowed rice. This happened in Japan shortly after World War II when a very large amount of rice was imported from various countries. *Penicillium citrinum*, a citrinin-producing fungus, and *Penicillium citreo-viride* were isolated together with *P. islandicum* as "yellowed rice fungus" at the time.[70)] It is interesting that studies of mycotoxins started in Japan a few years before aflatoxin was discovered in the United Kingdom to be the causative agent of Turkey X disease as well as a strong carcinogen.

The structures of rugulosin and luteoskyrin were first proposed to be partially reduced bisanthraquinones by Shibata and his colleagues, but revised structures have recently been presented based on further examination using nmr spectrometry and X-ray analysis.[71)] They have a very unusual cage-like structure in which two monomeric moieties are connected by three C–C linkages. Shibata *et al.*[72)] discussed the biosynthesis of these unusual modified bisanthraquinones based on an intermediate compound subjected to intramolecular Michael-type reaction *in vivo* (Fig. 2.27).

As mentioned at the beginning of this section, aflatoxin biosynthesis can occur through the acetate-malonate pathway, even though this mycotoxin contains a coumarin skeleton. As a general rule, coumarin compounds are formed through the aromatic pathway of biosynthesis involving phenylalanine, cinnamic acid and shikimic acid as intermediates. Administration of $^{14}CH_3$-methionine and [1-^{14}C]-acetate to *Aspergillus flavus* yielded radioactive aflatoxin B_1.[73)] Aflatoxin B_1 obtained from radioactive acetate as above was degraded and the distribution of radioactivity was determined by Büchi, who postulated a biosynthetic route for aflatoxin B_1 from acetate-malonate derived polyhydroxynaphthacene endoperoxide.[74)] However, the biosynthesis of aflatoxin B_1 from anthraquinone containing a linear C_6 chain probably derived from a single nonaketide such as norsoloric acid, has more recently been established using ^{13}C-nmr spectrometry.[75)] Averufin, previously isolated from *Aspergillus*

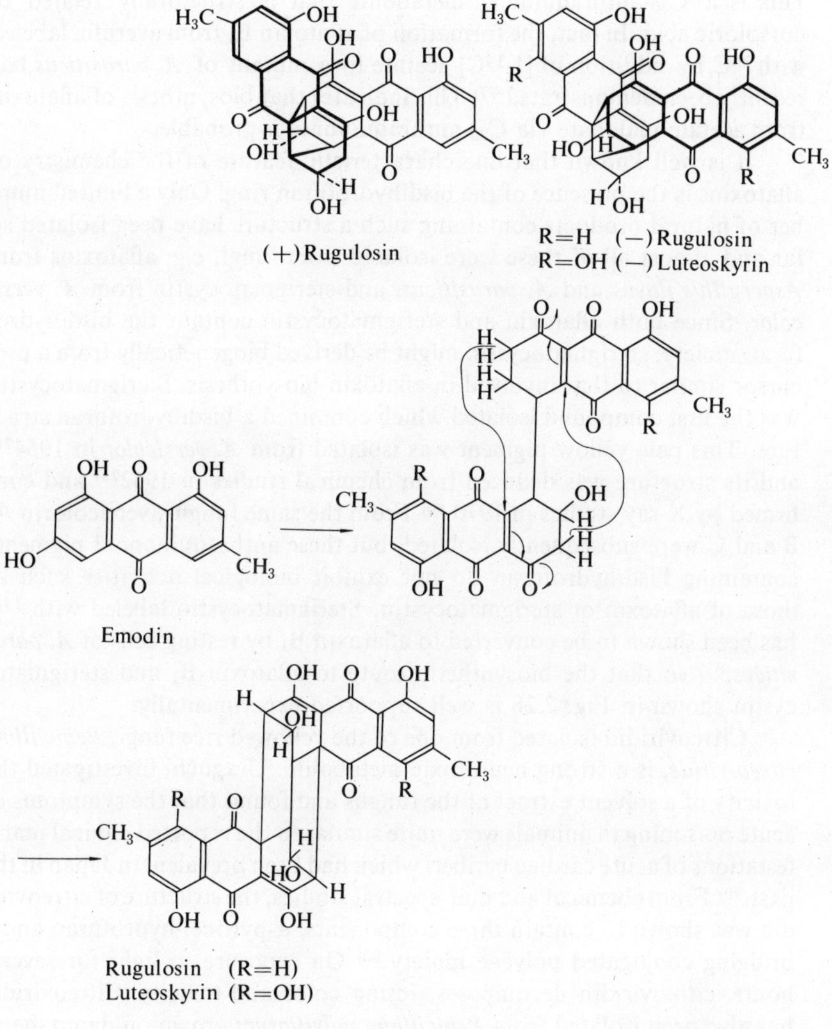

(+)Rugulosin

R=H (−)Rugulosin
R=OH (−)Luteoskyrin

Emodin

Rugulosin (R=H)
Luteoskyrin (R=OH)

Fig. 2.27. Structure and biosynthesis of rugulosin and luteoskyrin.

versicolor as a minor metabolite, has been shown to accumulate in a mutant of *Aspergillus parasiticus* with impaired aflatoxin production.[76] This is a C_{20}-anthraquinone metabolite that is structurally related to norsoloric acid. In fact, the formation of aflatoxin B_1 from averufin labeled with ^{14}C by addition of [1-^{14}C]-acetate to a mutant of *A. parasiticus* has recently been demonstrated.[77] This indicates that biosynthesis of aflatoxin from acetate-malonate via C_{20}-anthraquinones is probable.

It is well known that one characteristic feature of the chemistry of aflatoxins is the presence of the bisdihydrofuran ring. Only a limited number of natural products containing such a structure have been isolated so far and almost all of these were isolated from fungi, e.g. aflatoxins from *Aspergillus flavus* and *A. parasiticus*, and sterigmatocystin from *A. versicolor*. Since both aflatoxin and sterigmatocystin contain the bisdihydrofuran moiety, sterigmatocystin might be derived biogenetically from a precursor similar to that involved in aflatoxin biosynthesis. Sterigmatocystin was the first compound isolated which contained a bisdihydrofuran structure. This pale yellow pigment was isolated from *A. versicolor* in 1954[78] and its structure was deduced from chemical studies in 1962[79] and confirmed by X-ray studies in 1970.[80] From the same fungus, versicolorin A, B and C were subsequently isolated, but these anthraquinonoid pigments containing bisdihydrofuran do not exhibit biological activities such as those of aflatoxin or sterigmatocystin. Sterigmatocystin labeled with ^{14}C has been shown to be converted to aflatoxin B_1 by resting cells of *A. parasiticus*,[81] so that the biosynthetic route to aflatoxin B_1 and sterigmatocystin shown in Fig. 2.28 is well supported experimentally.

Citreoviridin isolated from one of the yellowed rice fungi, *Penicillium citreo-viride*, is a strong neurotoxic metabolite. Uraguchi investigated the toxicity of a solvent extract of the fungus and found that the symptoms of acute poisoning in animals were quite similar to the reported clinical manifestations of acute cardiac beriberi which had been prevalent in Japan in the past.[70] From chemical and nmr-spectral studies, the structure of citreoviridin was shown to contain three components, α-pyrone, hydrofuran and a bridging conjugated polyene moiety.[82] On exposure to light for several hours, citreoviridin decomposes, losing color and toxicity. Citreoviridin has also been isolated from *Penicillium pulvillorum* grown on damp maize meal as a toxic component.[83] Incorporation studies using $^{14}CH_3$-methionine and [2-^{14}C]-acetate have shown that citreoviridin can be formed by the condensation of nine acetate units, with methionine providing the C-methyl and O-methyl groups in *P. pulvillorum* as shown in Fig. 2.29.[84]

Zearalenone is an anabolic uterotrophic toxin responsible for vulvovaginitis in farm animals fed on moldy grain.[85] The structure of this estrogenic metabolite of *Gibberella zeae* was finally determined by Urry and

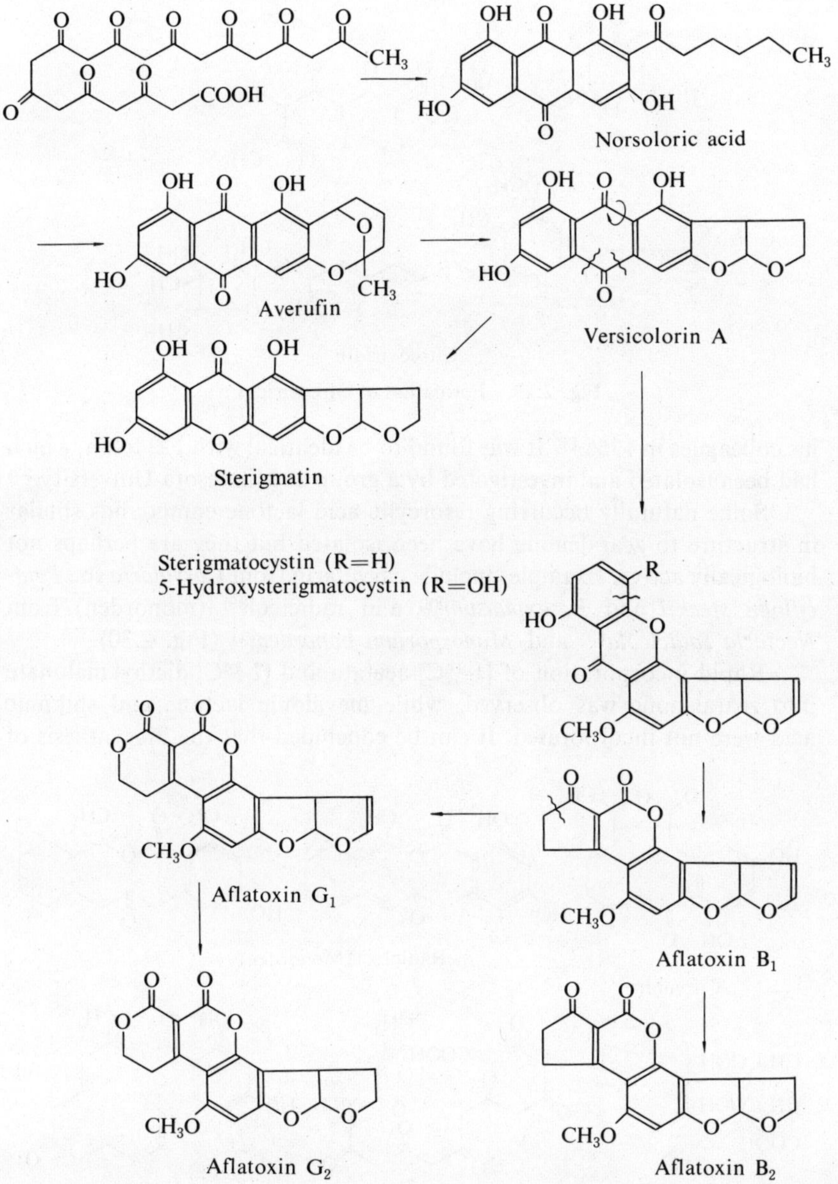

Norsoloric acid

Averufin

Versicolorin A

Sterigmatin

Sterigmatocystin (R=H)
5-Hydroxysterigmatocystin (R=OH)

Aflatoxin G_1

Aflatoxin B_1

Aflatoxin G_2

Aflatoxin B_2

Fig. 2.28. Biosynthesis of sterigmatocystin and aflatoxin.

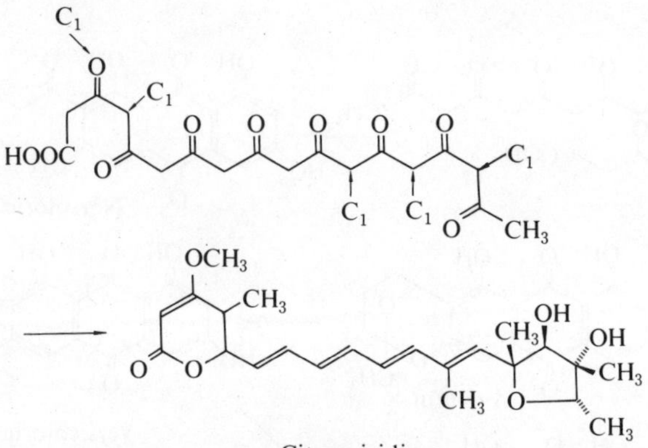

Citreoviridin

Fig. 2.29. Formation of citreoviridin.

his colleagues in 1966.[86] It was found to be identical with F-2 toxin, which had been isolated and investigated by a group at Minnesota University.[87]

Some naturally occurring resorcylic acid lactone compounds similar in structure to zearalenone have been isolated but they are perhaps not biologically active. Examples include curvularin from *Curvularia* sp., *Penicillium steckii* and *P. expansum*[88] and radicicol[89] (monorden) from *Nectoria radicicola*[90] and *Monosporium bonorden*[91] (Fig. 2.30).

Rapid incorporation of [1-^{14}C]-acetate and [2-^{14}C]-diethyl malonate into zearalenone was observed, while mevalonic lactone and shikimic acid were not incorporated. It can be concluded that the biosynthesis of

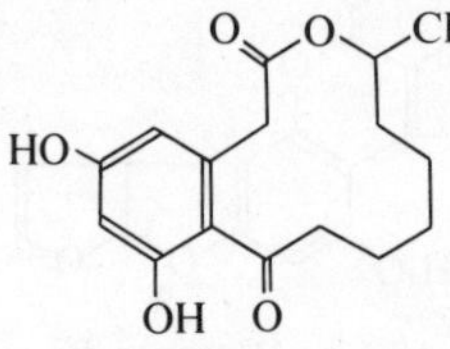

Curvularin

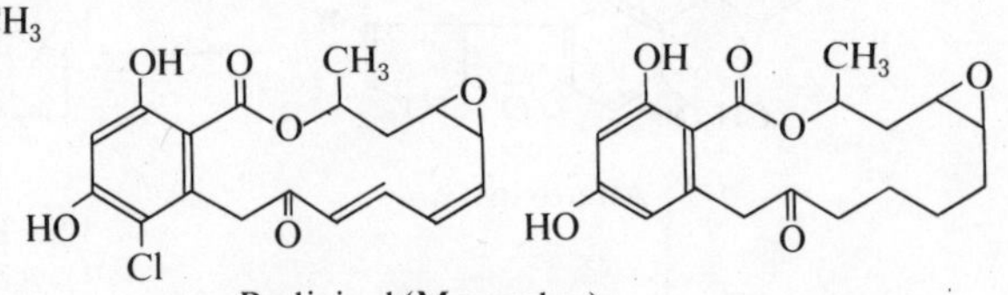

Radicicol (Monorden)

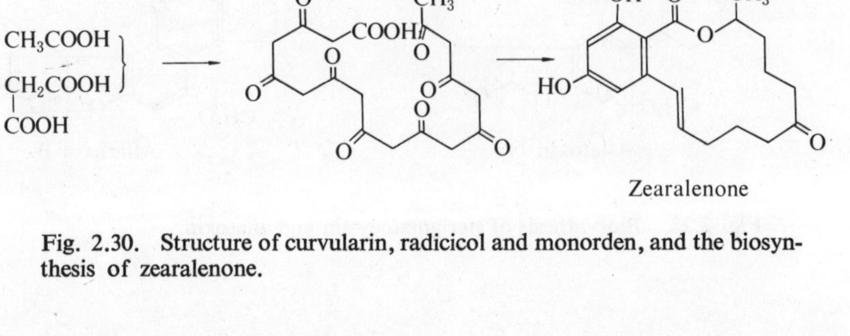

Zearalenone

Fig. 2.30. Structure of curvularin, radicicol and monorden, and the biosynthesis of zearalenone.

zearalenone involves the condensation of acetate and malonate units[92] (Fig. 2.30).

Patulin,[93] a lactonic compound, has been isolated under a variety of names from numerous species of *Penicillia* and *Aspergilli* and from *Byssochlamys nivea.* This compound has antibiotic activity. The toxicity of patulin in animals has been demonstrated by many investigators, and the LD_{50} for mice and rats is 0.3–0.7 mg/kg for i.v. administration. Dickens and his colleagues[94] have shown that patulin produces malignant tumors at the injection site on s.c. administration twice a week to rats for 15 months. A patulin-producing strain of *Penicillium urticae* was isolated from the feed responsible for lethal poisoning in dairy cattle in Japan in 1952.[95] In Germany, intoxication of cattle by germinated malt contaminated with *Aspergillus clavatus* was also assumed to be due to the toxic effects of patulin.[96]

The formation of patulin from an aromatic precursor was first postulated by Birkinshaw[97] and the conversion of 6-methylsalicylic acid to patulin in *Penicillium patulum* was actually proved. The sequence in Fig. 2.31 was proposed independently by Bu'Lock and Ryan[98] and Bassett and Tanenbaum.[99] The major biosynthetic route of patulin has been postulated on the basis of results obtained by kinetic pulse-labeling studies in which labeled acetate and other pertinent metabolites were fed to cultures of *P. urticae* and the kinetics of incorporation of the radioactivity into subsequent metabolites, including patulin, were examined.[100]

Penicillic acid[93] was first isolated in 1913 from a strain of *Penicillium*

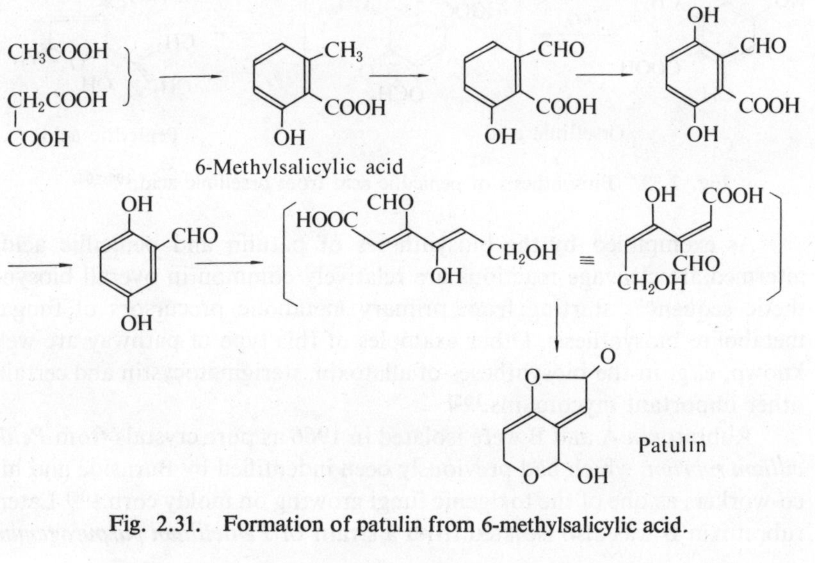

Fig. 2.31. Formation of patulin from 6-methylsalicylic acid.

puberulum and the structure was established in 1936 by Raistrick and co-workers, who obtained this compound from *Penicillium cyclopium.*[101] Dickens and his colleagues studied the carcinogenicity of penicillic acid and other lactones, including patulin, and demonstrated that penicillic acid produced transplantable tumors after 64 weeks in all rats receiving s.c. injection of 1.0 mg/kg twice weekly.[94]

Birch proposed the biosynthesis of penicillic acid from orsellinic acid by ring opening, as illustrated in Fig. 2.32.[102] However, Mosbach showed that cleavage of the aromatic ring did not occur as suggested by Birch, but as shown in Fig. 2.33.[103] Similar results were obtained by Bentley and Keil.[104]

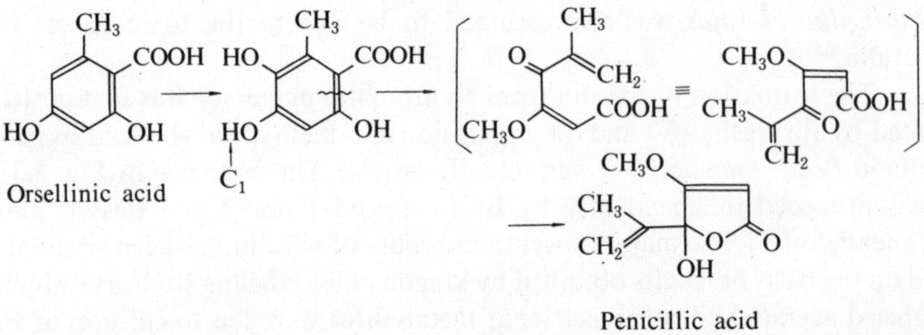

Fig. 2.32. Formation of penicillic acid from orsellinic acid, as proposed by Birch.[102]

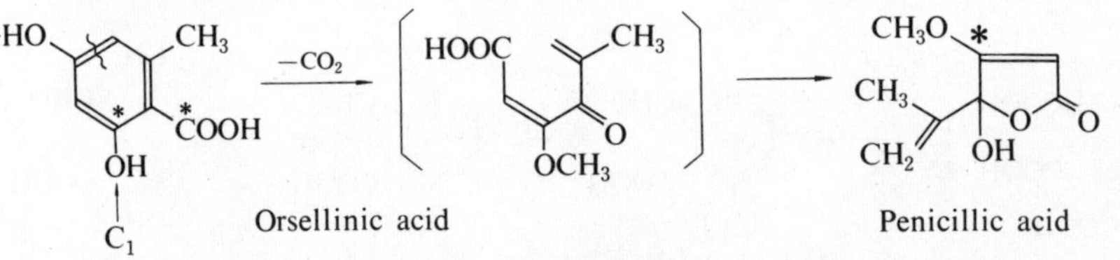

Fig. 2.33. Biosynthesis of penicillic acid from orsellinic acid.[103,104]

As exemplified by the biosyntheses of patulin and penicillic acid, intermediate cleavage reactions are relatively common in overall biosynthetic sequences starting from primary metabolic precursors of fungal metabolite biosynthesis. Other examples of this type of pathway are well known, e. g. in the biosyntheses of aflatoxin, sterigmatocystin and certain other important mycotoxins.[105]

Rubratoxin A and B were isolated in 1966 as pure crystals from *Penicillium rubrum*, which had previously been indentified by Burnside and his co-workers as one of the toxigenic fungi growing on moldy corn.[106] Later, rubratoxin B was also isolated from a strain of *Penicillium purpurogenum* which showed considerable toxicity to HeLa cells.[107] The structures of rubratoxin A and B were proposed by Moss *et al.* to be nonadrides with anhydride and lactone rings.[108] The structures of some other fungal nonadrides, glauconic,[109] glaucanic[110] and byssochlamic acids,[111] have been elucidated and the similarity between their structures and that of rubratoxins is clear (see Fig. 2.34). It may be postulated that rubratoxin is formed from two C_{13} units by head-to-tail condensation.[112] These units might be derived from decanoic acid and oxaloacetic acid. However, the course of rubratoxin biosynthesis is not clear as yet.

According to studies by Rose and Moss,[113] rubratoxin B is more toxic than any of its derivatives. It shows an LD_{50} of approximately 400 mg/kg on p.o. administration and 0.35 mg/kg for i.p. injection of a DMSO solution of this toxin.

Cytochalasins are a new class of fungal metabolite which produce unique effects on mammalian cells. Cytochalasin A and B are produced by *Helminthosporium dematioideum*[114] and *Phoma* sp.[115] and cytochalasin C and D were isolated from *Metarrhizium anisopliae*[116] and *Zygosporium masonii*.[117] More recently, the isolation of cytochalasin E from *Aspergillus*

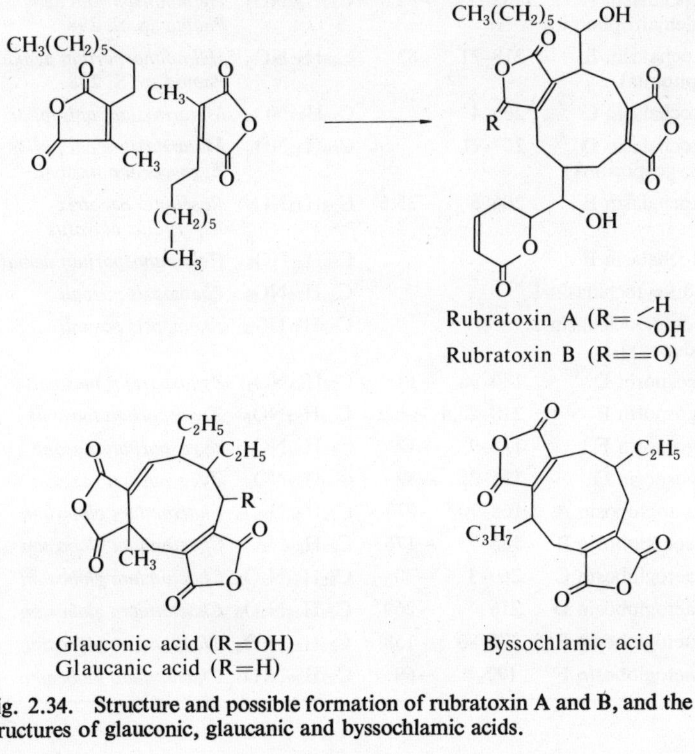

Fig. 2.34. Structure and possible formation of rubratoxin A and B, and the structures of glauconic, glaucanic and byssochlamic acids.

clavatus[118] and *Rosellinia necatrix*,[119] cytochalasin F from *H. dematioideum*,[119] and zygosporin E and F from *Z. masonii* together with zygosporin G[120] has been reported. During the course of screening for toxic fungal products, several chaetoglobosins have been isolated from *Chaetomium globosum*. The structural relationships between these cytochalasins, zygosporins and chaetoglobosins are shown in Table 2.6.[121]

The structures of the cytochalasins and zygosporins suggest that the macrolide portion of the cytochalasin molecule might be derived from polyketide and that the final formation of the structure might involve incorporation of phenylalanine. Cytochalasins containing a C_{11} ring, such as cytochalasin D, may be formed from octaketide and those containing a C_{13} ring, such as cytochalasin B, from a nonaketide with three and two

TABLE 2.6. Structure, physical properties and producing microorganisms of cytochalasins, zygosporins and chaetoglobosins

	mp(°C)	$[\alpha]_D$	Molecular formula	microorganism
Cytochalasin A (dehydrophomin)	182–5	+92	$C_{29}H_{35}NO_5$	*Helminthsporium dematioideum* *Phoma* sp. S. 298
Cytochalasin B (phomin)	218–21	+83	$C_{29}H_{37}NO_5$	*Helminthosporium dematioideum* *Phoma* sp. S. 298
Cytochalasin C	260–4		$C_{30}H_{37}NO_6$	*Metarrhizium anisopliae*
Cytochalasin D (zygosporin A)	267–71		$C_{30}H_{37}NO_6$	*Metarrhizium anisopliae* *Zygosporium masonii*
Cytochalasin E	206–8	−25.6	$C_{28}H_{33}NO_7$	*Rosellinia necatrix* *Aspergillus clavatus*
Cytochalasin F			$C_{29}H_{37}NO_5$	*Helminthosporium dematioideum*
Kodo-cytochalasin-1			$C_{30}H_{39}NO_5$	*Phomopsis paspali*
Kodo-cytochalasin-2 (deacetyl-1)			$C_{28}H_{27}NO_4$	*Phomopsis paspali*
Zygosporin D	180–90	−14.9	$C_{28}H_{35}NO_5$	*Zygosporium masonii*
Zygosporin E	218–23.5	+6.2	$C_{30}H_{37}NO_5$	*Zygosporium masonii*
Zygosporin F	126–9	−12	$C_{32}H_{39}NO_7$	*Zygosporium masonii*
Zygosporin G	115–25	−82	$C_{30}H_{37}NO_5$	*Zygosporium masonii*
Chaetoglobosin A	168–70	−270	$C_{32}H_{36}N_2O_5$	*Chaetomium globosum*
Chaetoglobosin B	186–7	−176	$C_{32}H_{36}N_2O_5$	*Chaetomium globosum*
Chaetoglobosin C	260–3	−30	$C_{32}H_{36}N_2O_5$	*Chaetomium globosum*
Chaetoglobosin D	216	−269	$C_{32}H_{36}N_2O_5$	*Chaetomium globosum*
Chaetoglobosin E	279–80	+158	$C_{32}H_{38}N_2O_5$	*Chaetomium globosim*
Chaetoglobosin F	177–8	−69	$C_{32}H_{38}N_2O_5$	*Chaetomium globosum*

TABLE 2.6—*Continued*

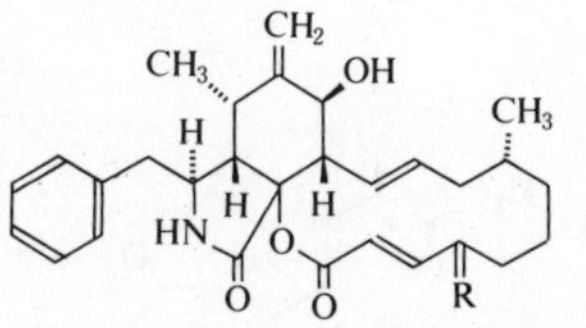

Cytochalasin A (R= =O)
Cytochalasin B (R=$<^{H}_{OH}$)

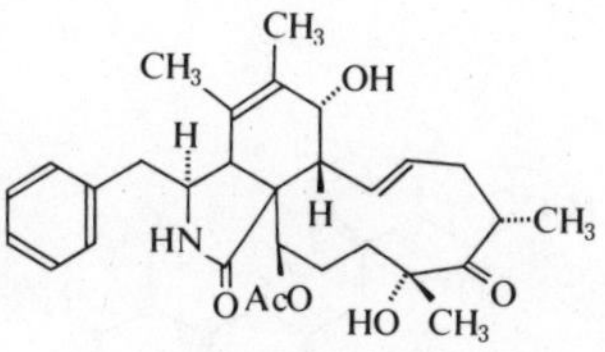

Cytochalasin C

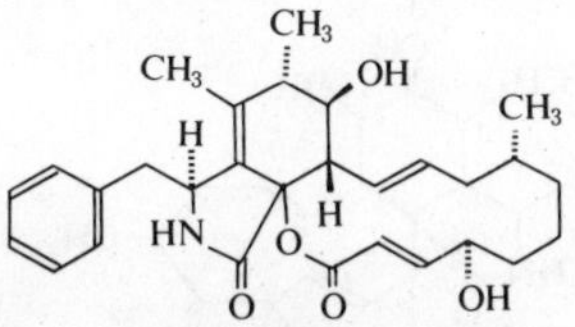

Cytochalasin F

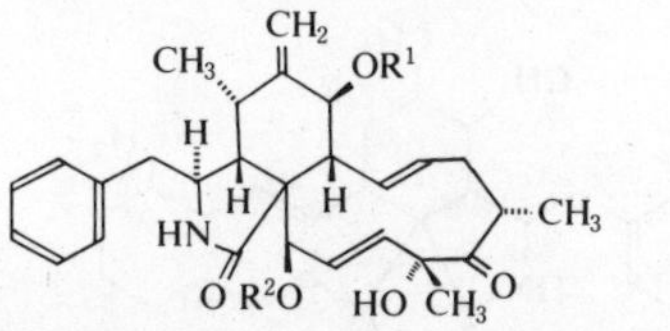

Cytochalasin D (R^1=H, R^2=Ac)
Zygosporin D (R^1=H, R^2=H)
Zygosporin F (R^1=R^2=Ac)

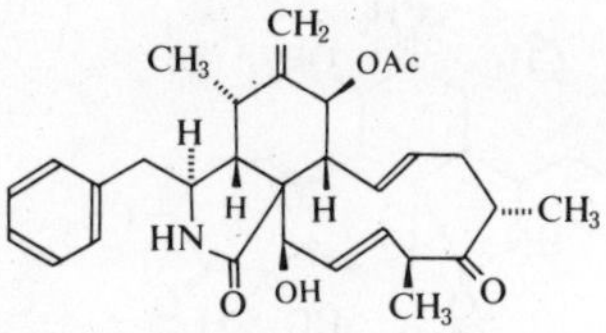

Zygosporin E

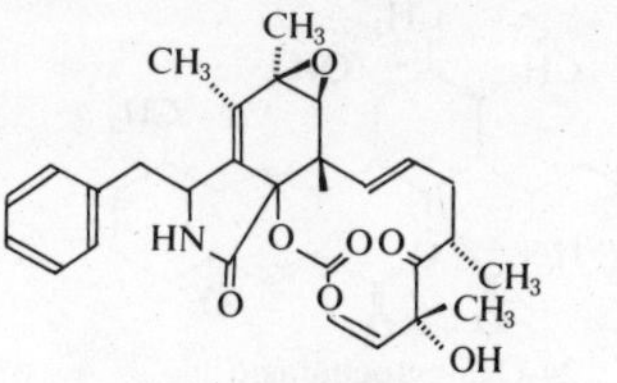

Cytochalasin E

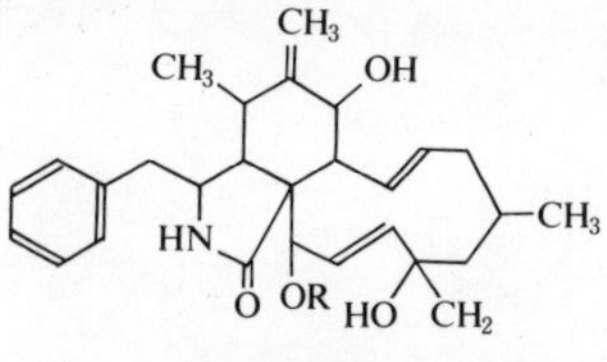

Kodo-cytochalasin-1 (R=Ac)
Kodo-cytochalasin-2 (R=H)

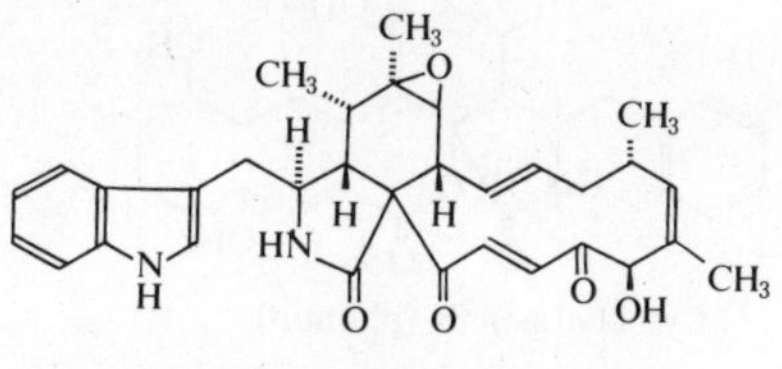

Chaetogloglobosin A

C_1 units incorporated from methionine, respectively. Tamm and his colleagues determined the distribution of ^{13}C in cytochalasin B obtained from *Phoma* sp. fed [2-^{13}C]-acetate and in cytochalasin D from *Z. masonii* after incorporation of [1-^{13}C]- and [2-^{13}C]-acetate and confirmed their proposed biosynthetic route, as shown in Fig. 2.35.[122)]

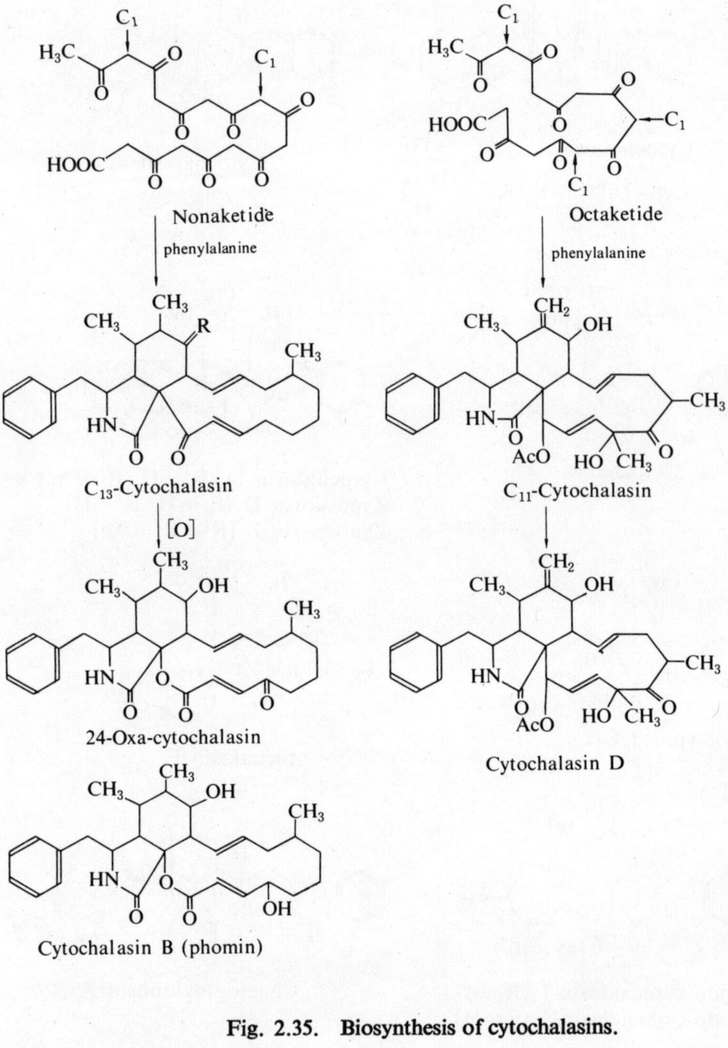

Nonaketide

phenylalanine

C_{13}-Cytochalasin

[O]

24-Oxa-cytochalasin

Cytochalasin B (phomin)

Octaketide

phenylalanine

C_{11}-Cytochalasin

Cytochalasin D

Fig. 2.35. Biosynthesis of cytochalasins.

2.5. MYCOTOXINS DERIVED FROM THE AROMATIC PATHWAY OF BIOSYNTHESIS (Shikimic Acid Pathway)

An important role of the shikimic acid pathway in secondary metabolism is to provide intermediates for the biosynthesis of aromatic compounds, including aromatic amino acids. Many simple phenolic compounds occur in fungi, but almost all of these compounds are derived through the acetate-malonate pathway, as described in the preceding section. The C_6–C_3 compounds derived through the aromatic pathway occur widely in higher plants but rarely in fungi. The shikimic acid pathway plays a relatively small part in secondary metabolism in fungi.[123]

Xanthocillin has been isolated from *Penicillium notatum*[124] and also from *Aspergilus chevalieri*.[125] It exhibits hepatotoxicity in experimental animals. This compound is an unusual isocyanide and is usually obtained as a mixture of xanthocillin X and Y. They both have antibiotic properties, but X is the predominant form. Grisebach *et al.*[126] showed that the radioactivity of [2-^{14}C]-tyrosine but not that of [1-^{14}C]-tyrosine is incorporated into xanthocillin (Fig. 2.36) An experiment using doubly-labeled tyrosine showed that the amino group of tyrosine did not participate in the formation of the isonitrile group in xanthocillins. Further, radioactive acetate and formate did not label the carbon atoms of the compound.

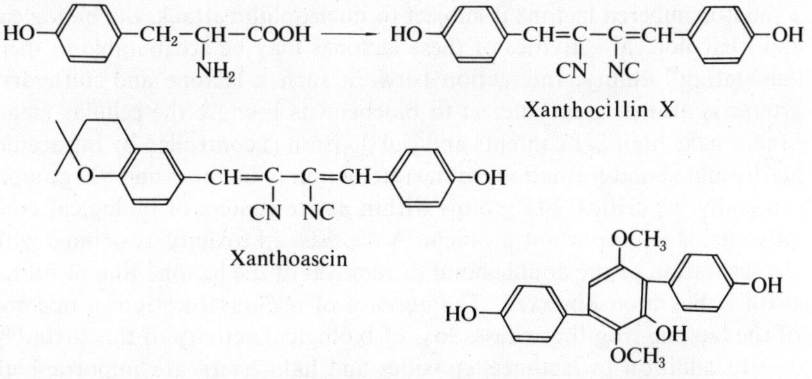

HO–C$_6$H$_4$–CH$_2$–CH(NH$_2$)–COOH ⟶ HO–C$_6$H$_4$–CH=C(CN)–C(NC)=CH–C$_6$H$_4$–OH

Xanthocillin X

Xanthoascin

Terphenyllin

Fig. 2.36. Formation of xanthocillin X from tyrosine, and the structures of terphenyllin (toxin A) and xanthoascin (toxin B).

Recently, a toxin-producing strain of *Aspergillus candidus* has been isolated and two new toxic metabolites tentatively named toxin A and B have been separated from the fungus by chromatographic methods monitored by cytotoxicity tests using HeLa cells.[127] The structure of toxin A has been elucidated, mainly by nmr spectrometry, as a terphenyl compound and identified with terphenyllin reported by Marchell and Vining.[128] Toxin B was subsequently investigated and its structure was found to be closely related to that of xanthocillin X. Toxin B has been named xanthoascin.[129]

2.6. SOME ASPECTS OF STRUCTURE-ACTIVITY RELATIONSHIPS IN MYCOTOXINS

Luteoskyrin and rugulosin, both containing an unusual cage-like structure, are able to chelate magnesium and calcium ions. This chelating property may play an important role in modifying the template activity of DNA and related biopolymers by binding through metal ion bridges.[130] Further, quinone-hydroquinone interconversion of these mycotoxins is also presumed to correlate with an uncoupling of mitochondrial respiration.[131] Both compounds are hepatotoxic and carcinogenic, with modified bisanthraquinone structures.

It has been established that compounds containing a lactone ring possess a wide range of cytotoxic and carcinogenic properties.[94,132] Since a small-membered lactone is subject to nucleophilic attack, the biological and toxicological activities of these lactones may be attributable to their "alkylating" ability. Interaction between such a lactone and sulfhydryl groups is also of great interest to biochemists because the cellular membranes have high SH contents and cell division is controlled by intracellular disulfide bond formation, so the interaction of lactones and SH groups, especially for critical SH groups within active centers of biological constituents, is an important problem. A decrease in toxicity associated with the saturation of the double bond or removal of the lactone ring of rubratoxin B has been observed. The absence of α,β-unsaturation or opening of the lactone ring thus causes loss of biological activity of this toxin.[133]

In addition to lactones, epoxides and halo-esters are important alkylating agents. As reviewed by van Duuren,[134] epoxides can be grouped into mono-, di and polyfunctional types. It is known that diepoxides are more frequently carcinogenic than monofunctional ones. It has been demonstrated that diepoxybutane reacts easily with guanine at the N-7 position *in vitro*. Like many fungal metabolites, trichothecenes possess an epoxy group. It is interesting that the biologically active trichothecenes lose their activity when they lose the epoxy group.[135]

Many unsaturated compounds can be transformed into epoxides which can react with cell constituents. This biological epoxidation can occur where a double bond is present in the molecule, and is affected by the stereospecificity of the enzyme involved (epoxidase in microsomes) and the stability of the resulting products. It is of interest to note that the LD_{50} of aflatoxin B_2, in which the terminal double bond of the bisfuran ring is saturated, is 10 times that of B_1 in 1-day-old ducklings.[136]

Wogan *et al.*[137] have investigated structure-activity relationships in the toxicity of aflatoxin analogs. They revealed that aflatoxin B_1 and G_1, in single doses, were lethal and had a similar relative potency to ducklings and rats (Table 2.7). Aflatoxin B_2 and G_2 were less potent in ducklings and were non-toxic to rats at doses of 200 mg/kg. Tetrahydrodesoxoaflatoxin B_1 and three synthetic compounds containing the substituted coumarin portion of the aflatoxin B_1 configuration (Fig. 2. 37) were non-toxic and non-carcinogenic at doses 100–200 times greater than aflatoxin B_1. Collectively, these data indicate that the furofuran moiety of the aflatoxin structure is essential for toxic and carcinogenic activity. Moreover, the presence of the double bond in the terminal furan is a vital determinant of potency, and the importance of the substituents on the lactone portion of the molecule is exemplified by the difference in potency between aflatoxin B_1 and aflatoxin G_1.

As for the tumorigenic activity of aflatoxin B_1, a total dose of 1 mg

TABLE 2.7. Single lethality of aflatoxins and aflatoxin analogs to rats (after Wogan *et al.*)[137]

Compound	I.p. dose (mg/kg)	Mortality at 14 days
Aflatoxin B_1	2.0–3.0	6/6
	1.0–2.0	29/46
	0.7–0.9	0/20
Aflatoxin B_2	12 –200	0/20
Aflatoxin G_1	1.0–1.5	0/19
	1.8–2.0	11/22
	3.0–10.0	54/54
Aflatoxin G_2	170 –200	0/4
Tetrahydrodesoxo B_1	32–128	0/6
Compound "11"	100–200	0/5
Compound "2"	200	0/5
Compound "8"	200	0/5

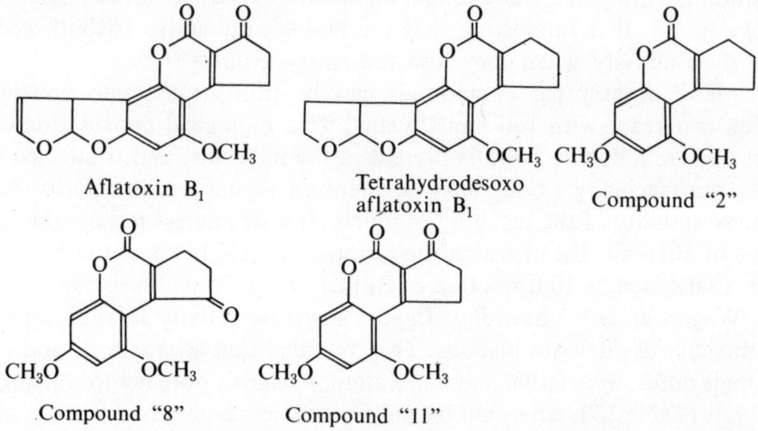

Fig. 2.37. **Structure of aflatoxin B_1, tetrahydrodesoxoaflatoxin B_1 and three related synthetic compounds.**

developed liver cancer in three out of 10 rats, whereas none developed in 10 rats receiving the same dose of aflatoxin B_2.[138] Data on the incorporation of ^3H-uridine into RNA of cultured liver cells of chicken embryos clearly demonstrate that aflatoxin B_1 strongly inhibits the biosynthesis of RNA, although the inhibition by aflatoxin B_2 was only moderate. In the specimen treated with tetrahydrodesoxoaflatoxin B_1, which was prepared from B_1 by catalytic hydrogenation (Fig. 2.38), no significant differences were observed between the intensities of ^3H-uridine-labeled cells in the control and in toxin-treated cells.[139] In the case of aflatoxins, the for-

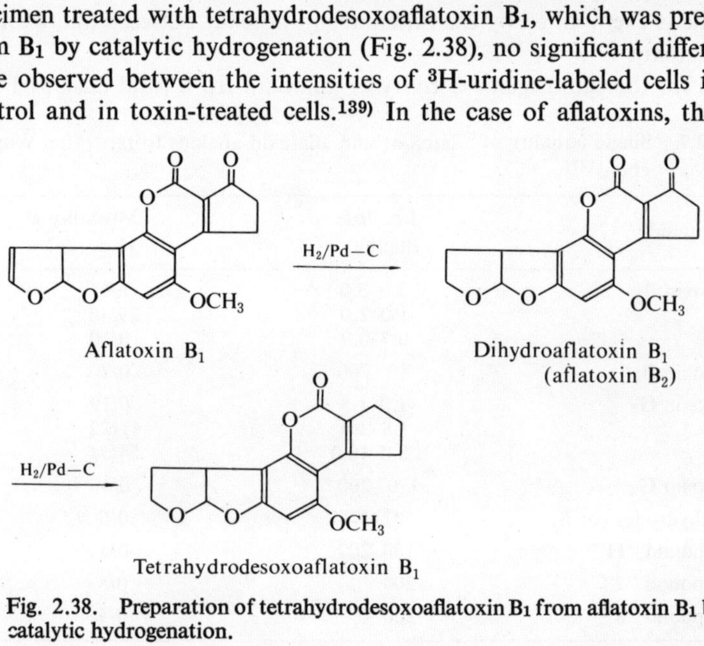

Fig. 2.38. **Preparation of tetrahydrodesoxoaflatoxin B_1 from aflatoxin B_1 by catalytic hydrogenation.**

mation of the epoxide in living organisms would thus be expected to produce biological activity. In fact, the formation of bound derivatives in hepatic DNA or RNA[140] or an epoxide-glutathione conjugate[141] upon incubation of aflatoxin B_1 with liver preparations has been demonstrated recently (Fig. 2.39), suggesting that aflatoxin B_1 forms an epoxide intermediate at the terminal double bond in the dihydrobisfuran ring.

Fig. 2.39. Formation of bound derivatives of aflatoxin B_1 with DNA or RNA via epoxide intermediate.

As in the case of aflatoxin B_1, the presence of carbonyl or alcohol groups sometimes affects the activity of other mycotoxins. When the carbonyl group at the 6′-position of zearalenone is reduced to alcohol, two stereoisomers are obtained. It is interesting that one (mp 178–180°C) is four times more active than the parent compound, while the other (mp 146–148°C) is only slightly more active. The reduction of the double bond between carbons 1' and 2' also enhances the activity of zearalenone, while the hydroxyl group on the benzene ring is essential for activity.[142]

Aspergillic and neoaspergillic acids are slightly toxic, having LD_{50} values of approximate 100 mg/kg (i.p.) to mice. However the deoxoderivatives of these pyrazine compounds have no toxicity.[143]

Numerous dipeptides forming 2,5-dioxopiperazine derivatives are produced by fungi. It is not clear whether they have any metabolic significance, and they are not pharmacologically active. However, certain dipeptides containing tryptophan are noteworthy. The tremorgenic properties of verruculogen[15] and fumitremorgin A and B[16] are remarkable. However, brevianamide and austamide are not tremorgenic even though they are similar dioxopiperazine compounds containing tryptophan and proline together with an isopentenyl group. The major difference between these and the tremorgenic toxins appears to be the lack of the methoxyl group at the 6-position of the indole ring and the hydroxyl groups on the side chain carbons of tryptophan in the latter.

Chlorinated mycotoxins such as ochratoxin A, griseofulvin and cyclochlorotine are highly hepatotoxic. Ochratoxin B, a dechloroderivative of ochratoxin A, has no toxicity.[46,144,145] Concerning the structure-activity relationships of ochratoxins, Chu *et al.*[145] employing a physicochemical method in their experiments, proposed that the phenolic group in the dissociated form is necessary for ochratoxin intoxication.

REFERENCES

1. *For ref.*: G. C. Ainsworth and A. S. Sussman (ed.), *The Fungi, An Advanced Treatise*, vol. 1, Academic Press (1965); J. H. Burnett, *Fundamentals of Mycology*, Edward Arnold (1968).
2. *For ref.*: M. W. Miller, *The Pfizer Handbook of Microbial Metabolites*, McGraw Hill (1961); S. Shibata, S. Natori and S. Udagawa, *List of Fungal Products*, Univ. of Tokyo Press (1964); J. D. BuLock, *The Biosynthesis of Natural Products*, McGraw Hill (1965); P. Bernfeld (ed.), *Biogenesis of Natural Compounds*, Pergamon Press (1967); H. Grisebach, *Biosynthetic Patterns in Microorganisms and Higher Plants*, John Wiley (1967); T. A. Geissman and D. H. G. Crout, *Organic Chemistry of Secondary Metabolism*, Freeman Cooper Co. (1969); M. Luckner, *Der Sekundärstoffwechsel in Pflanze und Tier*, VEB Gustav Fischer Verlag (1969); W. B. Turner, *Fungal Metabolites*, Academic Press (1971).
3. *For ref.*: G. N. Wogan(ed.), *Mycotoxins in Foodstuffs*, MIT Press (1965); R. I. Mateles and G. N. Wogan(ed.), *Biochemistry of Some Foodborne Microbial Toxins*, MIT Press (1967); A. Ciegler, S. Kadis and S. J. Ajl(ed.), *Microbial Toxins*, vol. VI–VIII, Academic Press (1971); I. F. H. Purchase(ed.), *Mycotoxins*, Elsevier (1974).
4. R. L. M. Synge and E. P. White, *Chem. Ind.*, 1546 (1959).
5. *For ref.*: L. G. Atherton, D. Brewer and A. Taylor, *Mycotoxins* (ed. I. F. H. Purchase), p. 29, Academic Press (1974).
6. N. R. Towers and D. E. Wright, *New Zealand J. Agr. Res.*, **12**, 275 (1969).
7. S. Wilkinson and J. F. Spilsbury, *Nature*, **206**, 619 (1965).
8. J. A. Winstead and R. J. Suhadolnik, *J. Am. Chem. Soc.*, **82**, 1644 (1960).
9. J. D. BuLock and A. P. Ryles, *Chem. Commun.*, 1404 (1970).
10. M. Nakagawa, T. Kaneko, H. Yamaguchi, T. Kawashima and T. Hino, *Tetrahedron*, **30**, 2591 (1974).
11. N. R. Towers, *New Zealand Agr. Res.*, **13**, 182 (1970).
12. M. S. Ali, J. S. Shannon and A. Taylor, *J. Chem. Soc.* (C), 2044 (1968).
13. J. C. McDonald, *J. Biol. Chem.*, **240**, 1692 (1965).
14. M. Lucker, *Eur. J. Biochem.*, **2**, 74 (1963); E. S. A. Aboutabl, A. E. Azzouny, K. Winter, M. Luckner, *Phytochem.*, **15**, 1925 (1976).
15. J. Foyes, D. Lockensgard, J. Clardy, R. J. Cole and J. W. Kirksey. *J. Am. Chem. Soc.*, **96**, 6785 (1974).
16. M. Yamazaki, H. Fujimoto and T. Kawasaki, *Tetr. Lett.*, 1241 (1975).
17. P. M. Scott, M.-A. Merrien and J. Polonsky, *Experientia*, **32**, 140 (1976).
18. S. D. Aust, *Biochem. Pharmacol.*, **19**, 427 (1970).
19. F. P. Guengerich, *Fed. Proc.*, **30**, 1067 (1971).
20. M. Sato and T. Tatsuno, *Chem. Pharm. Bull.*, **16**, 2182 (1968); H. Yoshioka, K. Nakatsu, M. Sato and T. Tatsuno, *Chem. Lett.* (*Tokyo*), 1319 (1973).
21. S. Marumo, *Bull. Agr. Chem. Soc. Japan*, **19**, 262 (1955).
22. K. Anzai and R. W. Curtis, *Phytochem.*, **4**, 263 (1965); S. Iriuchijima and R. W. Curtis, *ibid.*, **8**, 1397 (1969).
23. M. Yukioka and T. Winnick, *Biochim. Biophys. Acta*, **119**, 614 (1966); *J. Bact.*, **91**, 2237 (1966).
24. *For ref.*: J. H. Richards and J. B. Hendrickson, *The Biosynthesis of Steroids, Terpenes and Acetogenins*, W. A. Benjamin Inc. (1964).

25. E. B. Smalley and F. M. Strong, *Mycotoxins* (ed. I. F. H. Purchase), p. 199, Elsevier (1974).
26. S. Nozoe and Y. Machida, *Tetr. Lett.*, 2671 (1970).
27. S. Nozoe and Y. Machida, *Tetrahedron*, **28**, 5105 (1972).
28. Y. Machida and S. Nozoe, *Tetr. Lett.*, 1969 (1972).
29. Y. Machida and S. Nozoe, *ibid.*, 5113 (1972); *cf.* B. Achilladelis, P. M. Adams and J. R. Hanson, *Chem. Commun.*, 511 (1970).
30. R. E. Evans and J. R. Hanson, *Chem. Commun.*, 231 (1975).
31. M. Saito and K. Ohtsubo, *Mycotoxins* (ed. I. F. H. Purchase), p. 263, Elsevier (1974).
32. W. O. Godtfredsen, H. Lorck, E. E. van Tamelen, J. D. Willett and R. B. Clayton, *J. Am. Chem. Soc.*, **90**, 208 (1968); E. Capsi and L. J. Mulheirn, *ibid.*, **92**, 404 (1970).
33. S. Iwasaki, M. I. Sair, H. Igarashi and S. Okuda, *Chem. Commun.*, 1119 (1970).
34. T. S. Chou and E. J. Eisenbraun, *Tetr. Lett.*, 409 (1967).
35. W. O. Godtfredsen, S. Jahnsen, H. Lorck, K. Roholt and L. Tybring, *Nature*, **193**, 987 (1962).
36. J. N. Collie, *J. Chem. Soc.*, 1806 (1907).
37. *For ref.*: P. Bernfeld, (ed.) *Biogenesis of Natural Compounds*, Pergamon Press (1963); J. D. BuLock, *The Biosynthesis of Natural Products*, McGraw Hill (1965); S. W. Tanenbaum, *Biogenesis of Antibiotic Substances* (ed. Z. Vanek and Z. Hoštálek), p. 143, Czech. Acad. Sci. Prague (1965).
38. A. J. Birch and F. W. Donovan, *Australian J. Chem.*, **6**, 360 (1953).
39. R. Robinson, *Structural Relation of Natural Products*, Oxford (1955).
40. A. J. Birch, R. A. Massy-Westropp and C. P. Moye, *Chem. Ind.*, 683 (1955); *Australian J. Chem.*, **8**, 539 (1955).
41. A. C. Hetherington and H. Raistrick, *Phil. Trans. Roy. Soc. London*, **220B**, 269 (1931).
42. O. R. Rodig, *Chem. Commun.*, 1553 (1971).
43. E. Schwenk, G. J. Alexander, A. M. Gold and D. F. Stevens, *J. Biol. Chem.*, **223**, 1211 (1958).
44. R. F. Cartis, C. H. Hassal and M. Nazar, *J. Chem .Soc.* (C), 85 (1968).
45. O. R. Rodig, L. C. Ellis and I. T. Glover, *Biochemistry*, **5**, 2458 (1966).
46. K. J. van der Merwe, P. S. Steyn and I. Fourie, *J. Chem. Soc.* (C), 7083 (1965).
47. K. Nitta, Y. Yamamoto, T. Inoue and T. Hyodo, *Chem. Pharm. Bull.*, **14**, 363 (1966).
48. J. W. Searcy, N. D. Davis and U. L. Diener, *Appl. Microbiol.*, **18**, 622 (1969).
49. P. S. Steyn, C. W. Holzapfel and N. P. Ferreira, *Phytochem.*, **9**, 1977 (1970).
50. M. Yamazaki, Y. Maebayashi and K. Miyaki, *Tetr. Lett.*, 2301 (1971).
51. R. D. Wei, F. M. Strong and E. B. Smalley, *Appl. Microbiol.*, **22**, 276 (1971).
52. H. Iizuka and M. Iida, *Nature*, **196**, 681 (1962).
53. A. E. Oxford, H. Raistrick and P. Simonart, *Biochem. J.*, **33**, 240 (1939).
54. E. W. Hurst and G. E. Paget, *Brit. J. Derm.*, **75**, 105 (1963); S. S. Epstein, J. Andrea, S. Joshi and N. Mantel, *Cancer Res.*, **27**, 1900 (1967).
55. A. J. Birch, R. A. Massy-Westropp, R. W. Rickard and H. Smith, *J. Chem. Soc.*, 360 (1958); A. J. Birch, A. Cassera and R. W. Rickard, *Chem. Ind.*, 792 (1961).
56. A. Rhodes, G. A. Somerfield and M. P. McGonagle, *Biochem. J.*, **88**, 349 (1963).
57. W. J. McMaster, A. I. Scott and S. Trippett, *J. Chem Soc.*, 4628 (1960); A. Rhodes, B. Boothroyd, M. P. McGonagle and G. A. Somerfield, *Biochem. J.* **81**, 28 (1961).
58. C. H. Kuo, R. D. Hoffsommer, H. L. Slates, D. Taub and N. L. Wendler, *Chem. Ind.*, 1627 (1960); *Tetrahedron*, **19**, 1 (1963).
59. H. Raistrick and G. Smith, *Biochem. J.*, **30**, 1315 (1936); D. H. R. Barton and A. I. Scott, *J. Chem. Soc.*, 1767 (1967).
60. J. Balan, A. Kjaer, S. Kovac and R. H. Shapiro, *Acta Chem. Scand.*, **19**, 528 (1965); W. B. Turner, *J. Chem. Soc.*, 6658 (1965).
61. H. Fujimoto, H. Flasch and B. Franck, *Chem. Ber.*, **108**, 1224 (1975).
62. B. Franck and H. Flasch, *Fortschr. Chem. Org. Naturstoffe*, **30**, 151 (1973).

63. I. Yoshioka, T. Nakanishi, S. Izumi and I. Kitagawa, *Chem. Pharm. Bull.*, **16**, 2090 (1968).
64. P. S. Steyn, *Tetrahedron*, **26**, 51 (1970).
65. M. Yamazaki, Y, Maebayashi and K. Miyaki, *Chem. Pharm. Bull*, **19**, 199 (1971).
66. M. Harada, S. Yano, H. Watanabe, M. Yamazaki and K. Miyaki, *Chem. Pharm. Bull.*, **22**, 1600 (1974).
67. B. Franck and T. Reschke, *Chem. Ber.*, **93**, 347 (1960); B. Franck and I. Zimmer, *ibid.*, **98**, 1514 (1965).
68. D. Gröger, D. Erge, B. Franck, U. Oknsorge, H. Flasch and F. Hüper, *ibid.*, **101**, 1970 (1968).
69. S. Shibata, *Chemistry in Britain*, **3**, 110 (1967).
70. *For ref.*: M. Saito, M. Enomoto, T. Tatsunol and K. Uraguchi, *Microbial Toxins* (ed. A. Ciegler, S. Kadis and S. J. Ajl), vol. VI, p. 299, Academic Press (1971).
71. S. Shibata, *Pure Appl. Chem.*, **33**, 109 (1973).
72. S. Seo, U. Sankawa, Y. Ogiwara, Y. Iitaka and S. Shibata, *Tetrahedron*, **29**, 3703 (1973); S. Shibata, XXIth Intl. Congr. Pure Appl. Chem., vol. 2, p. 107 (Add. Publ. *Pure Appl. Chem.*), Butterworths (1974).
73. J. S. E. Holker and L. J. Mulheirn, *Chem. Commun.*, 1576 (1968).
74. M. Biollaz, G. Büchi and G. Milne, *J. Am. Chem. Soc.*, **90**, 5017, 5019 (1968); **92**, 1035 (1970).
75. P. S. Steyn, R. Vleggaar and P. L. Wessels. *Chem. Commun.*, 193 (1975); D. P. H. Hsieh, J. N. Sieber, C. A. Reece, D. L. Fitzell, G. N. LaMar, D. L. Budd and E. Motell, *Tetrahedron*, **31**, 661 (1975).
76. J. A. Donkersloot, R. I. Mateles and S. S. Yang, *Biochem. Biophys. Res. Commun.*, **47**, 1051 (1972).
77. M. T. Lin, D. P. H. Hsieh, R. C. Yao and J. A. Donkersloot, *Biochemistry*, **12**, 5167 (1973).
78. Y. Hatsuda and S. Kuyama, *Nippon Nogeikagaku Kaishi* (Japanese), **28**, 989 (1954).
79. J. C. Roberts and J. G. Underwood, *J. Chem. Soc.*, 2060 (1962).
80. N. Tanabe, Y. Katsube, Y. Hatsuda, T. Hamasaki and M. Ishida, *Bull. Chem. Soc. Japan*, **43**, 3635 (1970).
81. D. P. H. Hsieh, M. T. Lin and R. C. Tao, *Biochem. Biophys. Res. Commun.*, **52**, 992 (1973).
82. N. Sakabe, T. Goto and Y. Hirata, *Tetr. Lett.* 1825 (1964).
83. D. W. Nagel, P. S. Steyn and D. B. Scott, *Phytochem.*, **11**, 627 (1972).
84. D. W. Nagel, P. S. Steyn and R. P. Ferreira, *ibid.*, **11**, 3215 (1972).
85. M. Stob, P. S. Baldwin, J. Tuite, F. N. Andrew and K. G. Gillette, *Nature*, **196**, 1318 (1962).
86. W. H. Urry, H. L. Wehrmeister, E. B. Hodge and P. H. Hidy, *Tetr. Lett.*, 3109 (1966).
87. C. M. Christensen, G. H. Nelson and L. J. Mirocha, *Appl. Microbiol.*, **13**, 653 (1965).
88. O. C. Musgrave, *J. Chem. Soc.*, 4301 (1956); H. D. Munro, O. C. Musgrave and R. Templeton, *ibid.* (C), 947 (1967).
89. D. I. Fennel, *Chem. Ind.*, 1382 (1959); A. J. Birch, O. C. Musgrave, R. W. Rickards and H. Smith, *J. Chem. Soc.*, 3146 (1959).
90. R. N. Mirrington, E. Ritchie, C. W. Shopee, W. C. Taylor and S. Sternhell, *Tetr. Lett.*, 365 (1964).
91. F. McCapra, A. I. Scott, P. Delmotte, J. Delmotte-Plaquee and N. S. Bhacca, *ibid.*, 869 (1964).
92. J. A. Steele, J. R. Lieberman and C. J. Mirocha, *Can. J. Microbiol.*, **20**, 531 (1974).
93. *For ref.*: W. B. Turner, *Fungal Metabolites*, p. 106, Academic Press (1971); S. Shibata, S. Natori and S. Udagawa, *List of Fungal Products*, p. 41, Univ. of Tokyo Press (1964).
94. F. Dickens and H. E. H. Jones, *Brit. J. Cancer*, **15**, 85 (1961).
95. T. Yamamoto, *Yakugaku Zasshi* (Japanese), **74**, 810 (1954).

96. J. Schultz, R. Motz, M. Schäfer and W. Baumgart, *Monatsch. Vet. Med.*, **24**, 14 (1969).
97. J. H. Birkinshaw, *Ann. Rev. Biochem.*, **22**, 371 (1953).
98. J. D. BuLock and A. J. Ryan, *Proc. Chem. Soc.*, 222 (1958).
99. E. W. Bassett and S. W. Tanenbaum, *Biochim. Biophys. Acta*, **28**, 247 (1958).
100. R. I. Forrester and G. M. Gaucher, *Biochemistry*, **11**, 1102 (1972).
101. J. H. Birkinshaw, A. E. Oxford and H. Raistrick, *Biochem. J.*, **30**, 394 (1936).
102. A. J. Birch, G. E. Blanch and H. Smith, *J. Chem. Soc.*, 4582 (1958).
103. K. Mosbach, *Acta Chem. Scand.*, **14**, 457 (1960).
104. R. Bentley and J. G. Keil, *J. Biol. Chem.*, **237**, 867 (1962).
105. *For ref.*: R. Thomas, *Biogenesis of Antibiotic Substances* (ed. Z. Vaněk and Z. Hoštálek, p. 155, Czech. Acad. Sci. Praque (1965).
106. M. O. Moss, F. V. Robinson, A. B. Wood, H. M. Paisley and J. Feeney, *Nature*, **720**, 767 (1968).
107. S. Natori, S. Sakaki, H. Kurata, S. Udagawa, M. Ichinoe, M. Saito, M. Umeda and K. Ohtsubo, *Appl. Microbiol.*, **19**, 613 (1970).
108. M. O. Moss, F. V. Robinson and A. B. Wood, *Tetr. Lett.*, 367 (1969).
109. N. Wjkman, *Ann. Chem.*, **485**, 61 (1931).
110. D. H. R. Barton and J. K. Sutherland, *J. Chem. Soc.*, 1772 (1965).
111. H. Raistrick and G. Smith, *Biochem. J.*, **27**, 1814 (1933).
112. *For ref.*: J. K. Sutherland, *Proc. Chem. Org. Natl. Prod.*, **25**, 131 (1967); J. L. Bloomer, C. E. Moppett and J. K. Sutherland, *J. Chem. Soc.* (C), 588 (1968); R. K. Huff, C. E. Moppett and J. K. Sutherland, *ibid.* (C), 2584 (1972).
113. H. M. Rose and M. O. Moss, *Biochem. Pharmacol.*, **19**, 612 (1970).
114. D. C. Aldridge, J. J. Armstrong, R. N. Speake and W. B. Turner, *Chem. Commun.*, 26 (1967); *J. Chem. Soc.* (C), 1667 (1967).
115. W. Rothweiler and C. Tamm, *Helv. Chim. Acta*, **53**, 696 (1970).
116. D. C. Aldridge and W. B. Turner, *J. Chem. Soc.* (C), 923 (1969).
117. S. Hayakawa, T. Matsushima, T. Kimura, H. Minato and K. Katagiri, *J. Antibiotics*, **21**, 523 (1968); D. C. Aldridge and W. B. Turner, *ibid.*, **22**, 170 (1969); H. Minato and M. Matsumoto, *J. Chem. Soc.* (C), 38 (1970).
118. D. C. Aldridge, B. F. Burrows and W. B. Turner, *Chem. Commun.*, 148 (1972); G. Büchi, Y. Kitaura, S. S. Yuan, H. E. Wright, J. Clardy, A. L. Demain, T. Glinsukon, N. Hunt and G. N. Wogan, *J. Am. Chem. Soc.*, **95**, 5423 (1973).
119. D. C. Aldridge, B. F. Burrows and W. B. Turner, *Chem. Commun.*, 148 (1972).
120. H. Minato and T. Katayama, *J. Chem. Soc.* (C), 40 (1970).
121. *For ref.*: B. Binder and C. Tamm, *Angew. Chem. Intl. Ed.*, **12**, 370 (1973).
122. W. Graf, J.-L. Robert, J. Vederas, C. Tamm, P. H. Solomon, I. Miura and K. Nakanishi, *Helv. Chim. Acta*, **57**, 1801 (1974).
123. *For ref.*: W. B. Turner, *Fungal Metabolites*, p. 33, Academic Press (1971).
124. W. Rothe, *Pharmazie*, **5**, 190 (1950).
125. A. Takatsuki, S. Suzuki, K. Ando, G. Tamura and K. Arima, *J. Antibiotics*, **21**, 671 (1968).
126. H. Achenbach and H. Grisebach, *Z. Naturforsch.*, **20B**, 137 (1965).
127. C. Takahashi, K. Yoshihira, S. Natori, M. Umeda, K. Ohtsubo and M. Saito, *Experientia*, **30**, 529 (1974).
128. R. Marchelli and L. C. Vining, *Chem. Commun.*, 555 (1973).
129. C. Takahashi, K. Yoshihira, S. Natori, M. Umeda, *Chem. Pharm. Bull.*, **24**, 613 (1976); C. Takahashi, S. Sekita, K. Yoshihira, S. Natori, *ibid.*, 2317 (1976).
130. Y. Ueno, I. Ueno and K. Mizumoto, *Japan. J. Exptl. Med.*, **38**, 47 (1968).
131. I. Ueno, *Seikagaku* (Japanese), **38**, 741 (1966).
132. F. Dickens and H. E. H. Jones, *Brit. J. Cancer*, **19**, 392 (1965).
133. H. M. Rose and M. O. Moss, *Biochem. Pharmacol.*, **19**, 612 (1970).
134. B. L. van Duuren, *Mycotoxins in Foodstuffs* (ed. G. N. Wogan), p. 275, MIT Press (1965).

135. T. Tatsuno, *private communication.*
136. R. B. A. Carnaghan, R. D. Hartley and J. O'Kelly, *Nature*, **200**, 1101 (1963); S. B. Chang, M. M. Abdel-Kader, E. L. Wick and G. N. Wogan, *Science*, **142**, 1191 (1963).
137. G. N. Wogan, R. S. Edwards and P. M. Newberne, *Cancer Res.*, **31**, 1936 (1971).
138. W. H. Butler, M. Greenblatt and W. Lijinsky, *ibid.*, **29**, 2206 (1969).
139. K. Terao, M. Yamazaki and K. Miyaki, *Z. Krebsforsch.*, **78**, 303 (1972).
140. D. H. Swenson, E. C. Miller and J. A. Miller, *Biochem. Biophys. Res. Commun.*, **60**, 1036 (1974); R. C. Garner and C. M. Wright, *Chem. Biol. Interactions*, **11**, 123 (1975).
141. H. G. Raj, K. Santhaman, R. P. Gupta and T. A. Venkitasubramanian, *Chem. Biol. Interactions*, **11**, 301 (1975).
142. C. J. Mirocha and C. M. Christensen, *Mycotoxins* (ed. I. F. H. Purchase), p. 131, Elsevier (1974).
143. M. Sasaki, Y. Asao and T. Yokotsuka, *Nippon Nogeikagaku Kaishi* (Japanese), **42**, 351 (1968).
144. J. C. Peckham, B. Doupnik, Jr. and O. H. Jones, Jr., *Appl. Microbiol.*, **21**, 492 (1971).
145. F. S. Chu, I. Noh and C. C. Chang, *Life Sci.*, **11**, 503 (1972).

CHAPTER **3**

Toxicology and Biochemistry of Mycotoxins

Yoshio UENO
Tokyo University of Science, Ichigaya, Shinjuku-Ku, Tokyo 162, Japan
Ikuko UENO
The Institute of Medical Science, University of Tokyo, Shiroganedai, Minato-ku, Tokyo 108, Japan

3.1. GENERAL TOXICITY

3.1.1. Toxic Fungal Metabolites

Mycotoxins are secondary fungal metabolites capable of eliciting a toxic response, i.e. mycotoxicosis, in humans and animals. Over several tens of compounds have so far been reported as mycotoxins, although their toxicological features have been elucidated only in a limited number of cases.

Fungi which are capable of producing toxic metabolites are distributed over a large number of fungal families and no particular rule or principle exists concerning the ability to produce mycotoxins. Further, the chemical and biological nature of mycotoxins is highly diverse, and it is very difficult to distinguish clearly between types of fungal metabolites that are toxic and those that are non-toxic. Mycotoxins have therefore been classified here into *Penicillium* mycotoxins (Table 3.1), *Aspergillus* mycotoxins (Table 3.2), *Fusarium* mycotoxins (Table 3.3), and miscellaneous mycotoxins (Table 3.4). The LD_{50} values of the mycotoxins for mice, rats and other animals are also given. The acute toxicity, which is expressed as a lethal dose for animals, depends on the species, age and sex of the animals, and on the route of administration, as will be discussed later. Tentatively, the toxic potential will be classified according to the relative toxicity, as shown in Table 3.5.

3.1.2. Organ Specificity

Fungal toxic metabolites exhibit harmful effects on various organs and tissues of animals. From their different affinities to animal organs, mycotoxins may be grouped as hepatotoxins, nephrotoxins, dermotoxins, etc., as summarized elsewhere (see Table 4.10).

Since the discovery of the hepatotoxicity of luteoskyrin[1)] and aflatoxin,[2)] numerous surveys have demonstrated that the liver is the organ most frequently affected by many fungal metabolites. The pathological characteristics of the liver injuries indicate that different types of hepatotoxic mycotoxins occur. They may attack liver parenchymal cells or Kupffer's cells selectively, or may injure the central and/or peripheral areas of the liver lobules. Further, necrosis, fat accumulation and glycogen disappearance are indicative of biochemically different modes of hepatotoxic damage.

Reproductive organs are damaged by a limited number of mycotoxins. Zearalenone from *F. graminearum* affects the uterus of animals, causing

TABLE 3.1. Acute toxicity of *Penicillium* mycotoxins

Mycotoxin	Fungus	LD_{50} (mg/kg)		
		mice	rats	others
Citreoviridin	*P. citreo-viride*	8.2 (i.p.) 8.3 (s.c.) 29 (p.o.)	 3.6 (s.c.)	
Citrinin	*P. citrinum*	35 (s.c.)	50 (p.o)	19) i.p., rabbit)
Cyclochlorotine	*P. islandicum*	0.3 (i.v.) 0.45 (s.c.) 6.5 (p.o.)		
Cyclopiazonic acid	*P. cyclopium*		2.3 (i.p.) 36 (p.o.)	
Decumbin	*P. decumbens*		275 (p.o.)	
Erythroskyrine	*P. islandicum*	60 (i.p.)		
Gliotoxin	*P. terlikowskii*	45–65 (s.c.)		
Griseofulvin	*P. griseofulvum*		400 (i.v.)	
(−)Luteoskyrin	*P. islandicum*	6.6 (i.v.) 40.8 (i.p.) 147 (s.c.) 221 (p.o.)		
Mycophenolic acid	*P. brevi-compactum*	550 (i.v.) 2500 (p.o.)		
Patulin	*P. patulum*	10 (s.c.) 25 (p.o.)		170 (p.o., chick)
Penitrem A	*P. palitans*	1.1 (i.p.) 10 (p.o.)		
Penitrem B	*P. cyclopium*	5.8 (i.p.)		
Penitrem C	*P. crustosum*			
Penicillic acid	*P. puberulum*	250 (i.v.) 110 (s.c.) 530 (p.o.)		
PR-toxin	*P. roqueforti*	6 (i.p.) 20 (s.c.) >60 (p.o.)		
(+)Rugulosin	*P. rugulosum*	83 (i.p.)	44 (i.p.)	
Rubratoxin B	*P. rubrum*	2.6 (i.p.) 400 (p.o.)	0.35 (i.p.)	0.2 (i.p.,cat)
Rubratoxin A		6.3 (i.p.)		
Secalonic acid	*P. oxalicum*	42 (i.p.)		
Verruculogen (TR-1)	*P. verruculosum*	2.4		

TABLE 3.2. Acute toxicity of *Aspergillus* mycotoxins

Mycotoxin	Fungus	LD_{50} (mg/kg)		
		mice	rats	others
Aflatoxin B_1	*A. flavus*	9 (p.o.)	7.2 (p.o.)	see Table 3.6
Aspergillic acid	*A. flavus*	100 (i.p.)		
Fumagillin	*A. fumigatus*	800 (s.c.)		
Fumigatoxin	*A. fumigatus*	150 (LD_{100})		
Fumitremorgen A	*A. fumigatus*	5		
Helvonic acid	*A. fumigatus*	250 (i.v.) 400 (i.p.)		
Kojic acid	*A. flavus*, etc.	250 (i.p.)		
Malformin A	*A. nigar*	3.1 (i.p.)		
Maltoryzine	*A. oryzae* var. *microsporum*	3		
Nidultoxin	*A. nidulans*			1 μg/chick embryo
β-Nitropropionic acid	*A. flavus*	200 (p. o.)		
Ochratoxin A	*A. ochraceus*		20 (p.o.)	150μg/ducklling (p.o.)32–46 (monkey)
Ochratoxin B	*A. ochraceus*			
Palmotoxin B_0	*A. flavus*		6.4 (i.p.)	
Palmotoxin G_0	*A. flavus*			
Sterigmatocystin	*A. versicolor*	800 (p.o.)	60 (i.p.)	
Terreic acid	*A. terreus*	71–119 (i.v.)		
Terphenyllin	*A. candidus*			
Xanthoascin	*A. candidus*			
Viriditoxin	*A. viridi-nutans*	2.8 (i.p.)		
Xanthocillin-X	*A. chevalieri*	35 (i.p.) 25 (i.m.)	40 (p.o.)	60 (i.p., 150 s.c., guinea-pig)

TABLE 3.3. Acute toxicity of *Fusarium* mycotoxins

Mycotoxin	Fungus	LD_{50} (mg/kg)		
		mice	rats	others
Butenolide	*F. sporotrichioides*	43 (i.p.) 275 (p.0)		
Fusaric acid	*F. moniliforme*	100 (i.v.) 180 (p.o.)		180 (i.v., dog) 215 (i.v., rabbit)
Moniliformin	*F. moniliforme*			4 (o.p., cockerel)
Trichothecenes				
nivalenol	*F. nivale*	4.1 (i.p.)		
T-2 toxin	*F. tricinctum*	5.2 (i.p.)		
fusarenon-X	*F. nivale*	3.4 (s.c.)	4.4 (p.o.)	2 (s.c., cat)

TABLE 3.4. Acute toxicity of miscellaneous mycotoxins

Mycotoxin	Fungus	LD_{50} (mg/kg)		
		mice	rats	others
Chaetoglobosin A	*Chaetomium globusum*	6.5 (M, s.c.) 17.8 (F, s.c.)		
Cytochalasin D		1.85 (s.c.)		
Cochliodinol	*Chaetomium* sp.		<2 g (p.o.)	
Desmethoxyviridiol	*Nodulisporium hinnuleum*			4.2 (p.o., cockerel)
Diplodiatoxin	*Diplodia maydis*			
Oosponol	*Oospora astringens*			
Oosporein	*Chaetomium trilacterale*			6.12 (p.o., cockerel)
Orellanine	*Cortinarua orellanus*	8.3		4.9 (cat), 8.0 (guinea-pig)
Roridin A	*Myrothecium roridum*	1.0 (i.v.)		
Satratoxin	*Stachybotrys atra*			
Psoralen	*Pithomyces*			
Sporidesmin	*Pithomyces chartarum*	300	30	2–4 (guinea-pig), 1–2 (rabbit) 0.4 (s.c., sheep)
Stemphone	*Stemphylium sarcinaeforme*			100 μg/chick embryo
Trichotoxin A	*Trichoderma viride*			
Verrucarin A	*Myrothecium cerrucaria*	1.5 (i.v.)		

TABLE 3.5. Relative toxicity of mycotoxins

Range of toxicity	Lethal dose	Mycotoxins
Extremely toxic	below 1 mg/kg	cyclochlorotine, rubratoxin B
Very toxic	1–10 mg/kg	citreoviridin, luteoskyrin, aflatoxin B_1, maltoryzine, malformin, viridicatin, trichothecenes, cytochalasins
Toxic	10–100 mg/kg	others

vulvo-vaginitis and infertility,[3] and trichothecenes from *Fusarium* spp. cause cytological damage to the testis and ovary.[4]

Hematopoeitic organs, especially of the bone marrow, are severely attacked by trichothecenes. As will be discussed below, trichothecenes such as T-2 toxin and fusarenon-X induce a depression of leucopoiesis in acutely poisoned cats and mice, and from the evidence that subacute exposure of cats to T-2 toxin results in marked leucopenia and cellular destruction in the bone marrow, Ueno *et al.* have presented experimental proof for the etiological cause of alimentary toxic aleukia (ATA).[5]

Since Uraguchi[6] demonstrated ascending paralysis and respiratory arrest in several vertebrates intoxicated with citreoviridin from *P. citreoviride*, several kinds of neurotoxic mycotoxins have been identified among fungal metabolites. The pharmacologically active compounds include the tremorgenic mycotoxins such as fumitremorgens, verruculogen and peritrems,[7] all of which contain indole nuclei.

An interesting mycotoxin is xanthoascin from *A. candidus*. This attacks heart muscle, and in acutely poisoned mice, dilatation of the right and left ventricles and accumulation of myeloidal filaments in the heart muscle occur.[8]

Psoralens from *Sclerotinia sclerotiorum* and trichothecenes are dermotoxic. The photodynamic reaction of psoralens results in severe injury of the skin tissues. In the early literature, muconomycin A (verrucarin A) and muconomycin B, both macrocyclic trichothecenes, were reported as skin-inflammatory metabolites of *Myrothecium verrucaria*,[9] and subsequent research has shown T-2 toxin to be a highly potent skin-necrotizing metabolite of *F. tricinctum*.[10]

As mentioned above, many fungal toxins exhibit, in general, organ specificity. This specific nature may be attributable to (1) specific distribution of the mycotoxins into the target organs, (2) different susceptibility of the target cells according to cellular growth rate, (3) organ-specific activation of the mycotoxins to more toxic reagents, and (4) organ- or cell-specific permeability to mycotoxins.

Recent advances have indicated that microbial cells produce many pharmacologically important compounds such as hormone-like substances, enzyme inhibitors, histamine-sensitizing factors, all of which exhibit organ specificity. From the standpoint of mycotoxicology, however, safety evaluation of these metabolites should be made carefully.

3.1.3. Species and Strain Differences

In this section, examples of factors which introduce variation into the acute toxicity of mycotoxins are briefly considered. The experimental data quoted are either those previously reviewed in the literature or those obtained by Ueno *et al.* Summarized data are presented in Table 3.6.

TABLE 3.6. Species differences in the acute toxicity of mycotoxins

Species	LD_{50} (mg/kg)				
	aflatoxin B_1 (p.o.)	rubratoxin B (i.p.)	sporidesmin	fusarenon-X (s.c.)	citrinin (s.c.)
Mouse	9.0	0.27	200–300	4.2	35
Rat	5.5–17.9	0.35	20–30	4.0	67
Guinea-pig	1.4–2.0	0.48	2–4	0.5 (i.p.)	37
Hamster	10.2				
Rabbit	0.3–0.5		1–2		
Cat	0.55	0.2		*ca.* 2	
Dog	*ca.* 1.0	5.0			
Sheep	—		1–3		
Monkey	2.2–7.8				
Pig	0.62				
Duckling	0.3			2	

The extent of species variation in mycotoxin toxicity is appreciable. Animals of different species vary in their susceptibility to acute aflatoxin poisoning, with LD_{50} values ranging from 0.3 to 17.7 mg/kg. Rabbit, cat and pig are highly susceptible, followed by dog and monkey. Rat, mouse and hamster are resistant. Sheep and rabbit are sensitive to sporidesmin. Rubratoxin B exhibits a similar toxicity to mice, rat and guinea-pig, while fusarenon-X, one of the trichothecene mycotoxins, exhibits a high lethality to guinea-pig.

The strain of the experimental animal is one of the important factors influencing the toxic potential of mycotoxins. In feeding experiments of mice with *P. islandicum* Sopp-infected rice containing hepatocarcinogenic luteoskyrin and cyclochlorotine, definite differences in susceptibility among seven strains were observed.[11] C_{57}BL/6 mice proved very susceptible, while KK mice were the most resistant to these toxic metabolites, based on s-GOT and histopathological liver injuries. According to Saito *et al.*,[11] the liver lesions observed in these mouse strains could be divided into three types according to the site of the most remarkable cytological change in the liver lobules, as follows:

(1) periportal type—C_{57}BL/6, CFW, NC, and KK

(2) centrolobular type—C_3H/He
(3) centrolobular and periportal type—ddD, and CF 1.

These findings indicate that in the design of experiments, it is important to recognize the possible diversity of diagnostic pathology of cellular lesions occurring according to strain, even for a single mycotoxin.

3.1.4. Sex and Age Differences

Sex and age differences in acute toxicity are summarized in Table 3.7. In general, newborn and male animals are more susceptible to mycotoxins than adults and females. Exceptional cases are cyclochlorotine and citreoviridine, which are both more highly toxic to adult mice than newborns.

TABLE 3.7. Sex and age differences in the acute toxicity of mycotoxins

Mycotoxin	Animal	Route	LD_{50} (mg/kg)				
			newborn	weanling		adult	
				male	female	male	female
(−) Luteoskyrin	mouse	s.c.	7.2	27	22	154	2000
Aflatoxin B_1	rat	p.o.	0.56	5.5	7.4	7.2	17.9
Fusarenon-X	mouse	s.c.	0.2			4.2	
Rubratoxin B	rat	i.p.				0.36	0.36
Cyclochlorotine	mouse	s.c.	4.0			0.5	< 0.5
Citreoviridin	mouse	s.c.	128(♂), 106(♀)†				72†

†Lethal time (min), 500 mg/kg crude ether extract.

As a demonstration of the acute toxicity of (−) luteoskyrin to ddyS mice, the s.c. lethal dose of this mycotoxin is plotted against age in Fig. 3.1. The lethal doses to newborns and weanlings were very small compared to those for adults, and no sex differences were noted at this stage. However, above 3 weeks old, the lethal dose to female mice increased significantly. Biochemical tests such as s-GPT, and histopathological examinations revealed that the male adult mice were more highly sensitive to (−) luteoskyrin than the female adults. This sex difference in (−) luteoskyrin hepatotoxicity appears to be caused by specific accumulation of the toxicant in the male liver, as discussed in section 3.3.

3.1.5. Administration Routes and Vehicles

The lethal doses of seven mycotoxins were compared utilizing different routes of administration. As summarized in Table 3.8. the lethal doses were

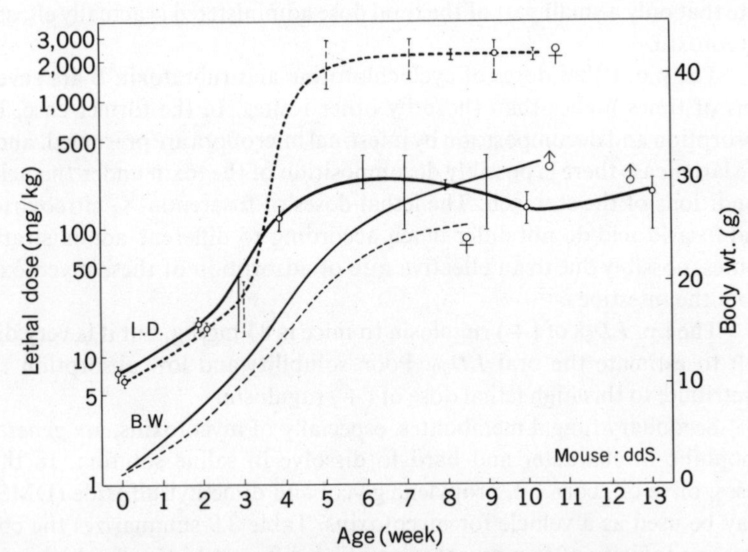

Fig. 3.1. Sex and age differences in luteoskyrin toxicity.

TABLE 3.8. Differences in lethal doses of mycotoxins according to administration route

Mycotoxin	Animal	*LD*$_{50}$ (mg/kg)			
		i.v.	i.p.	s.c.	p.o.
Citreoviridin	mouse	—	8.2	8.3	29
(—)Luteoskyrin	mouse	6.6	40.8	147	221
Cyclochlorotine	mouse	0.3	—	0.45	6.5
Rubratoxin B	rat	—	0.35	—	450
	duckling	—	5	—	60
PR-toxin	mouse	—	7	20	60
Fusaric acid	mouse	100	88	—	180
Sterigmatocystin	rat	—	60	—	166
(+)Rugulosin	mouse	—	83	—	<2000
Fusarenon-X	mouse	3.4	3.4	4.2	6

generally in the order, i.v. < i.p. < s.c. < p.o. For example, the single lethal dose (mg/kg) of (—) luteoskyrin dissolved in olive oil to male mice was i.p. 40.8, s.c. 147, p.o. 221, and the i.v. lethal dose of the mycotoxin dissolved in 0.9% NaCl solution was estimated at 6.6. The *LD*$_{50}$ values for these four different routes are thus in the ratio 1:6:22:33. These data indicate that only a small part of the total dose administered is actually effective mycotoxin.

The p.o. lethal doses of cyclochlorotine and rubratoxin B are several tens of times higher than those by other routes. In the former case, low absorption and decomposition by intestinal microflora are presumed, and in the latter case there is possibly decomposition of the toxin under the acidic conditions of the stomach. The lethal doses of fusarenon-X, citreoviridin and fusaric acid do not differ much according to different administration routes, possibly due to an effective rate of adsorption of these mycotoxins from the intestine.

The i.p. LD_{50} of (+) rugulosin to mice is 83 mg/kg, but it is very difficult to estimate the oral LD_{50}. Poor solubility and low absorption rate contribute to this high lethal dose of (+) rugulosin.

Secondary fungal metabolites, especially of mycotoxins, are generally lipophilic in character and hard to dissolve in saline solution. In these cases, olive oil, corn oil, propylene glycol and dimethylsulfoxide (DMSO) may be used as a vehicle for mycotoxins. Table 3.9 summarizes the comparative toxicity of two mycotoxins using different kinds of vehicles. (−) Luteoskyrin, one of the bis-anthraquinonoids, possesses a phenolic hydroxyl group in its structure. When dissolved in aqueous solutions by the addition of borax, sodium hydroxide or alkaline phosphate, the lethal dose decreases to about 1/10 of that when olive oil is used. Wogan *et al.*[12] have examined the acute toxicity of rubratoxin B. They showed that in rats, the LD_{50}'s of the mycotoxin dissolved in propylene glycol or DMSO are of the same order, while in mice the toxicity in DMSO solution is ten times that in propylene glycol solution.

It is well known that vehicles such as propylene glycol and DMSO are toxic to animals. Consideration must therefore be given to species differ-

TABLE 3.9. Solvent effects on the toxicity of mycotoxins

Mycotoxin	Animal	Solvent	Route	LD_{50}(mg/kg)
(−)Luteoskyrin	mouse (male)	olive oil	s.c.	147
		borax solution	s.c.	30
		phosphate buffer (pH 7)	s.c.	30
Rubratoxin B	rat (female)	propylene glycol	i.p.	0.36
		DMSO		0.35
	mouse (female)	propylene glycol	i.p.	2.6
		DMSO		0.27
	cat (male and female)	propylene glycol	i.p.	—
		DMSO		*ca.* 0.2

ences in the toxicity of such vehicles as well as the combined effects of mycotoxins and vehicles.

3.1.6. Combined Toxicity

Two or more mycotoxins may naturally occur simultaneously in foods and feedstuffs. Multiple fungal metabolites can accumulate in a given substrate through biosynthesis of several compounds by a single fungal species or by different species. *P. islandicum* Sopp which produces three different types of mycotoxins such as (−) luteoskyrin, cyclochlorotine and erythroskyrine, *A. ochraceus* which produces ochratoxin and penicillic acid, and *Fusarium* spp. which produce trichothecenes, zearalenone and butenolide, are examples of fungal species which can simultaneously elaborate multiple mycotoxins in grains.

In early biological and chemical experiments in search of toxic principles in yellowed rice infected with *P. islandicum*, the methanol extract of the cultured fungal mat was administered to mice and rats to observe possible acute toxic symptoms, liver malfunction by the BSP test (bromosulfophthalein-retention test), and histopathological changes.[13] The data obtained were divided into two different patterns, as illustrated in Fig. 3.2. This indicated that the liver injuries induced by the methanol extract were caused by two different toxicants. Using this as a working hypothesis for further chemical studies, two mycotoxins were isolated. One was a hydrophilic compound with rapid action, named Cl-peptide, and the other was a lipophilic compound with slow action, (−) luteoskyrin. Thus, crude toxin from *P. islandicum* exhibits a combined hepatotoxicity to animals.

In the experiments of Lindenfelser *et al.*,[14] ochratoxin (86% ochratoxin A, 14% ochratoxin B) and penicillic acid, alone or as mixtures of various proportions, were administered i.p. to female mice. The acute LD_{50} values obtained for ochratoxin and penicillic acid were 24 and 70 mg/kg, respectively. The LD_{50} values for the two mycotoxins alone and for various combinations were plotted by the method of Hewlett[15] to form an isobologram, as shown in Fig. 3.3 (A). The straight, dashed line shows the expected LD_{50} values assuming an additive response. The solid curve obtained by plotting the observed LD_{50} values signifies a synergistic effect for the two mycotoxins.

Similar experiments with aflatoxin B_1 and T-2 toxin revealed that the i.p. LD_{50} values to mice were 4 mg/kg for B_1 and 0.4 mg/kg for T-2 toxin. The isobologram indicated a synergistic effect, as illustrated in Fig. 3.3 (B). For example, at a dosage of 0.7 mg/kg B_1, the plot shows that a direct, additive response would require 0.3 mg/kg T-2 toxin for the combined LD_{50}

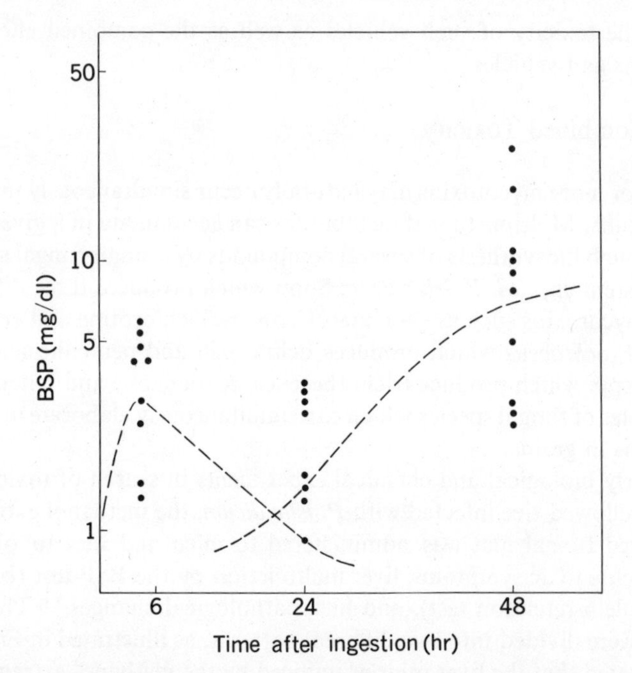

Fig. 3.2. BSP retention test in mice orally administered with methanol extract of *P. islandicum*.[13] Disturbed liver function occurred at two different periods in association with respective histological damage to the liver, indicating the presence of two toxins.

dose. However, the observed LD_{50} is 0.7 mg/kg B_1 plus 0.1 mg/kg T-2 toxin.[16] These data provide presumptive information regarding the enhanced acute toxicity that can be expected in commodities simultaneously contaminated with two or more mycotoxins.

Various tests measure toxic interactions, such as the animal acute toxicity test based on various combinations of toxicants and the initiator-promotor process in the development of skin tumors. The acute toxicity test permits an evaluation of toxin mixtures administered to animals in a single dose, and from the observed mortality the amount of toxicant that constitutes a lethal dose can be calculated. The distinctive feature of the skin test is the two-stage process in which a single application of an initiator compound followed by repeated application of croton oil produces tomorous changes in sensitive strains of test animals, while both the initiator and promotor alone are ineffective at similar dosages.

In mouse skin tumor tests, neither ochratoxin nor penicillic acid contributes to papilloma development in two-stage tumor induction on a base of 9, 10-dimethyl-1, 2-benz (*a*) anthrathene (DBA) as control initiation

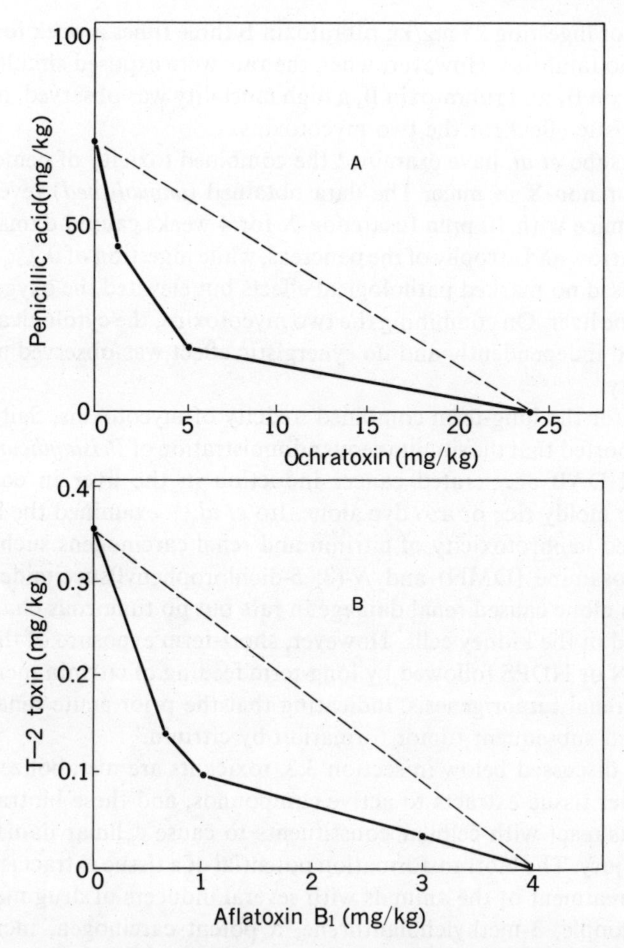

Fig. 3.3. Isobolograms for ochratoxin and penicillic acid (A) and for aflatoxin B_1 and T-2 toxin (B). Dashed lines represent expected LD_{50}'s, assuming an additive response; solid lines represent observed LD_{50}'s for various combinations of each pair of mycotoxins.

agent. Aflatoxin B_1 was found to be an initiator of mouse skin tumors but trichothecenes such as T-2 toxin and diacetoxyscirpenol were not, when croton oil served as the promotor. The trichothecenes acted as weak promotor substances on DBA-initiated mouse skin cells. Either DBA or aflatoxin B_1 initiation followed by T-2 toxin promotion caused extensive skin damage and subsequent tolerance of the treated skin to an elevated dose of trichothecene.

The combined effect of rubratoxin B and aflatoxin B_1 has been studied in rats.[12] Subacute dosing of rats either by feeding 0.2 ppm aflatoxin B_1 for

6 weeks or ingesting 25 mg/kg rubratoxin B three times a week for 5 weeks caused no fatalities. However, when the rats were exposed simultaneously to aflatoxin B_1 and rubratoxin B, a high mortality was observed, indicating a synergistic effect for the two mycotoxins.

Ohtsubo *et al.* have examined the combined toxicity of penicillic acid and fusarenon-X in mice. The data obtained (*unpublished*) revealed that feeding mice with 10 ppm fusarenon-X for 4 weeks caused damage to the bone marrow and atrophy of the pancreas, while ingestion of 0.2% penicillic acid caused no marked pathological effects but elevated the oxygenase system of the liver. On combining the two mycotoxins, the cytological changes appeared independently and no synergistic effect was observed as regards mortality.

As for the long-term combined toxicity of mycotoxins, Saito *et al.*[17] have reported that the simultaneous administration of *P. islandicum*-molded rice and DAB accelerated cancer induction in the liver in comparison with the moldy rice or azo dye alone. Ito *et al.*[18] examined the long-term combined nephrotoxicity of citrinin and renal carcinogens such as dimethylnitrosamine (DMN) and *N*-(3, 5-dichlorophenyl)succimide (NDPS). Citrinin alone caused renal damage in rats but no tumorous changes were observed in the kidney cells. However, short-term exposure of the animals to DMN or NDPS followed by long-term feeding of citrinin increased the rate of renal tumorigenesis, indicating that the prior acute renal damage enhanced subsequent tumor formation by citrinin.

As discussed below in section 3.3, toxicants are metabolized by liver and other tissue extracts to active compounds, and these biotransformed products react with cellular constituents to cause cellular damage or genetic injury. The biotransformation potential of a tissue extract is enhanced by pretreatment of the animals with several inducers of drug metabolism. For example, 3-methylcholanthrene, a potent carcinogen, increases the microsome-mediated binding of aflatoxin B_1 to DNA *in vitro*. This suggests that the DNA-modifying activity of aflatoxin B_1 is enhanced by other carcinogenic chemicals. In general, carcinogenic chemicals act as inducers of hepatic microsomal enzymes. It is therefore highly possible that a certain carcinogenic mycotoxin is capable of increasing the carcinogenic efficiency of other carcinogenic mycotoxins.

According to recent progress on the molecular mechanism of aflatoxin interaction with DNA, the 2,3-epoxide of aflatoxin is considered as an obligate intermediate. Epoxide derivatives may compete with aflatoxin epoxide for the specific hydrolase which is located in microsomes. Therefore, several kinds of epoxide analogs result in an increase in the cellular level of "active aflatoxin". This raises the important possibility that a mycotoxin which is capable of enhancing the microsomal oxygenase and depressing the microsomal epoxide hydrolase, may be able to promote the carcinogenic potency of aflatoxin.

Further research is expected to provide a key for understanding and evaluating the combined effects of mycotoxins on the basis of the molecular mechanisms of cytotoxicity as well as carcinogenicity. Full safety evaluation of each mycotoxin will ultimately be established by adopting this fundamental approach to the combined effects of mycotoxins themselves and of mycotoxins and other pollutants.

3.2. Comparative Toxicology

The toxicological and biological effects of mycotoxins are as varied as the toxins themselves and, because of diagnostic difficulties, usually only the acute and dramatic manifestations are observed. The most important aspects involve the exposure of man and farm animals to subchronic or chronic doses.

The broad scope of the mycotoxin problem is illustrated in this section by considering seven families of mycotoxins: luteoskyrin; aflatoxins and sterigmatocystin; trichothecenes; indole mycotoxins; penicillic acid and related lactones; ochratoxins and cyclochlorotine; and cytochalasins.

3.2.1. Anthraquinoid Mycotoxins

(−)Luteoskyrin ($C_{30}H_{22}O_{12}$, m.w. 574, m.p. 287°C (decomp.), $[\alpha]_D^{25}$ −880 (acetone)) and (+)rugulosin ($C_{30}H_{20}O_{10}$, m.w. 542, m.p. 290°C (decomp.), $[\alpha]_D^{25}$ +492 (dioxone)) are anthraquinoid in structure. Their solubility in water is very poor and they can be dissolved in saline only on addition of an alkali such as NaOH, Na_2CO_3, $NaHCO_3$ or Tris-aminomethane to the solution. Their maximum solubilities are around 1 mg/ml and 0.5 mg/ml, respectively. They can be dissolved or suspended in olive oil or corn oil.

These anthraquinoid mycotoxins possess a chelating potency for divalent cation such Mg^{2+}, Ca^{2+}, etc., forming insoluble complexes. Furthermore, they are photosensitive and change to a deep-colored precipitate on exposure to sunlight or a UV-lamp. The photoproduct of luteoskyrin, lumiluteoskyrin, is less toxic than the parent compound.

P. islandicum Sopp, *P. rugulosum*, and some other related fungi produce numerous kinds of anthraquinoid coloring pigments. Among these, (−)luteoskyrin has been extensively studied as the most toxic and carcinogenic agent. Comparative studies have revealed that (+)rugulosin is four times less toxic to mice and Ascites tumor cells than (−)luteoskyrin.[19] Recently, (−)rugulosin was isolated from *P. islandicum* Sopp and *Myrothecium verrucaria*. Due to the difficulty in obtaining large amounts of (−) rugulosin and related pigments, Ueno *et al.* compared their biological activity using the protozoan, *Tetrahymena pyriformis*,[20] and mutant cells of *E. coli*.[21] Umeda *et al.*[22] have also examined the cytotoxicity to cultured cells. As summarized in Table 3.10, (−)luteoskyrin, (−)rugulosin and (+)rugulosin are toxic to such cells, and in all cultured cells and *E. coli* Q-13, (−)rugulosin is about 4–10 times more toxic than (+)rugulosin.

TABLE 3.10. Comparative toxicity of anthraquinones

Anthraquinone	i.p. LD_{50} (mg/kg)		ED_{50} (mg/kg × 7)	ID_{50} (μg/ml)			
	mice[†1]	rats[†1]	Ascites tumors[†1]	HeLa[†2]	protozoa[†3]	*E. coli*[†4] F-11	Q-13
(−)Luteoskyrin	40.8		1–2	0.1	1	0.4	40
(−)Rugulosin	—			0.5		2.5	2.5
(+)Rugulosin	83.0	44	4–8	2	5	1.5	20
(−)Flavoskyrin				20	10	<48	48
Emodin				—	10	<50	50
Endocrocin				10–36[†5]			
(−)Rubroskyrin				20			
Iridoskyrin				< 32			
Dianhydrorugulosin				20			
Catenarin						136	136

†1 Y. Ueno *et al.* (1971).[19]
†2 M. Umeda *et al.* (1974).[22]
†3 Y. Ueno *et al.* (1968).[20]
†4 H. Yamakawa and Y. Ueno (1970).[21]
†5 T. Tatsuno *et al.* (1975).[23]

It has been demonstrated that (−)luteoskyrin can bind with DNA helix in the presence of magnesium ion.[24] This would be reasonable under the condition that the direction of linkage of two monomeric moieties in (−)luteoskyrin and (−)rugulosin could fit in the DNA helix more easily than the (+)compound.

Mice and rats administered with (−)luteoskyrin or (+)rugulosin usually became inactive, showing ruffled fur, loss of cutaneous tone and shallow respiration, and died after a prolonged comatose state. Liver function tests revealed an elevation of s-GOT and s-GPT that paralleled the dosage and extent of liver injury. Pathologically, acute intoxication is characterized by centrolobular necrosis. Coagulation necrosis of the liver cells, karyolysis, karyorrhexis, hyperchromatosis of the nuclear membrane, and fatty metamorphosis are conspicuous. Concerning changes in organs other than the liver, atrophy of the thymus, spleen and fat tissue are observed in mice dying of acute liver atrophy. Fatty degeneration of the tubular epithelium of the kidney also occurs.[1)]

Mice and rats are susceptible to anthraquinoids, and develop acute and chronic injuries including cirrhosis and hepatoma. Rhesus monkeys show acute lesions when fed with moldy rice.

From the s.c. and p.o. LD_{50} values of (−)luteoskyrin for mice (147 and 221 mg/kg, respectively), it appears that this compound is not highly toxic. However, when the compound was dissolved in normal saline, the i.v. LD_{50} obtained was very small (6.6 mg/kg), indicating a wide difference in LD_{50} values between the i.v. and other routes. Acute luteoskyrin toxicity usually appears comparatively slowly in animals when the mycotoxin is administered s.c. or p.o., yielding a relatively flat dose-response curve. However, as shown in Fig. 3.4, when injections are made at 24-hr intervals for 3 days, the total lethal dose decreases markedly and the dose-response curve becomes steep.

Tracer experiments and liver function tests have revealed that when luteoskyrin is administered s.c. for 3 consecutive days to mice, the liver luteoskyrin content and recovery curves, as observed 24 h after the first and second s.c. injections (5 mg/kg in each case), are consistent with curves for

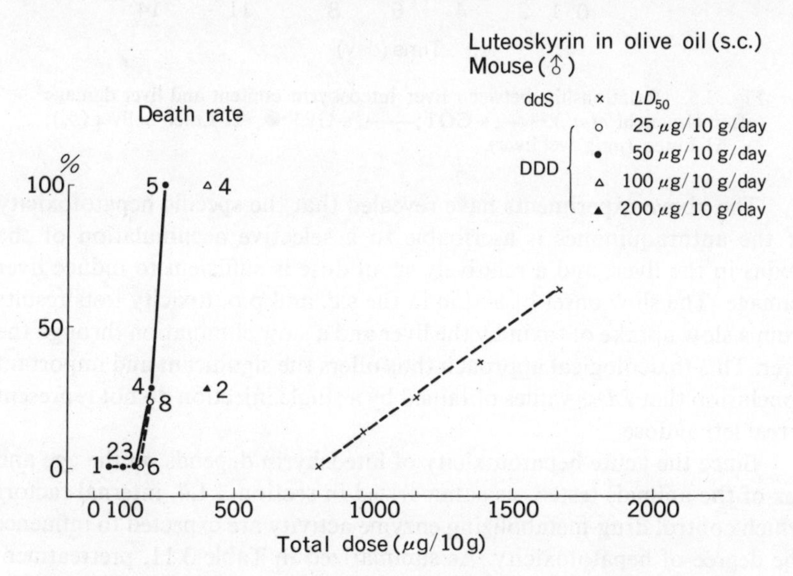

Fig. 3.4. Dose-response curves of (−)luteoskyrin for different injection times.

the respective enzyme activities, but that s-GOT and s-GPT reach maxima one day after the peak in luteoskyrin content (see Fig. 3.5). The presence of luteoskyrin in the liver must be considered to cause the induction of liver cell damage, and the time lag accounts for the slow hepatotoxicity of the compound.[25)]

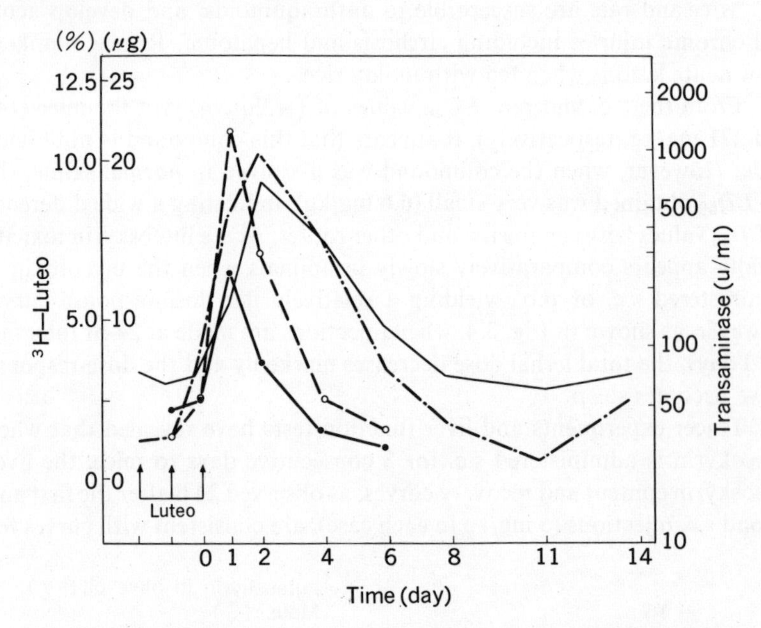

Fig. 3.5. Relationship between liver luteoskyrin content and liver damage (after Uraguchi *et al.*).[25] —, s-GOT;——, s-GPT;●, ^{3}H-Luteo in liver (%); ○, ^{3}H-Luteo (μg/g wet liver).

The above experiments have revealed that the specific hepatotoxicity of the anthraquinones is ascribable to a selective accumulation of the toxins in the liver, and a relatively small dose is sufficient to induce liver damage. The slow onset of action in the s.c. and p.o. toxicity tests results from a slow uptake of toxin by the liver and a slow elimination through the liver. This toxicological approach thus offers the significant and important conclusion that LD_{50} values obtained by a single injection do not represent a real lethal dose.

Since the acute hepatotoxicity of luteoskyrin depends on the age and sex of the animals tested, as summarized in section 3.1.4, internal factors which control drug-metabolizing enzyme activity are expected to influence the degree of hepatotoxicity. As summarized in Table 3.11, pretreatment with phenobarbital induces a decrease in the level of GPT in the serum and luteoskyrin content of the liver of mice intoxicated with luteoskyrin.

TABLE 3.11. Effect of phenobarbital on the hepatotoxicity of (−)luteoskyrin in mice (after I. Ueno *et al.*)[26]

Luteo-skyrin†1 (mg/kg)	Phenobar-bital†2 (mg/kg/day)	s-GPT (u/ml)	s-ALP (u/ml)	Microsome aniline hydroxylase (mμmoles/10 mg protein/10 min)	Microsome aminopyrine demethylase (mμmoles/10 mg protein/10 min)	Luteoskyrin in liver (μg/g)
—	—	28	8	38.7	0.213	—
25	—	865	7.4	34	0.204	10.7
50	—	884	8.0	66	0.278	11.6
50	80	45	8.2	64	0.323	0.7
25	80	45	8.8	46	0.296	undetermined
—	80	43	9.0	39	0.311	—

†1(−)Luteoskyrin dissolved in olive oil was administered p.o. to male 10–week-old mice of the DDD strain, 2 days before sacrifice.

†2Sodium phenobarbital was given i.p. for 3 days starting 2 days before the administration of (−)luteoskyrin.

Similar results have been obtained with 3-methylcholanthrene and promethazine (see Table 3.12). Pathological analysis also revealed decreased damage to liver cells in the pretreated mice.

These findings indicate that pretreatment of mice with a so-called "inducer" of drug-metabolizing enzymes may reduce the hepatotoxicity of luteoskyrin. Tracer experiments, however, have shown that the content of luteoskyrin is very closely related to the degree of severity of liver injury and that no detectable metabolite of luteoskyrin occurs in the liver. It is thus suggested that, in acute luteoskyrin toxicity in mice, an inducer such as phenobarbital causes an acceleration of luteoskyrin elimination from the liver cells.[26]

TABLE 3.12. Effect of pretreatments on the hepatotoxicity of (−)luteoskyrin in mice (after I. Ueno *et al.*)[26]

Luteoskyrin (mg/kg)	Pretreatment	s-GPT (μ/ml)	Luteoskyrin in liver (μg/g)
25	—	30	2.3
25	70 mg/kg methylcholanthrene, i.p. two days before luteo	58	1.4
50	—	195	6.4
50	100 mg/kg methylcholanthrene two days before luteo	56	3.4
50	—	958	4.8
50	25 mg/kg promethazine HCl, i.p. for 3 days from two days before luteo	64	0.6

The large number and variety of chemical agents that are capable of causing hepatic injury are best considered according to the presumed mechanisms and types of injury. The injuries caused may be an expression of the intrinsic toxicity of the respective hepatotoxins and of the particular susceptibility of the host. Table 3.13 gives summarized data on the comparative toxicology of four hepatotoxins; three mycotoxins and CCl_4. Clearly, the acute toxicity as well as biochemical changes are widely different among these four chemicals. (−)Luteoskyrin behaves in a similar manner to rubratoxin B in regard to age and sex differences and functional changes such as levels of blood sugar, liver lipid and s-GPT. Cyclochlorotine behaves in a similar manner to CCl_4 in regard to age differences and the marked elevation of s-GPT. The former two mycotoxins are presumed to injure the liver cells directly, and the latter two agents to affect well-organized or well-developed cellular membranes in the liver cells after metabolic conversion to reactive agents.

TABLE 3.13. Comparative toxicology of hepatotoxins (I. Ueno, *unpublished*)

		Hepatotoxic mycotoxins			CCl_4
		(−)luteoskyrin	Cl-peptide	rubratoxin B	
Acute toxicity	Age	new-born ≫ adult	new-born ≪ adult	new-born ≫ adult	new-born < adult
	Sex	male ≫ female	male < female	male > female	male ≈ female
	Castration male	↓			
	Castration female	—			
	Adrenoectomy	↓	↑	↑	↓
Functional changes	Blood sugar	↓	↑→↓↓	↓	↑→↓
	Liver glycogen	↓	↓↓↓	↓↓↓	↓
	Liver lipid	↑	↑	↑	↑↑↑
	Serum GPT	↑	↑↑↑	↑	↑↑↑

The above toxicological approach offers a key to understanding the characteristic features of the respective hepatotoxic chemicals at the molecular level. This may permit a correlation between the injurious changes in the whole animal and the cellular injuries.

3.2.2. Bisfuranoid Mycotoxins

Sterigmatocystin was the first known naturally occurring compound to contain the dihydrofurobenzofuran system, and after the elucidation of its structure a number of similar compounds were isolated from the secondary metabolites of fungi. One characteristic feature of this series is that all the metabolites contain either the unsaturated 7,8-dihydrofurano (2,

3-*b*)furan or the more fully reduced 2,3,7,8-tetrahydrofuro(2,3-*b*)furan. Aflatoxins, a family of closely related substances produced by *Aspergillus* flavus and *A. parasiticus*, contain the coumarin ring, and sterigmatocystin produced by *A. versicolor* and other fungi contains a substituted xanthone ring (Fig. 3.6). Further, there are fungal pigments in which a substituted anthraquinone is fused to the furofuran.

Toxicologically, aflatoxin may be regarded as a quadruple threat, as a potent toxin, a carcinogen, a teratrogen and a mutagen. The LD_{50} values of B_1 and other aflatoxins to various animal species are summarized in Table 3.14. There is a wide range in the acute lethal dose of aflatoxin B_1, varying from 0.3 mg/kg in ducklings to 16 mg/kg in mature female rats. Mature animals of a given species are generally more resistant than young ones, implying that aflatoxin biotransformation enzyme develops in the liver with age. From the lethal doses of aflatoxin B_1, the animal species may be divided into three groups: (1) extremely susceptible—duckling, rabbit, cat, pig, trout, (2) moderately susceptible—dog, sheep, guinea-pig, and (3) resistant—monkey, chicken, rat, mouse and hamster.

There is of course as yet no established toxic dose for humans. However, striking evidence from S.E. Asia, India and Africa indicates that aflatoxins have been involved in the occurrence of human fatalities, particularly among children. The characteristic features of the syndrome are as follows: fever, vomiting, diarrhea, coma, convulsions, fatty degenera-

TABLE 3.14. Acute toxicity of aflatoxins

Species	LD_{50} (mg/kg)							
	B_1	B_2	G_1	G_2	M_1	M_2	B_{2a}	P
Duckling	0.36	1.68	0.78	1.42	0.32	1.22	24	
Rabbit	0.3							
Cat	0.55							
Pig	0.62							
Trout†	0.81		1.90					
Dog	0.5–1.0							
Sheep	1.0–2.0							
Guinea-pig	1.40							
Monkey	2.2–7.8							
Chicken	6.3							
Rat	7.2(M)–16 (F)							
Mouse	9.0							
Hamster	10.2							
Chick embryo	0.025/embryo		1.0/embryo					< 150

†Intraperitoneal: all other values refer to the oral route.

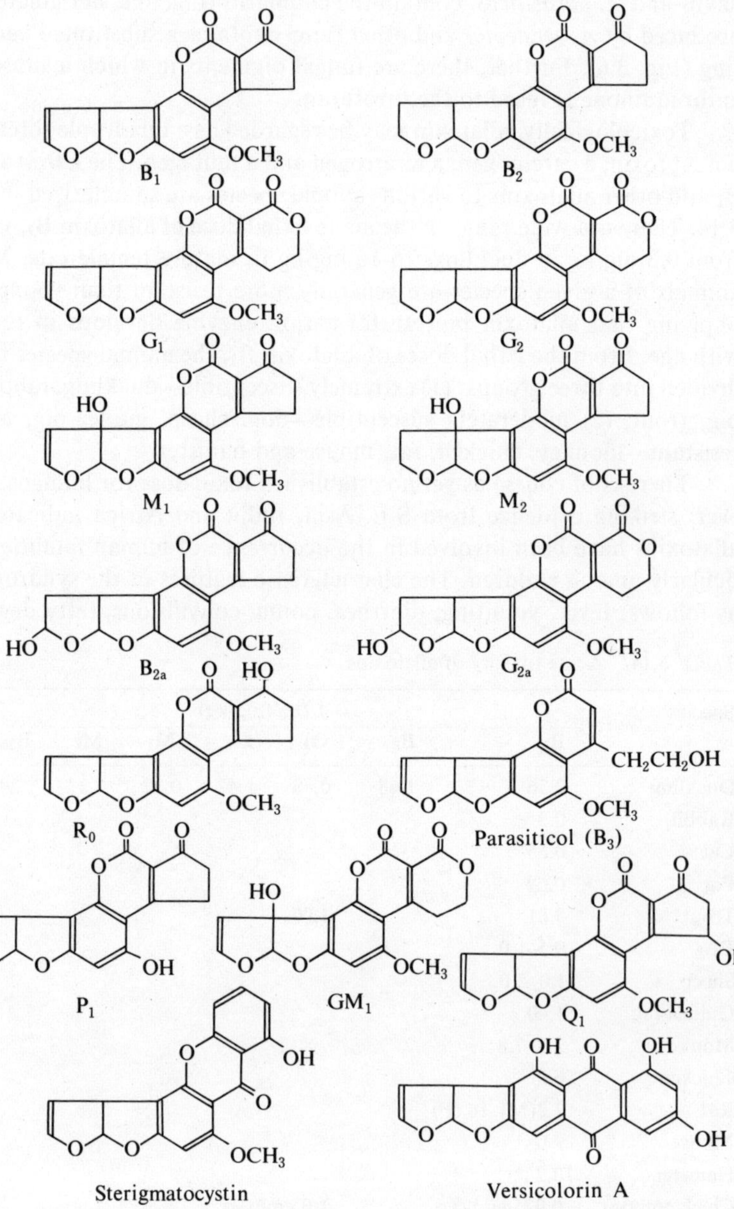

B_1 B_2

G_1 G_2

M_1 M_2

B_{2a} G_{2a}

R_0 Parasiticol (B_3)

P_1 GM_1 Q_1

Sterigmatocystin Versicolorin A

Fig. 3.6. Structure of aflatoxins, sterigmatocystin and versicolorin A.

tion of the liver, heart and kidneys, marked cerebral edemas with neural degeneration, and lymphocytolysis. This so-called Reye's syndrome as a child's disease[27] is therefore presumed to be related to the consumption of relatively large amounts of aflatoxin-contaminated foods, and aflatoxin was actually detected in the liver of 22 out of 23 cases. Similar acute fatal cases in man have developed in India through the intake of aflatoxin-contaminated corn.[28]

Recently, Salhab *et al.*[29] have compared the metabolic potential of human liver microsomes with that of the monkey. They demonstrated that the human liver can convert aflatoxin B_1 into its metabolites at a similar rate to monkey liver, implying that man may have a similar aflatoxin susceptibility to the monkey. This would mean that man should be classified under the resistant group as regards his acute aflatoxin toxicity.

Comparative studies on the mutagenicity of aflatoxin analogs to *Salmonella typhimurium* mutants have resulted in the following order of potency; $B_1 >$ aflatoxicol $> M_1 > G_1 > H_1 > Q_1$. This order is quite parallel to that for the toxic and carcinogenic activity.

In the case of sterigmatocystin, on the other hand, the acute toxicity to rats is very low compared to aflatoxin B_1. As shown in Table 3.15, the oral acute toxic dose of sterigmatocystin to rats is 166 mg/kg, which represents a toxicity 24 times less than that of aflatoxin B_1.[30] Chicken embryo assays demonstrated a sterigmatocystin toxicity 500 times less than that of aflatoxin B_1. Recent experiments by Kimura *et al.*[31] have revealed that the single lethal dose of sterigmatocystin to mature rats is about 800 mg/kg when olive oil is used as a vehicle. These data suggest that sterigmatocystin is very low in absorption. Fluorometric determinations indicate that over 70% of the p.o. injected dose is eliminated into the feces and only 0.1% of the administered sterigmatocystin was detected in the liver.

Upon *O*-acetylation of sterigmatocystin, the acute toxicity is markedly increased and is comparable to that of aflatoxin B_1 (see Table 3.15). Long-term feeding then induces liver tumors at comparatively lower doses and shorter exposures than with sterigmatocystin itself.[32]

TABLE 3.15. Acute toxicity of sterigmatocystins

Species and route	LD_{50} (mg/kg)	
	sterigmatocystin	*O*-acetylsterigmatocystin
Monkey p.o.	32	
Rat, male p.o.	166	
female p.o.	120	
male i.p.	60	11.3
Mouse, male p.o.	800	
Chicken embryo	14.9 μg/embryo	0.8 μg/embryo

Aflatoxin B_1 is a potent hepatotoxin and carcinogen, and many biochemical, toxicological and histological analyses have been performed to clarify the fundamental mechanisms of liver injury and hepatoma development. Toxicologically speaking, there are two approaches; one is the sequential analysis of alterations in liver function, and the other is the sequential analysis of aflatoxin biotransformation in the liver in relation to toxigenesis.

Using the former approach, Kalengayi *et al.*[33] examined the sequential histochemical changes in rat liver after single-dose aflatoxin B_1 intoxication. According to their systematic treatment, periportal cytoplasmic glycogen and RNA depletion occurred during the early period, and subsequently extended to the whole lobule. 5'-Nucleotidase, G-6-Pase, G-6-P dehydrogenase decreased or disappeared from the periportal area. On the other hand, alkaline phosphatase increased strikingly in the centrolobular area, while canalicular ATPase disappeared completely from the entire liver lobule. After necrosis, regenerative foci deficient in glycogen and RNA content, and in enzymes, appeared. This situation reflects the immaturity of the regenerating hepatocytes. Such early foci, however, subsequently disappeared and are thus considered irrelevant to hepatomagenesis.

In the case of rats exposed to a low dose of aflatoxin B_1 for long periods, no significant lesions occurred before 15 weeks. During this interval, the liver was histochemically unchanged except for periportal decreases in alkaline phosphatase and ATPase. These changes represent the toxic effect of aflatoxin B_1 and are not irrelevant to carcinogenesis. After 15 weeks, two types of liver cell hyperplasmic foci were detected as potential precursors of hepatocarcinomas. They were deficient in enzymes, especially ATPase, G-6-Pase and nucleases, although the question of whether a focal alteration of certain enzymes plays a fundamental role in the basic mechanism of hepatocarcinogenesis remains to be clarified.[34]

Using the second toxicological approach, Patterson *et al.*[35] investigated the sequential biotransformation of aflatoxin B_1 in susceptible cells. (The detailed mechanisms of aflatoxin metabolism will be discussed below in section 3.3.3.) They described the interaction of aflatoxin metabolites with cell organella, as shown in Fig. 3.7, which incorporates slight modifications by Ueno *et al.* Following transport across the cell membrane, the aflatoxin B_1 molecule may bind to nuclear DNA, causing nuclear damage, or bind to sex-determined sites on the endoplamsic reticulum. This latter binding may be related to injury of the endoplasmic reticulum system, including polysomal degradation and ribosomal detachment. Intracellular B_1 is reduced reversibly to aflatoxicol which may serve as an "aflatoxin

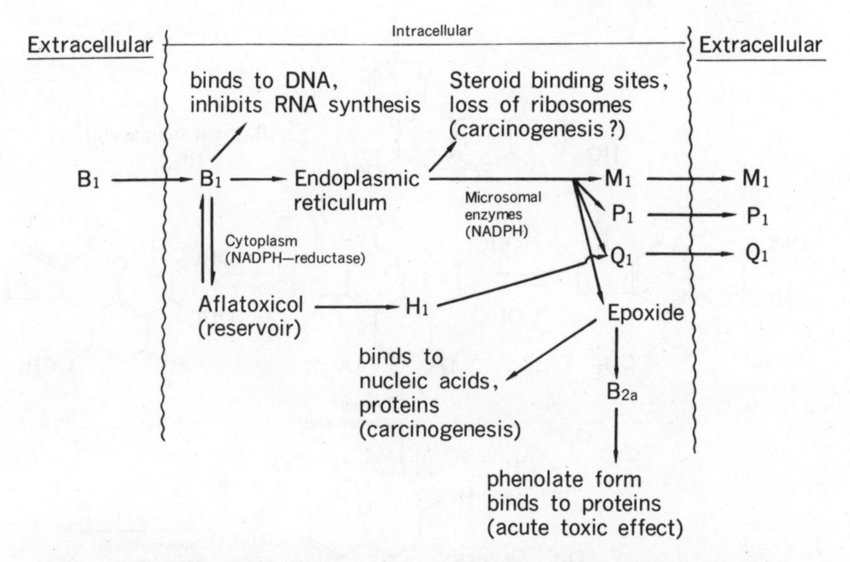

Fig. 3.7. Intracellular conversion of aflatoxin B_1 in relation to toxigenicity.

reservoir" in intracellular space. This means that the "active" or "toxic" aflatoxin level in the target organ may be determined by the cytoplasmic NADPH/NADP ratio and its catalyzing enzyme activity. The microsomal oxygenase system transforms aflatoxin B_1 into polar metabolites such as M_1, P_1 and Q_1, which can be eliminated from the liver cells. Aflatoxin B_{2a}, which is perhaps a hydrolytic product of aflatoxin epoxide, is relatively non-toxic, but at physiological pH (7.4) it is very unstable, probably due to its occurrence in phenolate form. The phenolate form of aflatoxin B_{2a} reacts with nucleic acids and proteins, probably as a result of Schiff base formation(Fig. 3.8).

Aflatoxins present enormous problems in the field of veterinary science.[36] Their toxic effects on domestic animals have recently been summarized by Ciegler,[37] as shown in Table 3.16. The most serious injury to animals after long-term exposure occurs in the liver. Undoubtedly, at subacute doses, feed efficiency and growth rate are affected. In poultry, increased fragility of the capillaries results in bruising of the birds during mechanical processing.

Sterigmatocystin has been shown to induce hepatomas in rats.[38,39] Zwicker *et al.*[40] reported that long-term administration of pure sterigmatocystin or rice cultures of *A. versicolor* to mice resulted in the appearance of pulmonary tumors. The tumors were classified as adenomas or adenocarcinomas, arising from the bronchial and alveolar epithelium. At the

Fig. 3.8. Resonance forms of the phenolate ion of aflatoxin B_{2a} and its reactivity at physiological pH.

TABLE 3.16. Dietary aflatoxin concentrations causing toxicosis in farm animals (after Ciegler)[37]

Species	Age	Aflatoxin content (ppm)	Duration of feeding	Effects
Calf	weanling	0.2–2.2	16 weeks	Stunting, death, liver damage
Steer	2 yr	0.2–0.7	20 weeks	Liver damage
Cow	2 yr	2.4	7 months	Liver damage
Pig	newborn	0.23	4 days	Stunting
Pig	2 weeks	0.17	23 days	Anorexia, stunting, jaundice
Pig	4–6 weeks	0.4–0.7	3–6 months	Stunting, liver damage
Chicken	1+ weeks	0.8	10 weeks	Stunting, liver damage
Duck	unknown	0.3	6 weeks	Liver damage, death

present time, no definite evidence is available to explain this increased incidence of pulmonary neoplasms, although it is quite possible that there is an interaction between an oncogenic virus in the pulmonary tissues of the test mice and sterigmatocystin.

Bisfuranoid mycotoxins are hepatotoxic and tumorigenic to various species of animals, and their toxicity, acutely or chronically, is thus enhanced or depressed by a number of internal and environmental factors. As summarized in Table 3.17, ethionine and CCl_4 have been reported as accelerators of liver cirrhosis in aflatoxin-administered animals, while hypophysectomy and administration of phenobarbital decrease the hepatotoxic and carcinogenic potential of aflatoxin. The content of vitamins and fatty acids in the feed may result in some modification of the liver injuries.

TABLE 3.17. Factors affecting the toxicity of aflatoxin B_1

Increase:	ethionine (cirrhosis)
	CCl_4 (cirrhosis)
	lipotope deficiency
	methionine
Decrease:	hypophysectomy
	phenobarbital
	choline deficiency
	low dietary vitamin A
	dodecan

3.2.3. Epoxide Mycotoxins

Naturally occurring epoxide compounds are widely distributed in the metabolites of animals, insects, plants and microbes (Table 3.18). Some epoxides of industrial origin are carcinogenic to animals (Table 3.19). In general, the epoxide ring is presumed to be an essential and functional group of these compounds. Among the fungal metabolites, most are reported as antibiotic to fungi, bacteria and protozoa, and in connection with human and animal intoxication, the trichothecene compounds are the most important mycotoxins.

The trichothecenes are a chemically related group of biologically active fungal metabolites produced by various species of *Fusarium*, *Myrothecium*, *Trichoderma*, *Cephalosporium*, *Verticimonosporium* and *Stachybotrys* (Fig. 3.9). The first trichothecene compounds, verrucarins, were isolated during a search for antibiotics in 1946, and diacetoxyscirpenol represents the earliest isolation of a phytotoxic trichothecene. The above trichothecene-producing fungi invade various agricultural products and a wide range of plants as plant pathogens or plant parasites, and their biological characteristics govern to a large extent the distribution and development of serious intoxications of man and animals. Ueno *et al.* have sum-

TABLE 3.18. Naturally occurring epoxides

Plant hormones	abscisic acid
	graphinone
	haliangine
	pyrethrosine
Phytotoxins	prolactone A,B,C,D
	inumakilactone A,B,C
	phyllosinol
	diacetoxyscirpenol
Insect hormomes	
juvenile hormones	
sex pheromones	
aggregation hormones	
Antibiotics	fumagillin
	illudin
	pentalenolactone
	corriolin A,B,C
	trichothecin
	crotocin
	radicicol (monorden)
	terreic acid
	streptolydigin
	hirsutic acid
	cerulenin
	cervicarcin
	carbomycin
	oleandomycin
	chalcomycin,
	neutramycin
	pimaricin
Others	nepetaefulin
	cleroden (vermifuge)
	jatrophone (antitumor agent)
	daphnetoxin (plant toxin)

TABLE 3.19. Carcinogenic epoxides of industrial origin

Monoepoxides	glycidaldehyde
	1,2 - epoxybutene - 3
	styrene oxide
	1,2 - epoxyhexadecane
Bifunctional epoxides	1,2,5,6 - diepoxyhexane
	1,2,7,8 - diepoxyoctane
	1- ethyleneoxy - 3, 4 - epoxycyclohexane

(A)

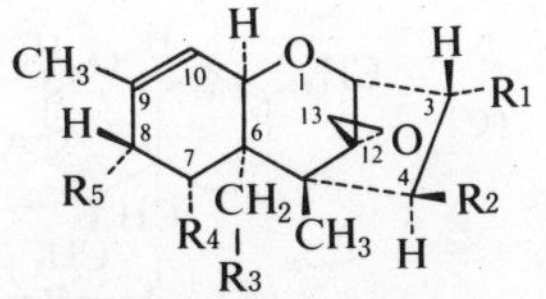

	R_1	R_2	R_3	R_4	R_5	†
Trichothecene	H	H	H	H	H	d
Trichodermol (roridin C)	H	OH	H	H	H	a
Trichodermin	H	OAc	H	H	H	a
Dihydroxytrichothecene	H	OH	H	H	OH	d
Verrucarol	H	OH	OH	H	H	b
Scirpentriol	OH	OH	OH	H	H	c
T-2 tetraol	OH	OH	OH	H	OH	c
Monoacetoxyscirpenol	OH	OH	OAc	H	H	c
Diacetoxyscirpenol	OH	OAc	OAc	H	H	c
8-Hydroxydiacetoxyscirpenol (Neosolaniol)	OH	OAc	OAc	H	OH	c
8-Acetyldiacetoxyscirpenol (Acetylneosolaniol)	OH	OAc	OAc	H	OAc	c
7-Hydroxydiacetoxyscirpenol	OH	OAc	OAc	OH	H	c
7,8-Dihydroxydiacetoxyscirpenol	OH	OAc	OAc	OH	OH	c
HT-2 toxin	OH	OH	OAc	H	$OCOCH_2CH(CH_3)_2$	c
T-2 toxin	OH	OAc	OAc	H	$OCOCH_2CH(CH_3)_2$	c
Acetyl T-2 toxin	OAc	OAc	OAc	H	$OCOCH_2CH(CH_3)_2$	c
Calonectrin	OAc	H	OAc	H	H	c
Deacetylcalonectrin	OAc	H	OH	H	H	c

(B)

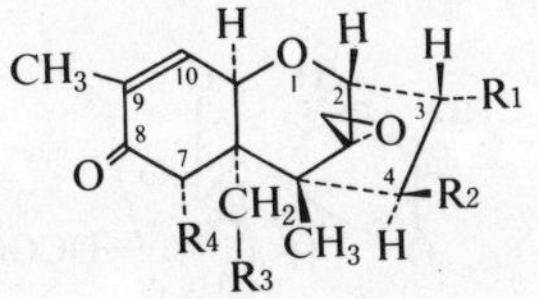

	R_1	R_2	R_3	R_4	†
Nivalenol	OH	OH	OH	OH	c
Monoacetylnivalenol (fusarenon-X)	OH	OAc	OH	OH	c
Diacetylnivalenol	OH	OAc	OAc	OH	c
Deoxynivalenol (Rd-toxin)	OH	H	OH	OH	c
Deoxynivalenol-monoacetate (Rc-toxin)	OAc	H	OH	OH	c
Trichothecin	H	$OCOCH{=}CHCH_3$	H	H	d

Fig. 3. 9. Naturally occurring trichothecenes.

(C)

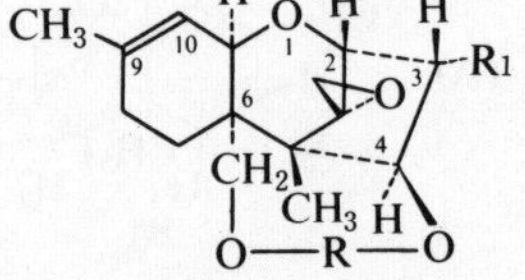

	R	†
Verrucarin A	–C(=O)CH(OH)CH(CH$_3$)CH$_2$CH$_2$OC(=O)CH=CHCH=CHC(=O)–	b
Roridin A	–C(=O)CH(OH)CH(CH$_3$)CH$_2$CH$_2$OCH(CH$_3$CHOH)CH=CHCH=CHC(=O)–	b
Satratoxin H	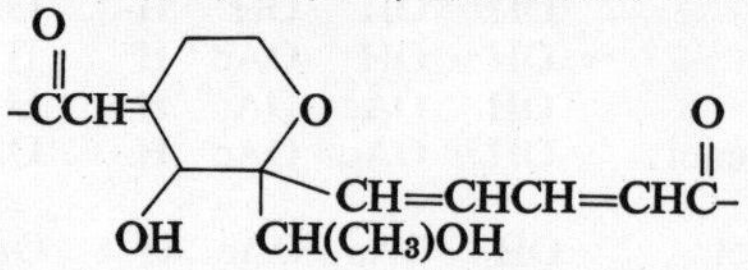 –C(=O)CH= [ring] O; OH; CH(CH$_3$)OH; CH=CHCH=CHC(=O)–	f
Vertisporin	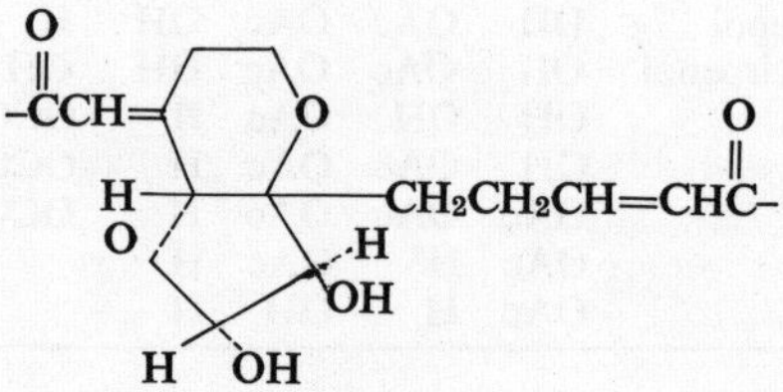 –C(=O)CH= [ring] O; H–; O; H; OH; H OH; CH$_2$CH$_2$CH=CHC(=O)–	g

(D)

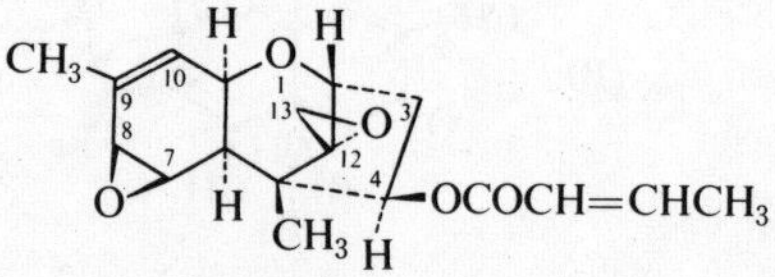

Crotocin

†Isolated originally from (a) *Trichoderma viride,* (b) *Myrothecium verrucaria or M. roridum,* (c) *Fusarium* sp., (d) *Trichothecium roseum,* (e) *Cephalosporium crotocigenum* (f) *Stachybotrys atra,* (g) *Verticimonosporium diffractum.*

Fig 3.9—*Continued*

marized past cases of serious intoxications due to foodstuffs contaminated with trichothecene mycotoxin-producing fungi, as shown in Table 3.20.

The lethal toxicity of the trichothecene family to mice and rats is summarized in Table 3.21(A) and (B), respectively. Macrocyclic trichothecenes such as verrucarin A and roridin A are the most toxic agents, while trichothecin, trichodermin and crotocin are very weak in comparison with the other toxins. No marked differences in lethal trichothecene dose are observable according to different routes of administration, indicating that both the orally and intraperitoneally administered toxins are quite smoothly absorbable within these animals. In acutely poisoned mice and rats, no remarkable syndrome appears, although there is diarrhea and marked erosion and hemorrhage in the intestine and lung.

In avians, cats, dogs and swine, vomiting is an early response to trichothecene intoxication. In ducklings, 0.1 mg/kg of T-2 toxin or 1.0 mg/kg of fusarenon-X is capable of inducing vomiting 5–10 min after s.c. or p.o. administration. This characteristic sympton of ducklings has been utilized for bioassay of the "vomiting factor" in toxic corn samples.[41,42] In cats, emesis and vomiting occur about 30 min after s.c. dosing with 0.1 mg/kg of T-2 toxin, and these symptoms recur for several hours at intervals of 0.5 to 1 h.[5] As summarized in Table 3.22, vomiting doses of type A trichothecenes are about ten times less than those of type B trichothecenes.

As has been reported in many papers dealing with *Fusarium* toxicosis, the skin test is widely used for the screening of toxic strains and identification of toxic agents, and in fact, T-2 toxin, diacetoxyscirpenol, and butenolide were discovered in this way. Based on this historical background, Ueno *et al.* have re-examined the skin-necrotizing effect of trichothecenes, as summarized in Table 3.23.[46] All the trichothecenes possess skin-irritant or skin-necrotizing activity for animals. The type-A toxins (T-2 toxin, HT-2 toxin, etc.) are about ten times more toxic than the type-B toxins (nivalenol, fusarenon-X, etc.), and the skin of guinea-pig is notably more sensitive than that of other animals. Dendrochine, a mycotoxin from *Dendrodochium toxicum*, is also reported to be a skin-irritant.[47] Since this fungus is presumed to be synonymous with *Myrothecium roridum*, it is highly possible that dendrochine is one of the trichothecene family.

One notable finding in this connection was that oral lesions characterized by swelling and necrosis of the oral cavity, were inducible in poultry by feeding with T-2 toxin This disease syndrome is identical to actual cases of avian fusariotoxocosis, as reported by Wyatt *et al.*[48] In addition, according to Ohtsubo *et al.* (*unpublished*), long-term feeding of rats with a T-2 toxin diet resulted in the induction of papillomas on the surface of the stomach.

Besides the trichothecene mycotoxins, the following fungal metabolites

TABLE 3.20. Major symptoms and suspected fungi in food-borne diseases of humans and animals

Toxicosis	District	Fungus	Symptoms	Affected species
"Taumelgetreide" toxicosis	Siberia (U.S.S.R)	*G. saubinetti*	headaches, vertigo, shivering chills, nausea, vomiting, visual disturbance	man, horse, pig, fowl
Red-mold toxicosis	Japan	*G. graminearum*	vomiting, diarrhea, hemorrhage, refusal of feed	man, horse, cow, pig
	Europe	*F. roseum*	weakness, vertigo, headaches, nausea, vomiting	man, horse
Moldy-corn toxicosis	U.S.A.	*F. tricinctum*	vomiting, emesis, refusal of feed	cow, pig
"Alimentary toxic aleukia"	U.S.S.R.	*F. sporotrichioides* *F. poae*	vomiting, inflammation, diarrhea, leukopenia, multiple hemorrhage, necrotic angina, septis	man, domestic animals
Stachybotryotoxicosis	U.S.S.R. Europe	*S. atra*	shock, somatitis, dermal necrosis, thrombocytopenia, leukopenia, nervous disorders, hemorrhage	man, horse, pig, calf, poultry
Bean-hull poisoning	Japan	*F. solani*, etc.	convulsions, cyclic movements, respiratory failure, decrease of heart rate and reflexes	horse
Dendrodochiotoxicosis	U.S.S.R.	*Dendrodochium toxicum* (*Myrothecium roridum*)	acute death, hemorrhage, cyanosis, leucocytosis, leucopenia, dilatation of micro-blood capillaries	horse, pig, sheep, man

TABLE 3.21. Acute toxicity of trichothecencs to mice and rats

(A) Mice

Type[†1]	Trichothecene	LD_{50} (mg/kg) i.v.	s.c.	i.p.	p.o.
A	T-2 toxin			5.2	7.0
	HT-2 toxin			9.0	
	Diacetoxyscirpenol	10.0		23.0	7.3
	Neosolaniol			14.5	
	7-Hydroxydiacetoxyscirpenol			*ca.* 4.0	
	7,8-Dihydroxydiacetoxyscirpenol			*ca.* 6.0	
	Trichodermin				1000
B	Nivalenol			4.1	
	Monoacetylnivalenol (fusarenon-X)	3.4	4.2	3.4	4.5
	Diacetylnivalenol			9.6	
	Dihydronivalenol			15.0	
	Deoxynivalenol (Rd-toxin)		(27)[†2]	70.0	
	Deoxynivalenol-monoacetate (Rc-toxin)		(37)[†2]	46.9	
	Trichothecin	300			
C	Roridin A	1.0		0.5–0.7	
	Verrucarin A	1.5			
	Verrucarin B	7.0			
D	Crotocin	700	500		1000

[†1]For structures, see section 2.3.
[†2]Duckling.

(B) Rats

Trichothecene	LD_{50} (mg/kg) i.v.	s.c.	i.p.	p.o.
T-2 toxin				3.8
Diacetoxyscirpenol			0.75	7.3
Trichodermin				
Monoacetylnivalenol (fusarenon-X)				4.4
Trichothecin		250		
Crotocin		100		

TABLE 3.22. Vomiting doses of trichothecenes

Type	Trichothecene	Vomiting doses (s.c. mg/kg)				
		Duckling[†1]	Pigeon[†3]	Cat[†1]	Dog[†2]	Pig[†4]
A	T-2 toxin	0.1	0.72	0.1–0.2		
	HT-2 toxin	0.1	*ca.* 0.7			
	Diacetoxyscirpenol	0.2				
	Neosolaniol	0.1				
B	Nivalenol	1.0				
	Monoacetylnivalenol (fusarenon-X)	0.4		1–2		
	Diacetynivalenol	0.4				
	Deoxynivalenol (Rd-toxin)	10			0.1–0.2	(positive)
	Deoxynivalenol-monoacetate (Rc-toxin)	10			0.1–0.2	

[†1] Y. Ueno *et al.* (1974),[41] and N. Sato *et al.* (1975).[5]
[†2] S. Morooka *et al.* (1975).[43]
[†3] R. A. Ellison *et al.* (1973).[44]
[†4] R. F. Vesonder *et al.* (1973).[45]

TABLE 3.23. Skin toxicity of trichothecenes (after Ueno *et al.*)[46]

Type	Trichothecene	Skin-necrotizing dose (μg/spot)		
		mouse	guinea-pig	rabbit
A	T-2 toxin	1	0.2	
	HT-2 toxin	1	0.2	
	Diacetoxyscirpenol	10	0.2	1
	Neosolaniol	10	1	
B	Nivalenol	100	10	10
	Monoacetylnivalenol (fusarenon-X)	10	1	10
	Diacetylnivalenol	10	1	

have been reported as inflammatory to skin tissues; terreic acid from *Aspergillus terreus*, fumagillin from *A. fumigatus*, 8-methoxypsoralen and 4,5′,8-trimethylpsoralen from *Sclerotinia sclerotiorum*, and patulin from *A. clavatus*.

Of note are the hematological changes occurring in animals acutely or chronically intoxicated with trichothecene mycotoxins. When mice were intraperitoneally administered with 2–5 mg/kg of fusarenon-X or 5–20 mg/kg of neosolaniol, a temporary elevation of circulating white blood cells was induced 1–3 h after administration. In cats, similar leucocytosis occurred 0.5–1 h after s.c. dosing of 0.1 mg/kg of T-2 toxin, as shown in Fig. 3.10. When cats were administered with 0.05–0.1 mg/kg of T-2 toxin for several days or continuously (see Fig. 3.11), the number of white blood cells decreased markedly, and the cats died from leucopenia. During such subchronic intoxication of cats, hemorrhage of the lung, intestines, and destruction of the bone marrow were remarkable. An important sign was the meningeal hemorrhage of the brain, as reported by Sato *et al.*[5] These

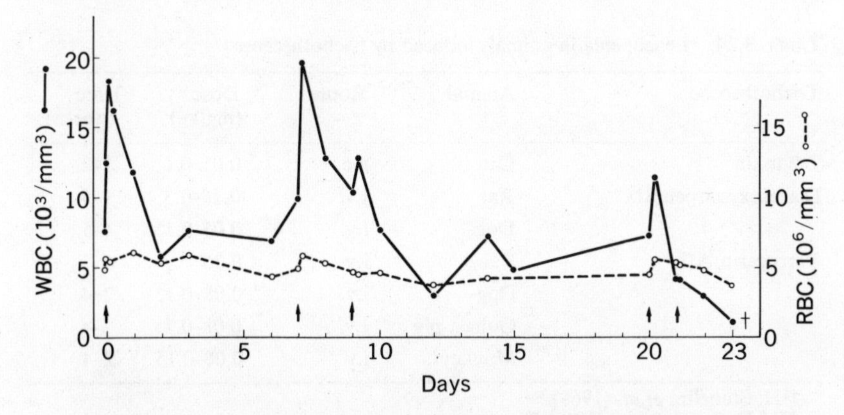

Fig. 3.10. Leucocytosis in cats induced by T-2 toxin (after Sato *et al.*).[5] A female cat was administered 0.1 mg/kg T-2 toxin subcutaneously at the times indicated by the arrows.

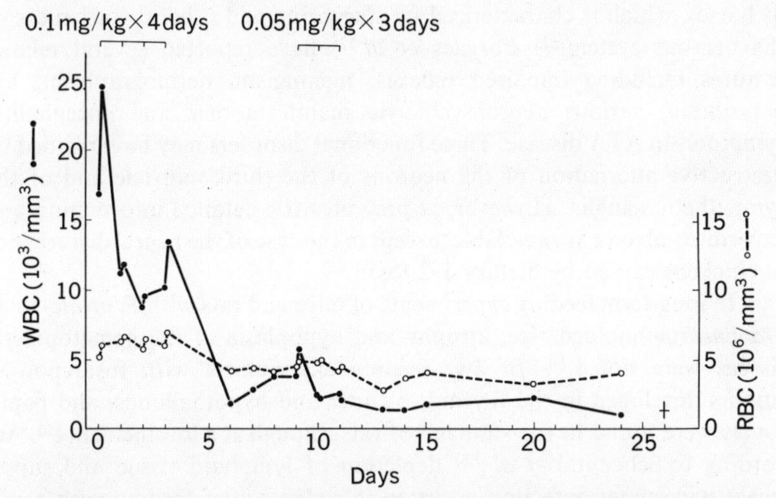

Fig. 3.11. Leucopenia in cats induced by T-2 toxin (after Sato *et al.*).[5] A male cat was administered with 0.1 or 0.05 mg/kg T-2 toxin for one week simultaneously.

toxicological features of T-2 toxin intoxicated cats are quite similar to actual human cases that suffered from the effects of overwintered millet in the U.S.S.R. Since leucopenia is inducible in experimental animals with other kinds of trichothecenes such as diacetoxyscirpenol[49] and verrucarin A,[50] as summarized in Table 3.24, the above hematological changes are considered to be general toxicological characteristics of trichothecene compounds.

TABLE 3.24. Leucopenia in animals induced by trichothecenes

Trichothecene	Animal	Route	Dose (mg/kg)	Time (weeks)
T-2 toxin	Cat	s.c.	0.05–0.1	1–2
Diacetoxyscirpenol†1	Rat	i.v.	0.15–0.3	1–5
	Dog	i.v.	0.05–0.15	4–5
Verrucarin A†2	Rat	i.v.	0.25	4
	Dog	i.v.	0.08–0.15	2–4
	Guinea-pig	i.v.	0.08–0.15	2–4
	Monkey	i.v.	0.08–0.15	2–4

†1H. Stahelin. *et al.* (1968).[49]
†2M.E. Rusch. *et al.* (1965).[50]

Nervous disturbance represents another problem in farm animals, and sometimes in humans. It has been implicated in the "bean-hull poisoning" of horses, which is characterized by staggering and other disturbances of the nervous system.[51] Forgacs *et al.*[52] have reported several related features including impaired reflexes, meningism, dermographism, hyperesthesia, various neuropsychiatric manifestations, and encephalitic symptoms in ATA disease. These functional disorders may be explained by destructive alternation of the neurons of the third ventricle and of the sympathetic ganglia. However, at present, little detailed information and experimental data are available, except in the case of the neural disturbance in chickens caused by dietary T-2 toxin.[53]

In long-term feeding experiments of mice and rats with *F. nivale*- or *F. graminearum*-molded rice, atrophy and hypoplasia of the hematopoietic tissues were noted.[54] In 2-yr feeding experiments with fusarenon-X, tumors developed in the thyroid, bladder and hypothalamus, and papillomas were found in the stomach of rats, though at a low incidence[2]. According to Schoental *et al.*,[55] depletion of lymphoid tissue and subsequent widespread infection occurred in rodents after feeding with crude toxins from *F. poae* and *F. sporotrichioides*, the authentic isolates from overwintered grains. These results indicate that the crude toxins act as immunosuppressive agents. On small-sample dosing to the skin, the esophagus and stomach show local cytotoxic effects which are followed by regeneration and basal cell hyperplasia of the squamous epithelium, suggesting that the use of moldy cereals contaminated with trichothecenes or skin-irritant metabolites may play a role in the development of tumorous changes of the digestive tract. Long-term, low-dose feeding experiments with rainbow trout and rats, as well as the mouse papilloma induction studies of Marasas *et al.*,[56] have indicated that T-2 toxin is not carcinogenic.

As indicated above when summarizing the toxicological characteristics of the trichothecene family, all trichothecene mycotoxins are capable of inducing skin-necrotization, vomiting, leucocytosis, and leucopenia in experimental animals. Such experimental findings lend strong support to the idea that all syndromes observed in *Fusarium* toxicosis and related intoxications originate from the ingestion of fungal trichothecenes. On the other hand, detailed observations indicate that the toxicity and other biological activities of trichothecenes differ markedly between type-A toxins (T-2 toxin, etc.) and type-B toxins (nivalenol, etc.). As summarized in Table 3.25, the former toxins are far more active than the latter as regards their cytotoxicity, skin toxicity and induced vomiting, although in the case of lethal toxicity and hematologic changes, there are no marked differences between them.

These general features suggest an important role for C-8 ketone in the expression of biological activity in the trichothecene family, although of course the side chains and number of hydroxyl groups also exert a significant effect on the activity.

Table 3.25. Comparative toxicology of Type A and Type B toxins

Trichothecene group:	T-2 toxin (A)		Nivalenol (B)
Cytotoxicity Skin toxicity Chicken embryo toxicity Vomiting	A	$\gg$	B
Mice lethal toxicity Leucocytosis Leucopenia	A	$\approx$	B
Inhibition of protein synthesis			
whole cell	A	$\gg$	B
cell-free	A	$\approx$	B

3.2.4. Indole Mycotoxins

Since Wilson *et al.*[57)] reported a tremorgen-convulsant in metabolites of *A. flavus*, several tremorgens have been isolated from various species of *Penicillium* and *Aspergillus* (Table 3.26). Wilson *et al.*[58)] isolated penitrem A ($C_{37}H_{44}O_6NCl$) from a strain of *P. cyclopium* causing disease in sheep and horses. The compound produced neurological and renal effects in mice, rats, chickens, rabbits and guinea-pigs. In mice, a small oral dose of 0.25 mg/kg induced perceptible tremors which persisted for several hours, and larger doses of penitrem A caused increased irritability, an inability to grasp, and marked tremors. At high doses, the tremors progressed to clonic or tetanic convulsions and often culminated in death.

TABLE 3.26. Indole mycotoxins†1

Mycotoxin†2	Fungus	LD_{50} (mg/kg) mouse	rat
Penitrem A	*P. cyclopium* *P. crustosum* *P. palitans*	1.05	
Penitrem B (dechloro-penitrem A)		5.84	
Paxilline	*P. paxilli*	150	
Cyclopiazonic acid	*P. cyclopium*		2.3 (i.p.), 36 (p.o.M) 63 (p.o.F)
Fumitremorgin A	*A. fumigatus*		
Fumitremorgin B	*A. fumigatus*		
Verruculogen TR_1	*P. verruculosum, A. caespitosus*		20 (p.o.), 1 day, cockerel
Verruculogen TR_2	*P. verruculosum, A. caespitosus*		
Brevianamide	*P. brevicompactum*		
Austamide	*A. ustus*		
Oxaline	*P. oxalicum*		
Roquefortine	*P. roqueforti*		15–20 (i.p.)
Tryptoquivaline	*A. clavatus*		
Tryptoquivalone	*A. clavatus*		

†1For further details, see section 1.3.10.
†2For structures, of selected tremorgenic mycotoxins, see Fig. 2.7.

According to Wilson *et al.*,[59)] the toxin caused increases in resting potential, end-plate potential, end-plate duration and miniature-end potential in rat phrenic nerve-diaphragma preparations. Since the toxin exerts no effect on diaphragma cholinesterase activity, it presumed to cause these changes by influencing pre-synaptic transmitter release. Also,

Stern[60] has suggested that the toxin is related to the activity of α-motor cells of the anterior horn.

Concerning the diuretic effects of the toxin, Wilson[61] has indicated that reabsportion in the kidney tubulus is prevented.

Holzapfel[62] isolated cyclopiazonic acid ($C_{20}H_{20}N_2O_3$) as a principal toxin from several strains of *P. cyclopium*. This compound is toxic to mice, rats, ducklings and calves. Rats receiving 8 mg/kg of the toxin i.p. exhibited exterior spasms with cyanotic mucous membranes and died within 2 h. Rats receiving 2–4 mg/kg of the toxin died within 1–3 days after dosing. In the case of oral administration, the absorption rate was rather slow and mortality occurred up to 6 days after dosing.

Histological examinations have revealed cytological lesions in various organs of rats receiving the toxin, i.e. single cell necrosis of hepatocytes, enlargement of bile duct lining cells, necrosis at the cortico-medullar junction of the kidneys, and degeneration of the islets of Langerhans. Such cellular damage is presumed to induce nervous symptoms, although no detailed studies on the central nervous system have yet been reported.

Yamazaki *et al.*[63] have investigated the strong tremorgenic effect induced in mice intraperitoneally administered with crude toxins from *A. fumigatus*. Subsequent fractionation led to the isolation of two tremorgenic indole derivatives, fumitremorgin A ($C_{32}H_{41}N_3O_7$) and fumitremorgin B ($C_{27}H_{33}N_3O_5$),[64] which are presumed to affect the activity of motor neurons of the central nervous system.

Cole *et al.*[65,66] isolated and determined the structure of verruculogen ($C_{27}H_{33}N_3O_7$) from *P. verruculosum*. This mycotoxin induces tremors, hypersensitivity to sound, tetanis spasmus and ataxia in mice, cocktails and chickens. Hortujac *et al.*[67] have indicated that a decrease in γ-aminobutyric acid level in the central nervous system is responsible for the induction of tremors.

3.2.5. Lactones

Among certain fungal metabolites, α,β-unsaturated lactone plays a functionally important role in the expression of biological activity, e.g. in carcinogenicity, mutagenicity, antimicrobial activity, etc. This lactone moiety is found, for example, in aflatoxins, ochratoxins, β-propiolactone, butenolide, citreoviridin, penicillic acid, patulin, etc. As reviewed by Dickens,[68] many lactone compounds are carcinogenic to animals, and their chemical structure other than the lactone moiety largely determines their relative activity, as shown in Fig. 3.12. In this section, patulin, citreoviridin and other related lactones are discussed.

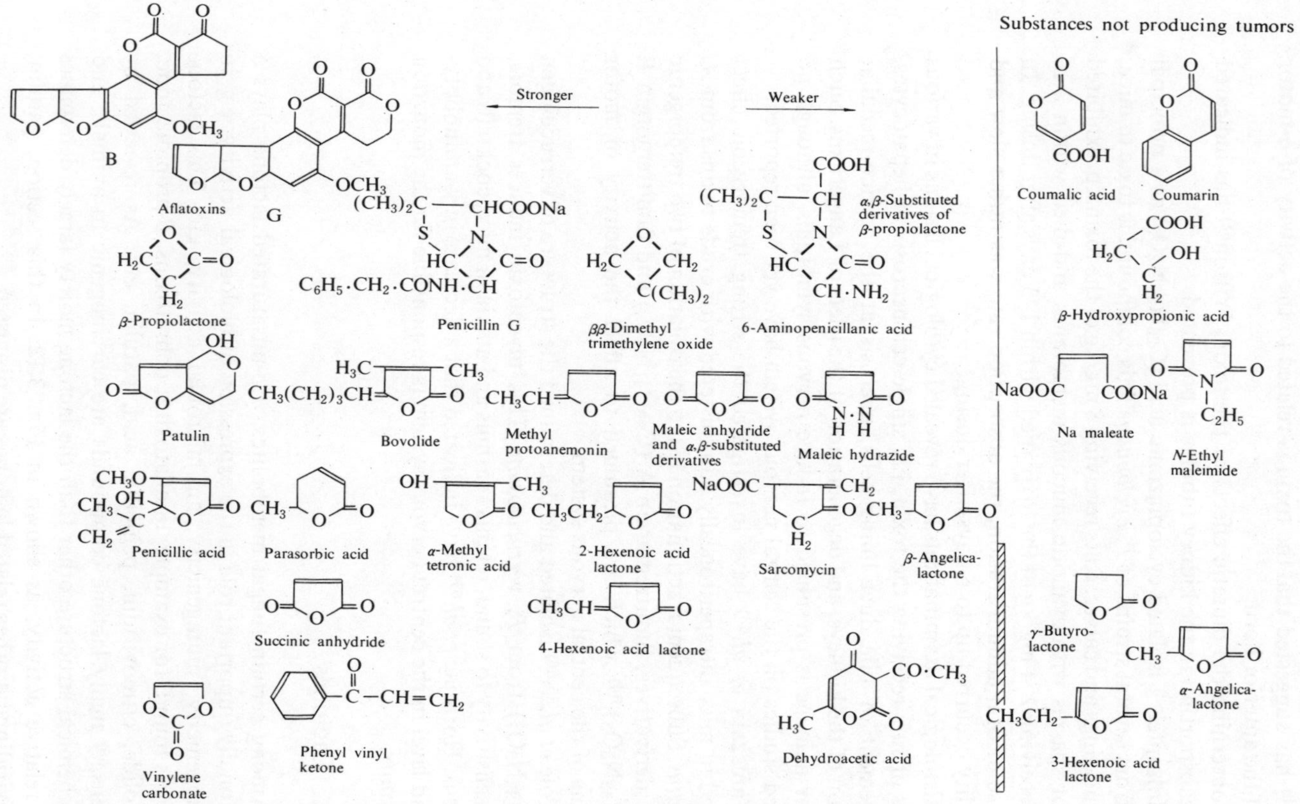

Substances not producing tumors

Stronger

Weaker

B

Aflatoxins

G

β-Propiolactone

Penicillin G

ββ-Dimethyl trimethylene oxide

6-Aminopenicillanic acid

α,β-Substituted derivatives of β-propiolactone

Coumalic acid

Coumarin

β-Hydroxypropionic acid

Patulin

Bovolide

Methyl protoanemonin

Maleic anhydride and α,β-substituted derivatives

Maleic hydrazide

Na maleate

N-Ethyl maleimide

Penicillic acid

Parasorbic acid

α-Methyl tetronic acid

2-Hexenoic acid lactone

Sarcomycin

β-Angelica-lactone

Succinic anhydride

4-Hexenoic acid lactone

γ-Butyro-lactone

α-Angelica-lactone

Vinylene carbonate

Phenyl vinyl ketone

Dehydroacetic acid

3-Hexenoic acid lactone

Fig. 3.12. Relative potential of carcinogenic lactones. (Source: ref. 68. Reproduced by kind permission of Springer-Verlag K.G., West Germany.)

Patulin ($C_7H_6O_4$), 4-hydroxy-4*H*-furo[3,2-*c*]pyran-2(6*H*)-one, produced by numerous species of *Aspergillus* and *Penicillium*, was initially reported as an antibiotic, and various names were give to it, i.e. clavacin, claviformin, expansin, penicidin. Recently, two chemical analogs, ascradiol and desoxypatulinic acid, have been isolated from fungal metabolites (Fig. 3.13). The former toxin is 1/4 less toxic than the parent compound.

Patulin

Desoxypatulinic acid

Ascradiol

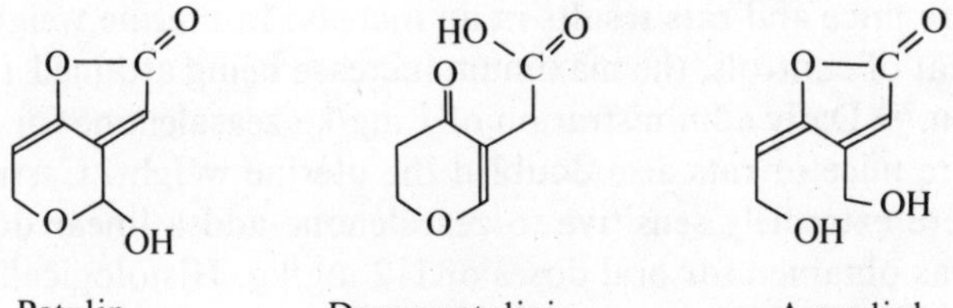

Fig. 3.13. Structure of patulin and its derivatives.

According to Dickens *et al.*,[69] following s.c. injection of 0.2 mg patulin in arachis oil twice a week to male rats, sarcomas were induced at the site of infection in 6 of 8 rats after 64–69 weeks. Long-term ingestion of patulin in mice and rats, however, revealed no evidence of carcinogenic changes, indicating an apparent resistance in these animals to oral dosing of the mycotoxin.

According to Lovett,[70] the oral LD_{50} of patulin to White Leghorn cockerels, averaging 191 g, was 170 mg/kg. The acute symptoms were extensive hemorrhage along the entire digestive tract, particularly the preventriculus, gizzard, and intestines. Also, Mintzlaff *et al.*[71] have demonstrated liver lesions in chicks fed 0.2 mg patulin daily for 6 weeks.

Patulin is reactive with SH-compounds, and the resulting SH-adducts are less toxic than the parent compound to mice, chick embryos and rabbit skin.[72]

In vitro experiments have revealed that patulin non-competitively inhibits aldolase with a K_i of 1.3×10^{-5}M, and less inhibition was observed with a cysteine adduct of patulin.[73] Chemical analysis of the inactivated enzyme revealed an interaction of patulin with the –SH and $-NH_2$ groups of the protein molecule.

Patulin is known to inhibit K^+ transport in erythrocytes,[74] and Ueno *et al.* have demonstrated that it inhibits Na^+-dependent transport of glycine in reticulocytes.[75] These inhibitory effects are presumed to be based on a potential for inactivating SH-group in the active center of membrane transport systems.

An estrogenic response observed in swine consuming moldy corn has been attributed to compounds produced by *Gibberella zeae* (*Fusarium roseum* and *F. graminearum*). One of these compounds has been isolated and characterized as zearalenone ($C_{18}H_{21}O_5$). Substances structurally similar but different as regards biological activity are produced by several other fungi, e.g. curvularin (*P. expansum*) and radicicol (*Nectria radicicola*).

Zearalenone can be quite properly regarded as an oestrogen for various kinds of animals. A single oral ingestion of 1–10 mg/kg zearalenone to immature mice and rats results in an increase in uterine weight to several times that of controls, the maximum increase being attained 18–24 h after ingestion.[76] Daily administration of 1 mg/kg zearalenone for one week to immature mice or rats also doubled the uterine weight. Castrated female mice were extremely sensitive to zearalenone and a linear dose-response curve was obtained for oral doses of 1–2 mg/kg. Histologically, proliferation and mitosis of the uterine muscle were observed.

Swine are very sensitive to zearalenone. Five mg/day of zearalenone for 5 days resulted in enlarged vulvas, mammae and nipples, prolapse of the vagina, and atrophy of the ovaries.[77]

Meronuck *et al.*[78] have investigated the activity of zearalenone in birds. Turkey poults fed rations containing *F. roseum*-invaded corn, developed swollen vents and prolapsed cloacae. According to Sherwood *et al.*,[79] 40 ppm zearalenone fed to chicks for 10 days had little effect on body weight but caused an enlargement of the testis and comb.

According to Mirocha *et al.*,[80] zearalenone occurs naturally in feed. Recently, Sugimoto *et al.*[81] detected heavy pollution of imported maize by the mycotoxin. Since some zearalenone-producing strains of *Fusarium* are capable of producing trichothecene mycotoxins such as T-2 toxin, etc., Ueno *et al.* have examined the combined effects of zearalenone and trichothecenes.[82]

Johnston *et al.*[83] and others have examined the structure-activity relationships of zearelenone analogs, and pointed out that optimum uterotropic activity is obtained when the 6′ carboxyl is reduced to an OH group (see also Chapter 2). Synthetic zearalanol (6′-reduced zearalanone) is already available commercially under the brand name "Ralgro" for use as an additive to livestock fodder to accelerate growth and fattening in young animals. From a toxicological viewpoint, however, no safety evaluation of zearalenone and its analogs has yet been made. As pointed in section 3.3.5, a wide distribution for zearalenone in animal tissues has been indicated, and further, the supernatant enzyme from rat liver converts zearalenone into a hydroxylated derivative which still exhibits estrogenic activity. It is thus quite possible that zearalenone or its metabolite(s) may be excreted into milk or accumulated in tissues.

Schoental[84] has pointed out a possible role for dietary zearalenone in the induction of spontaneous tumors in laboratory animals, and Drane *et al.*[85] have attempted to detect contaminant zearalenone and estrogenic factor in laboratory rat cake. The combined results appear to establish the possible contamination of certain experimental animal feeds with zearalenone, which may itself have originated from low-grade cereals infected with zearalenone-producing fungi.

Concerning the long-term toxicity of zearalenone, no detailed information is yet available. However, a DNA-attacking ability for zearalenone has been detected in a recombination-less mutant of *Bacillus subtilis*,[86] and long-term feeding of mice with zearalenone-containing feed resulted in the production of severe changes in several tissues (Y. Ueno, *unpublished data*).

Citreoviridin ($C_{23}H_{30}O_6$) produced by *P. citreo-viride*, *P. ochrasalmoneum*, *P. fellatum* and *P. pulvillorum*, is composed of three moieties: α-pyrone chromophore, conjugated polyene and a hydrofuran ring. Photochemical reaction of citreoviridin in the presence of iodine yields isocitreoviridin, and under UV-light the mycotoxin exhibits brilliant yellow fluorescence.

According to Uraguchi,[87,88] ethanol extracts of *P. citreo-viride*-molded rice typically induce acute poisoning in cats, dogs, and other vertebrates. Early symptoms are progressive paralysis of the hind legs, vomiting and convulsions. Respiratory distress appears gradually, and at an advanced stage of intoxication, cardiovascular disorders and hypothermia are marked. In the final stage, dyspnea, gasping and Cheyne-Stokes respiration are followed by respiratory arrest. These characteristic symptoms develop strongly in higher animals such as dogs and cats. Based on the electrophysiological approach, the following intoxication process was envisaged: (1) spinal and medullary depression are responsible for progressive paralysis, (2) the mycotoxin selectively inhibits motor neurons on internuncial neurones along the spinal cord and motor nerve cells in the medulla oblongata, and (3) the attack on the respiratory center is the cause of death.

With the pure mycotoxin, similar symptoms appear in experimental animals. The LD_{50}'s for male mice are 8.3 (s.c.), 8.2 (i.p.) and 29 (p.o.) mg/kg, and in female rats, the s.c. LD_{50} is 3.6 mg/kg. Acute symptoms are lameness of the posterior extremities, impairment of voluntary movement, tremors, paralysis, coma, gasping and convulsions.

Citreoviridin is capable of directly impairing ATPase activity at a low concentration (Y. Ueno, *unpublished*), and according to Roberton *et al.*,[89] aurovertin B_1, which possesses a similar structure to citreoviridin (Fig. 3.14), is a potent inhibitor of ATP synthesis and ATP hydrolysis catalyzed by mitochondria. Beechey *et al.*[90] have also reported that citreoviridin diacetate is an inhibitor of energy-linked processes occurring in intact mitochondria.

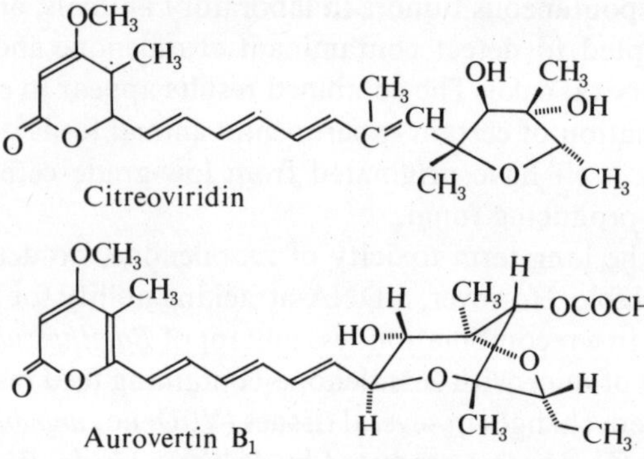

Citreoviridin

Aurovertin B_1

Fig. 3.14. Structure of citreoviridin and aurovertin B_1.

Rubratoxin A ($C_{26}H_{32}O_{11}$) and rubratoxin B ($C_{26}H_{30}O_{11}$) are toxic metabolites of *P. rubrum* and *P. purpurogenum*. Moss[91] and Newberne[92] have given extensive reviews of the chemical and biological properties of rubratoxins. Toxicological signs of intoxication are available for dogs and rabbits. The characteristic symptoms are anorexia, dehydration, somnolence, diarrhea and jaundice. Hemorrhage is also observable throughout a wide range of organs and tissues. The major site of injury is generally the liver.

Hayes *et al.*[93] have investigated the embryocidal and teratogenic effects of rubratoxin B on rats. This toxin caused increased levels (27–61 %) of prenatal mortality at an oral dose of 50 mg/kg at day 8 of gestation. However, the prenatal growth rate was unaffected and only a few gross abnormalities were noted. On the other hand, according to Tanaka (*personal communication*), administration of 0.5 mg/kg rubratoxin B to pregnant mice on day 7 of gestation caused marked malformations such as umbilical hernia, exencephaly, open eyes and anophthalmia, and bent ribs.

3.2.6. Halogens

Halogen-containing compounds are distributed to some extent in many biologically active fungal metabolites. In particular, chlorine is present in the structure of mycotoxins as well as antimicrobial agents, as shown in Table 3.27. It should be noted that chlorine-containing compounds such as fungal toxins (cyclochlorotine of *P. islandicum* and ochratoxins of *A. ochraceus*) as well as organic reagents (CCl_4 and chloroform) are established as hepatoxic chemicals. Furthermore, many organic halogens (cyclochlorotine and insecticides such as BHC) are hepatocarcinogens.

TABLE 3.27. Halogen-containing fungal metabolites

Halogen	Group	Metabolite	Biological activity
Cl	Mycotoxins	cyclochlorotine	hepatotoxic, carcinogenic
		ochratoxin A	hepatotoxic
		sporidesmin A, B, C	facial eczema
		penitrem A	tremorgenic
	Antimicrobials	caldoriomycin	antibacterial
		chloramphenicol	antibacterial
		grisan antibiotics	antifungal
		griseofulvin	
		erdin	
		geodin	
		geodoxin	
		mollisin	antibacterial, antifungal
		chlorflavonin	antifungal
		pyoluteorin	antibacterial, antiprotozoal
		chlortetracycline	antibacterial
Br	Antimicrobials	aerothionin A	antibacterial

Among halogen-containing antimicrobial agents, there are many known cases of so-called "side-effects." The most outstanding example is chloramphenicol, which is able to induce aplastic anemia. Chlortetracycline has been reported to be hepatotoxic. Thus, by causing toxicological disturbance in animal cells, certain antibiotic fungal metabolites exhibit mycotoxic characteristics.

"Chlorine-containing peptide" (cyclochlorotine) was first isolated by Tatsuno *et al.*[94] from culture filtrates of *P. islandicum* Sopp, a causal fungus of yellowed rice. This compound ($C_{25}H_{36}N_5O_8Cl_2$) is a cyclic peptapeptide [L-ser–dichloropro-L-β-aminobutyric acid–L-ser–L-β-phe].[95] On alkaline hydrolysis, it is converted into a non-toxic peptide which lacks the two chlorine atoms and cyclic structure.

Toxicologically, this mycotoxin is characterized by the induction of hepatic damage[1]. In mice and rats, the liver appears pale and anemic 5 min after administration of the lethal dose, and gradually turns red 30 min later. From the reduced volume of circulating blood in the liver, it is presumed that the peptide causes an increase in the permeability of the liver microcapillaries. In dogs, similar changes are also observed in the skin on s.c. injection of 100 μg of the peptide.

Other characteristic changes occur in the blood sugar level and glycogen content of the liver. On s.c. injection of cyclochlorotine at the LD_{50}

dose, the blood glucose shows an abrupt increase shortly after administration and then decreases 30–60 min later. In adrenoectomized rats or ergotamine-pretreated mice, the blood sugar is reduced by cyclochlorotine without any initial elevation. These findings indicate that the initial elevation is related to adrenal function. The content of liver glycogen is rapidly reduced with 0.2–0.5 mg/kg cyclochlorotine. Concerning functional changes in the liver of mice and rats intoxicated with cyclochlorotine, Ueno *et al.* have established that this mycotoxin induces accelerated glycogen catabolism and inhibits glycogen neogenesis. Specific accumulation of the peptide into the liver tissues is responsible for alteration of these biochemical and membrane functions.[96)]

Ochratoxin A ($C_{20}H_{18}ClNO_6$), B (dechloro-OA), and C (4-hydroxy-OA) are isolated from metabolites of *A. ochraceus*, *P. viridicatum* and other fungi. Ochratoxin A contains 7-carboxy-5-chloro-3,4-dihydro-8-hydroxy-3*R*-methyl-isocoumarin which is linked by an amide bond over the 7-carboxyl group of L-*β*-phenylalanine. Ochratoxin B lacks the 5-chlorine atom of ochratoxin A.

Ochratoxins are very toxic to experimental animals including ducklings, mice, rats, hens, dogs and trout. According to Choudhury *et al.*,[97)] 1–2 ppm ochratoxin A fed to hens for 14 weeks reduced the hatchability of fertile eggs and depressed the subsequent performance of the progeny for the first two weeks of their life. At 2–4 ppm ochratoxins, delayed sexual maturity and lower egg production rates were noted. Peckham *et al.*[98)] demonstrated suppression of hematopoiesis in the bone marrow and depletion of lymphoid elements from the spleen of chicks administered with ochratoxin A.

Szczeck *et al.*[99)] have demonstrated the toxicity of ochratoxin to beagle dogs. The sensitivity to ochratoxin A was high, and renal injuries characterized by necrosis of the epithelial cells of the proximal tubulus were induced.

Doster *et al.*[100)] investigated the acute toxicity of five derivatives of ochratoxins to rainbow trout. The 10-day i.p. LD_{50} values for ochratoxin A and ochratoxin B methylester were 3.0 and 13.0 mg/kg, respectively. No lethal effect was observed at a dose of 66.7 mg/kg ochratoxin B. Histologically, the liver and kidneys showed necrotic changes after ochratoxin A and C administration.

In rats intubated near the oral LD_{50} of ochratoxin A (20–22 mg/kg), the liver and kidneys showed marked necrotic injuries. In the kidneys, the distal convoluted tubules were injured at low levels of ochratoxin A. Concerning the liver injuries caused by ochratoxins, many reports have dealt with the biochemical modification of carbohydrate metabolism. Suzuki *et al.*[101)] have demonstrated that impairment of active transport of glucose is responsible for the marked depletion of glycogen as well as the depression of glycogen neogenesis, although no molecular analysis has yet been performed.

3.2.7. Macrolides

Macrocyclic structures are widely distributed in naturally occurring toxic metabolites such as trichothecenes (verrucarins and satratoxins), zearalenone, cytochalasins and chaetoglobosins. In this section, the latter two toxins will be considered, since these metabolites exhibit a characteristic toxic effect on animal tissues.

Cytochalasins, phomins and chaetoglobosins possess a common structure, perhydro-isoindole linked with a macrocyclic ring, and are produced by a wide range of fungi, as summarized in Chapter 2.

Up to the present, over 20 cytochalasins and related toxins have been identified. They induce common, distinctive effects, i.e. they all inhibit cytoplasmic cleavage in cultured mammalian cells, yielding polynucleate cells, and impair cell movement. Nuclear extrusion is one of their most curious effects. According to Umeda *et al.*,[102] no clear structure-activity relationships were observed among 12 compounds (8 chaetoglobosins and 4 cytochalasins). Wessells *et al.*[103] have classified biological processes which are sensitive/insensitive to cytochalasin B. As shown in Table 3.28, the effects of cytochalasin can be thought of as resulting from contractile activity of cellular organella. Reversible disorganization of microfilaments is presumed to be involved in the primitive contractile process. Thus, cytochalasins represent important biological reagents for elucidating the biological machinery of cellular organella, especially the microfilaments.

In spite of the large amount of data on the *in vitro* cytotoxity of cytochalasins, detailed knowledge on their *in vivo* toxicity is quite limited. Hayakawa *et al.*[104] have reported LD_{50} values for cytochalasin D in mice, i.e. 18.5 mg/kg (s.c.) and 36.0 mg/kg (p.o.). According to Buchi *et al.*,[105] cytochalasin E killed rats within a few hours of dosing. The LD_{50} values were 2.6 or 9.1 mg/kg after i.p. or p.o. administration of a single dose, respectively. Death was due to circulatory collapse caused by massive extravascular effusion of plasma (Table 3.29).

Glinsukon *et al.*[106] have also examined the toxicological features of cytochalasin E in rats. The LD_{50} values obtained for a single i.p. administration were 0.98 mg/kg for one-day-old rats and 2.60 mg/kg for adolescent rats. In the case of p.o. administration, the LD_{50} for adolescent rats was 9.10 mg/kg, but it became 1.30 mg/kg when the toxin was administered intrathoracically. The rats became ataxic and drowsy, and severe hematocytes developed at the infection site. Cyanosis, then sudden coma, follow-

TABLE 3.28. Cytochalasin B-sensitive and -insensitive processes (after Wessells *et al.*)[103]

A. Sensitive	1. Cytokinesis (cultured cells, lymphocytes, eggs)
	2. Single cell movement (L and heart fibroblasts)
	3. Axonal growth cone activity (nerve outgrowth)
	4. Tubular gland formation (oviduct)
	5. Morphogenesis in salivary epithelia (oviduct, salivary, lung)
	6. Premitotic migration of nuclei in epithelia (oviduct, salivary)
	7. Cytoplasmic streaming (Nitella, Avena)
	8. Blood clot retraction (platelets)
	9. Tail resorption in tunicate metamorphosis (*Distaplia*)
	10. Invagination during gastrulation (*Urechis*)
	11. Smooth muscle contraction (gut peristalsis)
	12. Cardiac muscle cell contraction
	13. Calcium-induced cortical contraction in eggs (*Xenopus*)
	14. Calcium-mediated cortical healing in eggs (*Xenopus*)
B. Insensitive	1. Karyokinesis (spindle function and microtubule integrity)
	2. Sperm tail function (*Urechis*)
	3. Ciliary function (*Paramecium, Urechis*)
	4. Cytoplasmic microtubule integrity (nerve, salivary)
	5. Filament (100 Å) integrity (salivary, oviduct)
	6. Neurofilament (100 Å) integrity (nerve)
	7. Microvillious core filament integrity (oviduct)
	8. Protein synthesis (salivary, glial, nerve)
	9. Striated muscle actomyosin superprecipitation
	10. *E. coli* cell division, flagellar movement, and mating
	11. Contractile vacuole function (*Paramecium*)

TABLE 3.29. Acute toxicity of cytochalasins

Compound	Animal	Route	LD_{50} (mg/kg)
Cytochalasin D	mouse	s.c.	1.85
		p.o.	36
Cytochalasin E	mouse	i.p.	4.60
	1-day-old rats	i.p.	0.98
	adolescent rats	i. thor.	1.30
	adolescent rats	i.p.	2.60
	adolescent rats	p.o.	9.10
	guinea-pig	i.p.	0.5–1.5
Kodo-cytochalasin-1	mouse		2 (lethal at 45 min)
Chaetoglobosin A	mouse	s.c.	6.5 (male)
			17.8 (female)

ed. Rats receiving a fatal i.p. dose of cytochalasin E died within 2–18 h with 2–3 ml of fluid in the peritoneal cavity. Histologically, congestive degenerative changes, necrosis of the liver, kidney, spleen and small intestine, brian edemas, pulmonary hemorrhage, and injury to the vascular wall were evident. Further examination of the fluid balance in intoxicated rats[107)] revealed that a single lethal dose of cytochalasin E (3.50 mg/kg, i.p.) produced a rapid decrease in plasma concentration in adult male rats within 20 min. The reduction in plasma volume was paralleled by the appearance of fluid at the site of injection. The plasma glucose concentration increased during the first hour after toxin treatment but this was followed by mild hypoglycemia. The liver glycogen content fell to 9% of the control value within 60 min after dosing. Based on analytical results for the fluid components, cytochalasin E was presumed to cause direct impairment of blood capillaries, so permitting the extravascular effusion of plasma fluid, albumin and globulin. The apparent cause of death was shock. Since antihistaminics and sympathomimetic drugs gave no protection to rats or guinea-pigs administered with a lethal dose of cytochalasin E, the possibility that this type of mycotoxin liberates bound histamine *in vivo* is excluded.

The cytochalasin E used in the above experiments represents one of the toxic principles obtained from a strain of *A. clavatus*. It was isolated from cooked rice eaten by a Thai child who subsequently died with Reye's syndrome. However, the lesions observed in animal tissues treated with a single lethal dose of cytochalasin E differ from those of autopsy specimens taken from fatal cases in Thailand. It thus appears that cytochalasin E may not be a factor in the etiology of Reye's syndrome.

In the case of kodo-cytochasin-1 and -2, no precise toxicological features are reported at the present. However, according to Patwardhan *et al.*,[108)] Kodo millet (*Paspalum scrobiculatum* Linn., syn. *P. commersonii* Lam; Sanskrit, *kodrava*; Hindi, *kodo*) represents a minor grain crop that has, since ancient times, been cultivated in India, and the grains have often been reported to cause poisoning in man and animals when used as food. Also, ancient texts describe the use of this material as a poison for tigers. The chief symptoms of kodrava poisoning are unconsciousness, delirium with violent tremors of the voluntary muscles, vomiting, and difficulty in swallowing. The toxic grains are often contaminated with fungi, among which *Phomopsis paspalli* is predominant. Subsequent analysis has revealed kodo-cytochalasin-1 and -2 as toxic agents of this type of fungus. Toxicological experiments with higher animals are expected to provide information on the possible relationships between kodrava poisoning and these kinds of cytochalasins.

3.3. FATE AND METABOLISM

3.3.1. (—)Luteoskyrin and (+)Rugulosin

(−)Luteoskyrin and (+)rugulosin act selectively on the liver of animals and only a small proportion of the administered dose is sufficient to cause functional and morphological damage. Acute toxicity usually develops rather slowly in animals and a flat dose-response curve is obtained. Uraguchi *et al.*,[25] Ueno *et al.*[109,110] and Sato[111] have performed pharmacokinetic investigations on the relationship between the distribution and toxicity of hepatotoxic anthraquinones. Male mice received 3 daily s.c. injections of an olive oil suspension of ^{3}H-luteoskyrin at a daily dose of 5 mg/kg for 3 consecutive days. As shown in Fig. 3.15, elimination of the compound from the body occurred very slowly and more radioactivity was excreted in the feces than in the urine. The total amount excreted within 18 days after injection was about 25% of the administered dose (19% in feces, 6% in urine). The concentration of ^{3}H in the blood

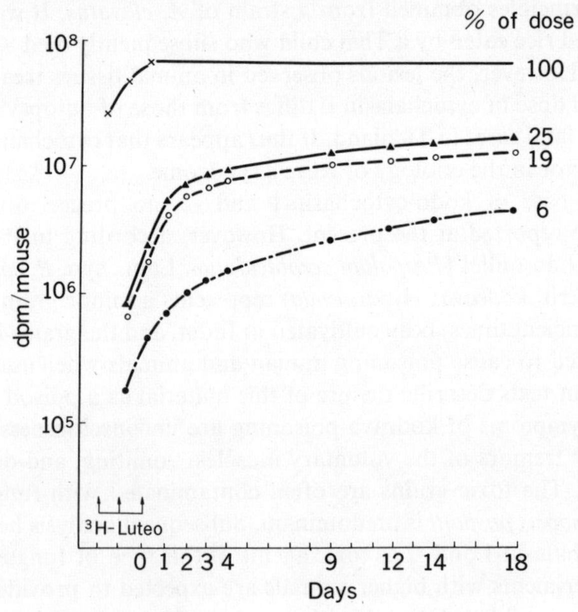

Fig. 3.15. Cumulative excretion of ^{3}H-luteoskyrin in the feces and urine of male mice.[25] ×, Dose administered; ○, recovery in the feces; ●, recovery in the urine; ▲, total recovery.

increased gradually with time and the maximum level of ^{3}H-luteoskyrin was 0.13 μg/ml. Among several organs, the liver showed the highest level of radioactivity, corresponding to 6.5% of the administered dose or 22μg ^{3}H-luteoskyrin/g liver. This concentration was about 700 times that in the blood cells. ^{3}H-Luteoskyrin was also detected in the kidneys, lung and brain, but the radioactivity levels in these organs were much lower than that in the liver. These findings suggest that (−)luteoskyrin is preferentially concentrated in the liver. Chemical analysis revealed that over 80% of the radioactivity in the liver derived from (−)luteoskyrin itself. A high affinity of (−)luteoskyrin for liver mitochondria may result in the functional and morphological damage in the liver, and the observed sex and age susceptibility differences in mice may derive from these pharmacokinetic features of the mycotoxin.

According to Sato,[111] (+)rugulosin behaves in a similar manner. Male mice were injected s.c. with 100 mg/kg of ^{3}H-rugulosin daily for 3 consecutive days, and 3 days after the last injection, when the s-GOT activity reached a maximum, the distribution pattern of radioactivity was analyzed. About 20% of the total dose was recovered from the organs and injected site; 11.2% occurred in the liver, which corresponded to 100 μg (+)rugulosin/g liver. A large portion (80% of the total recovery) was found at the injected site. These findings indicate that (+)rugulosin injected s.c. is distributed to the liver and remains for a long time at the injection site.

As mentioned above, (−)luteoskyrin exhibits deleterious effects on the function and morphology of liver mitochondria, and its toxicity varies with the species, strain, sex, and age of the experimental animals. The susceptibility of mice is higher than that of rats, and in mice, (−)luteoskyrin is more toxic to males and infants than to females and adults. As regards age differences, the ^{3}H-luteoskyrin levels in the liver of suckling mice were 1.62% of the administered dose in males and 1.43% in females, while the corresponding figures for adult mice were 0.26 and 0.19%, respectively (Table 3.30). Further pharmacokinetic studies demonstrated that ^{3}H-luteoskyrin and ^{3}H-rugulosin, when administered s.c. to mice, accumulated preferentially in the mitochondrial fraction of the liver (Tables 3.31 and 3.32), and the levels of accumulation were higher in animals showing higher susceptibility to the mycotoxins.

3.3.2. Cyclochlorotine

According to the authors' unpublished date for s.c. administration of 1 mg/kg of ^{3}H-labeled cyclochlorotine to male mice, 64% of the total radioactivity was detected in the liver at 20 min after injection, 54% at 1 h, and 25% at 4 h. For the kidneys, the corresponding figures were 28, 13,

TABLE 3.30. Effect of age on the accumulation and intracellular distribution of ^{3}H-luteoskyrin in the liver of mice (after I. Ueno *et al.*)[109]

Age (week)	Sex	^{3}H-Luteo in liver (μg/g)	^{3}H-Luteo in liver fractions (dpm/mg protein)			
			mitochondria	light mitochondria	microsomes	supernatant
2	male	1.2	820	590	240	447
	female	1.0	660	390	202	306
10	male	0.6	185	78	55	49
	female	0.4	79	52	42	37

TABLE 3.31. Sex differences in the subcellular distribution of ^{3}H-luteoskyrin in the liver of mice (after I. Ueno *et al.*)[109]

Fractions	Specific radioactivity (dpm/mg protein) Time after ^{3}H-luteoskyrin injection					
	15 h		23 h		1 week	
	male	female	male	female	male	female
Nuclei	1645	972	605	790	832	136
Mitochondria	5298	1460	6027	1428	3553	363
Light mitochondria	1132	235	737	323	830	140
Microsomes	780	818	1185	817	488	185
Supernatant	278	462	165	238	315	135
Cell debris	1898	657	752	258	793	132

TABLE 3.32. Intracellular distribution of ^{3}H-rugulosin in the liver of mice (after Sato)[111]

Fractions	^{3}H-Rugulosin	
	dpm/mg protein $\times 10^{-3}$	dpm/g liver $\times 10^{-3}$
1000g $\times$ 10 min precipitates	3.8	0.4
6000g $\times$ 10 min precipitates (heavy mitochondria)	12.2	56
15,000g $\times$ 10 min precipitates (light mitochondria)	10.6	24
105,000g $\times$ 60 min precipitates (microsomes)	6.2	12
105,000g $\times$ 60 min supernatant	7.2	64

and 12%, respectively. Charcoal and Dowex-50 adsorption techniques revealed that the peptide structure remained intact. These findings indicate that the peptide is quickly absorbed and transported to the liver cells, and excreted from the kidneys without marked modification.

After incubation of cyclochlorotine with various kinds of pure proteases or tissue homogenates *in vitro*, its lethal toxicity was assayed by injecting the incubation mixture into mice. Among the proteases, pronase and nagase, which are both known to cleave cyclic peptides, reduced the lethal toxicity of cyclochlorotine, while trypsin, chymotrypsin and papain had no observable effect. Among the tissue homogenates of mice, only brain homogenates caused appreciable inactivation. The peptide thus seems to resist conventional proteolytic enzymes as well as tissue proteolytic enzymes, and only those proteases having an ability to hydrolyze cyclic peptides can degrade cyclochlorotine. Another possibility for biological detoxication is dehalogenation. In fact, treatment of cyclochlorotine with ammonia or alkali results in a non-toxic dechloropeptide.

3.3.3. Aflatoxins and Sterigmatocystin

Knowledge of the fate and metabolism of mycotoxins in experimental animals is important for understanding the factors determining the biological responses and also for extrapolating animal data to assess the human risks. Aflatoxins are the most potent hepatocarcinogens among numerous naturally occurring and synthetic chemicals. Therefore, precise knowledge of their fate and metabolism, as well as of the biological activity of each metabolite, is urgently required to prevent human hazards arising from possible aflatoxin pollution. The metabolism of aflatoxin B_1 in animals is illustrated schematically in Fig. 3.16.

A. Demethylation: Aflatoxin P_1

Wogan *et al.*[112] prepared ring-labeled and *O*-methyl-labeled aflatoxin B_1 by submerged culture of *Aspergillus flavus* in liquid media containing either [1-C^{14}]-acetate or methyl-^{14}C-L-methionine, and administered the labeled toxins i.p. to Fischer rats. The major excretory route of the ring-labeled aflatoxin B_1 was biliary excretion into the feces. This accounted for nearly 60% of the administered ^{14}C. A further 20% was excreted into the urine. In contrast, about 25% of the administered *O*-methyl-labeled aflatoxin B_1 appeared in the respiratory CO_2, with a concominant decrease in the feces level. This indicates that *O*-demethylation probably represents a significant metabolic pathway of aflatoxin B_1 in rats.

Dalezois *et al.*[113,114] administered ^{14}C-ring-labeled aflatoxin B_1 to

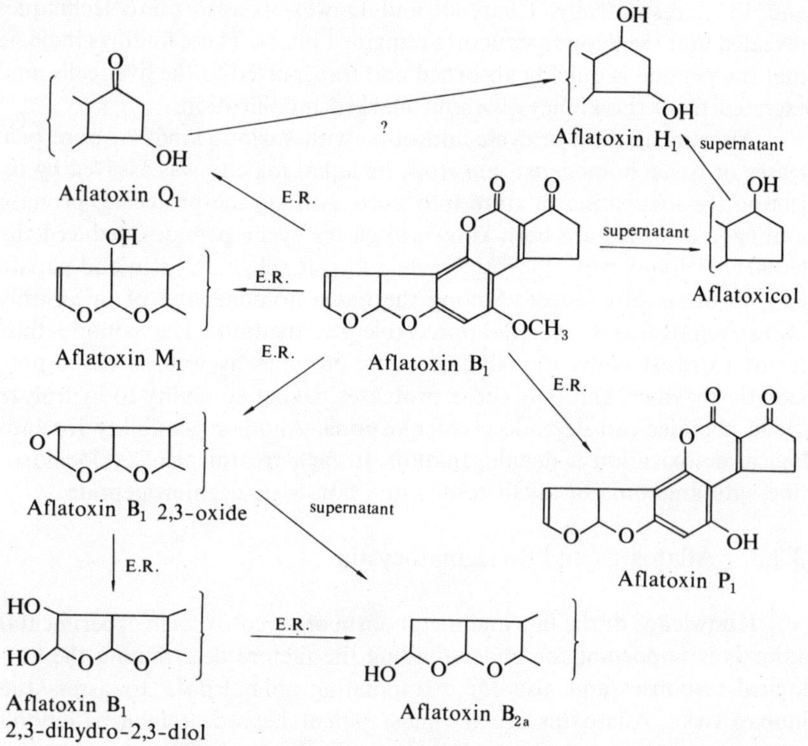

Aflatoxin Q_1

Aflatoxin H_1

Aflatoxicol

Aflatoxin M_1

Aflatoxin B_1

Aflatoxin B_1 2,3-oxide

Aflatoxin P_1

Aflatoxin B_1 2,3-dihydro-2,3-diol

Aflatoxin B_{2a}

Fig. 3.16. Metabolic conversion of aflatoxin B_1.

monkeys, and demonstrated that an *O*-demethylation product of the toxin, aflatoxin P_1, in the urine represented about 20% of the administered dose, 17% as the glucuronide, 3% as the sulfate and 1% as the unconjugated phenol. In this experiment, they found that aflatoxin B_1 and M_1 in the urine accounted for 2.3 and 0.01–0.1% of the dose, respectively. If aflatoxin P_1 is a major urinary metabolite of aflatoxin B_1 in man, as in the monkey, it may be feasible to utilize tests for aflatoxin P_1 in epidemiological surveys.

As for the direct and indirect effects of aflatoxin P_1 on animals, no detailed information is yet available. Büchi *et al.*[115] reported that aflatoxin P_1 prepared by treatment of aflatoxin B_1 with lithium *t*-butylmercaptide in hexamethylphosphamide, caused some degree of mortality in newborn mice at an i.p. dose of 150 mg/kg, whereas aflatoxin B_1 showed an i.p. LD_{50} of 9.5 mg/kg under comparable conditions. Since aflatoxin P_1 retains the bisfuran moiety, which is presumed to represent the active center of aflatoxin carcinogenesis, further studies on its biological activity and metabolic modification are required.

B. Hydroxylation: Aflatoxin M_1, B_{2a}, Q_1 and H_1

Numerous reports demonstrate that aflatoxin M_1, a 4-hydroxylated derivative of B_1, can be detected by fluorescence techniques as a toxic metabolite in the milk and urine of cows, sheep and rats after experimental administration of aflatoxin B_1. This compound has also been found in the urine of humans ingesting aflatoxin-polluted foods. In animals and man, about 5% of the ingested aflatoxin B_1 is excreted as M_1 in the urine.

Hydroxylation of aflatoxin B_1 is mediated by mixed function oxygenase of the endoplasmic reticulum (or microsomal enzymes). Indeed, in the presence of fortified NADPH, crude and isolated microsomal preparations from the liver of many animal species have been found to transform aflatoxin B_1 to its 4-hydroxy derivative (aflatoxin M_1) to some extent. An analogous compound, GM_1, was formed from aflatoxin G_1.

In vitro assays of aflatoxin-metabolizing activity have demonstrated a second pathway of aflatoxin hydroxylation, i.e. NADPH-dependent microsomal enzyme hydrates the vinyl ether double bond of aflatoxin B_1 and M_1 to form the 2-hydroxyl derivatives, aflatoxin B_{2a} and G_{2a}, respectively.[116] Aflatoxin B_{2a} is the same compound as hydroxydihydro-aflatoxin B_1 and aflatoxin B_1-W. Since aflatoxin B_1-2,3-oxide has been identified as an active intermediate, as will be described in section D, B_{2a} is presumed to represent a hydrolyzed product of the oxide. Ciegler *et al.*[117] demonstrated that aflatoxin B_{2a} is a detoxified metabolite of aflatoxin B_1 by acid-forming mold cultures. It exhibited a lower toxicity than the parent compound in duckling assays: feeding of 55μg B_{2a} per duckling caused no deaths, whereas the LD_{50} of B_1 was 40 μg per duckling. Bile duct hyperplasia was not noted, in contrast to the case of B_1, although there was some fatty metamorphosis in the liver.

A third pathway of hydroxylation occurs at the terminal cyclopentane ring of aflatoxin B_1. Hsieh *et al.*[118] and Krieger *et al.*[119] demonstrated that an hepatic microsomal oxidase system from monkeys catalyzed the hydroxylation of aflatoxin B_1 to aflatoxin M_1 and Q_1. The oxidase system requires NADPH and molecular oxygen. Following incubation of aflatoxin B_1 with the microsomal fraction, 52.9% was recovered as Q_1, 6.1% as M_1 and 5.5% as water-soluble derivatives. The optimum pH for Q_1 production was 7.4 and that for M_1 was 8.4. The apparent K_m and V_{max} of Q_1 formation were 0.166 mM and 7.7 μmol/mg protein/min, respectively, and those for M_1 formation were 0.012 mM and 0.44 μmol/mg protein/min respectively. In a CO-air atmosphere ($CO:O_2 = 5:1$), the rates of formation of M_1 and Q_1 were reduced to 33 and 66% respectively, and SKF 252-A and the methylenedioxyphenoxy compound, sesamez, were strikingly more effective in inhibiting Q_1 formation than M_1 formation. These data support the assumption that biotransformation of aflatoxin B_1 to M_1 and Q_1 requires the participation of cytochrome P-450, the hemoprotein terminal oxidase, and that there is a fundamental difference between the two microsomal aflatoxin B_1 hydroxylase systems.

Concerning the biological activity of aflatoxin Q_1, it is only 5.5% as toxic as the parent compound B_1. It is not mutagenic to *Salmonella typhimurium*.

Salhab *et al.*[29] demonstrated that human and monkey liver preparations modified aflatoxin B_1 to a derivative with the ketone carbonyl on the cyclopentane ring reduced to a secondary alcohol, and a hydroxyl group introduced onto the carbon β to the alcohol group. This metabolite is formed from aflatoxin B_1 at levels similar to those for aflatoxin M_1. Formation of this newly identified metabolite, named aflatoxin H_1, required both the microsomal hydroxylase and cytoplasmic reductase systems. The following two possible pathways are considered:

(1) Aflatoxin $B_1 \xrightarrow[\text{microsomes}]{} $ aflatoxin $Q_1 \xrightarrow[\text{supernatant}]{} $ aflatoxin H_1

(2) Aflatoxin $B_1 \xrightarrow[\text{supernatant}]{} $ aflatoxicol $\xrightarrow[\text{microsomes}]{} $ aflatoxin H_1

Bioassays using chicken embryos and a mutant of *S. typhimurium* revealed no toxicity.

C. Reduction of Cyclopentanone: Aflatoxicol

According to Patterson *et al.*,[120] soluble fractions from chicken, duck, turkey and rabbit, were capable of reducing the carbonyl group of the terminal cyclopentane of aflatoxin B_1 and B_2 to the corresponding aflatoxicol and dihydro-aflatoxicol. Little or no such activity was present in soluble fractions from guinea-pig, mouse and rat. This metabolic activity is located in the 105,000*g* supernatant fraction of liver homogenates and requires NADPH.

Since 17-ketosteroid sex hormones inhibit the cytoplasmic reduction reaction, it has been proposed that a soluble NADPH-linked 17-hydroxysteroid dehydrogenase may be involved in the transformation.[116] In this respect, the biotransformation differs fundamentally from the abovementioned microsomal oxidase system.

D. Epoxidation: Aflatoxin B_1-2,3-epoxide

In 1950, it was suggested by Boyland[121] that epoxides are intermediates in the metabolism of carcinogenic hydrocarbons to dihydrodiols, phenols and glutathione conjugates. In fact, the K-region epoxides of dibenzo (*a*)anthracene and related compounds are converted by rat liver microsomes to the corresponding phenols and dihydrodiol.[122] It has further been shown that epoxides are obligatory intermediates in the oxidation of benzene and naphthalene by the microsomal oxidase.

Based on these observations, Schoental[123] proposed that the vinyl ether double bond of aflatoxins is susceptible to the metabolic activation system in the same way as in the case of the K-region of polycyclic aromatic hydrocarbons. Garner *et al.*[124] demonstrated that rodent and fish liver microsomes converted aflatoxin B_1 to derivatives which are highly toxic to certain bacteria, although the toxic derivatives could not be isolated. They obtained an RNA-aflatoxin B_1 adduct, however, on addition of RNA to the incubation mixture. Swenson *et al.*[125] showed that the RNA of phenobarbital-induced microsomes from hamsters and rats bound aflatoxin B_1, but not the much less carcinogenic B_2, *in vitro* in the presence of an NAD PH system. On mild acid hydrolysis, the RNA-aflatoxin B_1 adduct yielded 2,3-dihydro-2,3-dihydroxy-aflatoxin B_1, and they assumed that aflatoxin B_1-2,3-epoxide was a possible reactive intermediate derived from the parent B_1. This might account for the observed mutagenic and carcinogenic properties of the mycotoxin, as will be discussed in a later section.

It is important to note that the microsomal or cytoplasmic conversion of aflatoxins to P, Q, H and aflatoxicol usually yields less toxic metabolites than the parent compound, whereas the above-mentioned epoxidation yields more reactive toxicants. Thus, modifications which are toxicologically opposing occur in the same liver microsomal fractions.

E. Sterigmatocystin

Thiel and Steyn[126] prepared ^{14}C-labeled sterigmatocystin in *Aspergillus versicolor* cultures using a liquid medium containing [1-^{14}C]-acetic acid and administered the labeled toxin orally to a male vervet monkey at a dose of 100 mg/5.4 kg. Amberlite XAD-2, DEAE-Sephadex 25 and TLC analysis revealed that more than 50% of the urinary metabolites were in the form of glucuronide conjugate of sterigmatocystin. Unlike aflatoxin B_1 which undergoes *O*-demethylation followed by conjugation with glucuronic acid in Rhesus monkeys, sterigmatocystin possesses a free phenolic hydroxyl group which is capable of conjugation. The methyl group of sterigmatocystin remains intact in the major urinary product (Fig. 3.17).

Sterigmatocystin shows a lower toxicity than aflatoxin B_1. On *O*-acetylation, the resulting *O*-acetylsterigmatocystin exhibits an acute cytotoxicity comparable to that of aflatoxin B_1. Yamazaki *et al.*[127] prepared *O*-acetyl(^{14}C-ring-labeled)-sterigmatocystin and administered it orally to rats. Analytical data on the excreta revealed that the absorption rate of

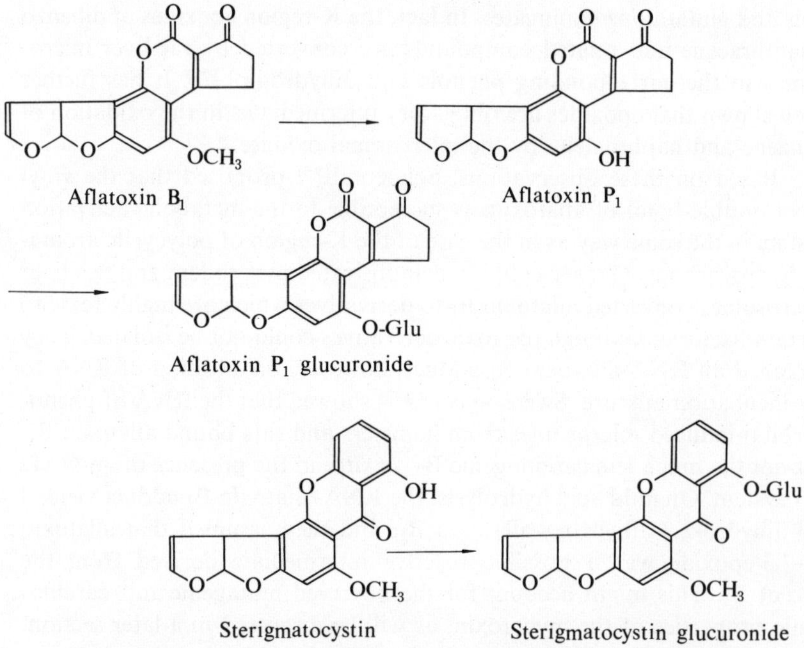

Aflatoxin B_1 → Aflatoxin P_1 → Aflatoxin P_1 glucuronide

Sterigmatocystin → Sterigmatocystin glucuronide

Fig. 3.17. Urinary metabolites of aflatoxin B_1 and sterigmatocystin.

O-acetylsterigmatocystin was higher than that of the parent compound. The question of whether the metabolites of *O*-acetylsterigmatocystin in the excreta are in the free or conjugated form remains to be clarified. Comparative studies on the relative absorbabilities of aflatoxin B_1, sterigmatocystin and its *O*-acetyl derivative, would provide further valuable data for assessing the relative toxicities of these bisfuran mycotoxins.

3.3.4. Trichothecenes

In spite of the potent cytotoxic properties of the trichothecene mycotoxins, details of their fate and metabolism are not yet clear. Ueno *et al.*[128] examined the fate and tissue distribution of ^{3}H-fusarenon-X in mice. ^{3}H-Labeled fusarenon-X was administered s.c. to male mice (3.6×10^5 cpm/animal), and its excretion into the urine and feces and the distribution of radioactivity in the organs were examined at 0.5, 3, 12 and 24 h after injection. As summarized in Table 3.33, at 30 min the radioactivity was highest in the liver, followed by the kidneys, small intestine, and large intestine. At 3 h, radioactivity was detected only in the kidneys among the organs tested, and the greater proportion was found in the urine, in which

TABLE 3.33. Distribution and excretion of ^{3}H-fusarenon-X in mice (after Ueno *et al.*)[128]

No. of mouse		1	2	3	4
Time of examination (h)		0.5	3	12	24
Tissues	liver	9000 cpm	0 cpm	0 cpm	0 cpm
	kidneys	3500	1200	0	0
	small intestine	2900	0	0	0
	large intestine	2300	0	0	0
	stomach	750	0	0	0
	spleen	250	50	0	0
	bile	100	100	0	0
	heart	0	0	0	0
	brain	0	0	0	0
	plasma	300	0	0	0
Excreta	urine	980	16,300	98,300	89,000
	feces	2980	3540	6060	3490

maximum radioactivity was observed at 12 h after injection. No detectable radioactivity was found in the heart or brain.

Based on these data, it was concluded that fusarenon-X was eliminated rapidly via the kidneys into the urine. Furthermore, the charcoal adsorption method as well as TLC analysis revealed that the major part of the radioactivity in the urine represented a compound more polar than the parent trichothecene, fusarenon-X.

Recently, Ueno *et al.*[129] developed a technique for the GLC analysis of trichothecene compounds in fungal metabolities, and adopted this technique to analyze trichothecenes present in urine and tissue samples. The data demonstrated that urine collected from rats and horses which had been administered fusarenon-X or T-2 toxin, contained nivalenol, HT-2 toxin, and neosolaniol. All studies so far (Ueno *et al.*, *unpublished*) suggest that the acetyl group at C-4 of T-2 toxin and fusarenon-X is selectively deacylated to the corresponding HT-2 toxin and nivalenol.

In order to examine whether these trichothecenes are deacetylated by cell-free liver systems, *in vitro* experiments with homogenates and subcellular fractions of rat liver were performed. Ohta and Ueno[130] demonstrated that the S-9 fraction (9000*g* supernatant) of rat liver homogenate deacetylated stereospecifically the C-4 acetyl group of T-2 toxin and fusarenon-X to HT-2 toxin and nivalenol, respectively (Fig. 3.18), and that this deacetylating activity was localized in the microsomal fraction of the liver (Table 3.34). Addition of NADPH or an NADH-generating system to the *in vitro* system did not enhance the deacetylation activity, and the

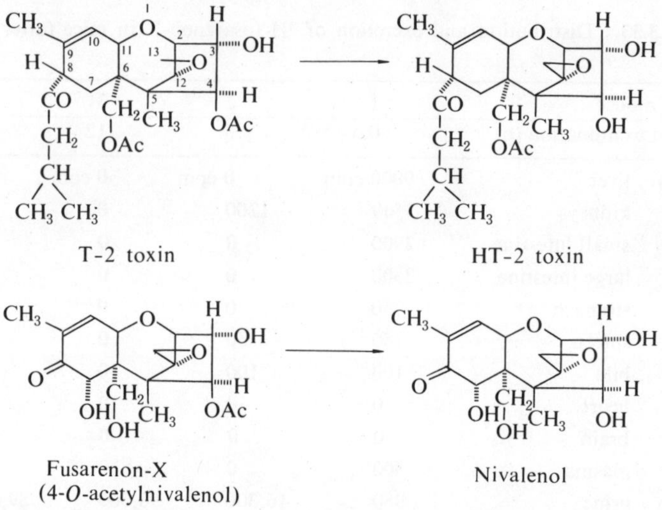

T-2 toxin → HT-2 toxin

Fusarenon-X (4-*O*-acetylnivalenol) → Nivalenol

Fig. 3.18. Deacetylation of trichothecene mycotoxins.

TABLE 3.34. Enzymatic deacetylation of T-2 toxin by rat liver (after Ohta and Ueno)[130]

Enzyme fraction	HT-2 toxin formed (μmol/mg protein/10 min)
S-9 fraction	0.030
Microsomal fraction	0.076
Supernatant	0

administration of inducers such as phenobarbital, 3-methylcholanthrene and PCB (Kanechlor-400) to the animals elevated the total liver microsomal deacetylase activity. It is probable therefore that the trichothecenes, when administered to animals, are first deacetylated at C-4 by the stereospecific esterase of liver microsomes, and that the resulting products such as nivalenol and HT-2 toxin are eliminated via the kidneys into the urine.

During the course of investigation of the emetic effect of T-2 toxin, Ellison *et al.*[44] demonstrated the *in vitro* conversion of T-2 toxin to HT-2 toxin by human and bovine liver homogenates.

Other routes of metabolic alteration of trichothecene mycotoxins such as demethylation at C-9 and hydrolytic cleavage of the 12, 13 epoxide remain to be clarified. If it can be clearly established that HT-2 toxin neosolaniol and nivalenol are urinary metabolites of T-2 toxin and fusarenon-X in animals and man and that their excretion is quantitatively related to the trichothecene uptake, then the exposure of humans as well as veterinary animals to these mycotoxins can be estimated by screening urine samples.

3.3.5. Zearalenone

As discussed above, zearalenone exhibits a uterotrophic effect in mice and rats irrespective of the administration route, although the greatest effect is seen after oral administration. This finding suggests that zearalenone is metabolized to an active compound after ingestion. In order to examine this possibility, experiments with the labeled compound were recently performed by Ueno *et al.*[131,132] Two female rats weighing 100–120 g were orally administered 10 mg/kg of ^{3}H-zearalenone, and its fate was analyzed (Fig. 3.19). Most of the fecal excretion of the ^{3}H-compound(s)

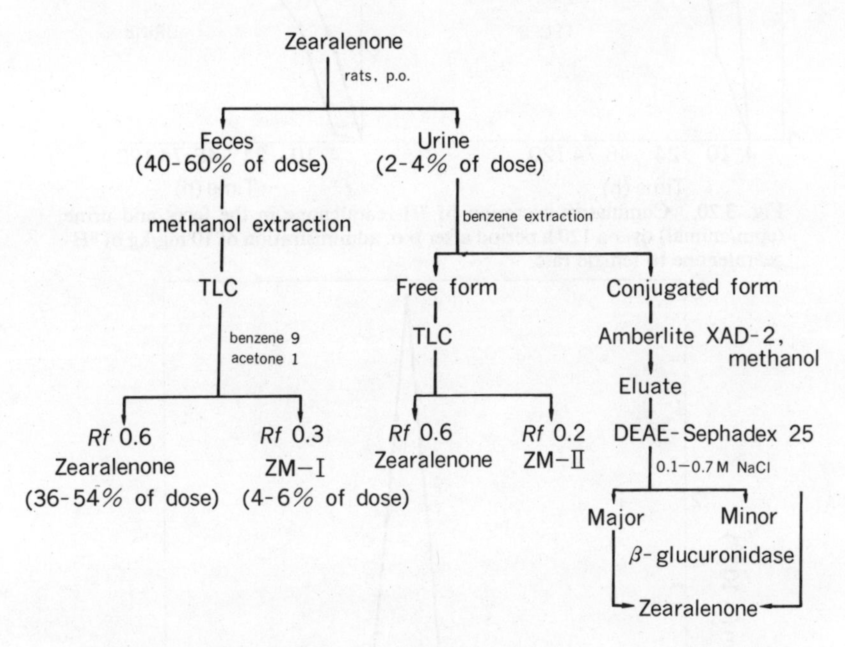

Fig. 3.19. Distribution pattern of ^{3}H-zearalenone in rat excreta (after Ueno *et al.*).[131]

occurred during the first 4–10 h and a total of 40–60% of the administered dose was recovered in the feces within 24 h. Excretion into the urine was rather slow compared to the fecal excretion, and a total of 2–4% of the dose was recovered in the urine collected over the first 120 h (Fig. 3.20).

All of the radioactivity in the feces was extractable with methanol. On TLC (Kieselgel G, benzene-acetone 9:1), the radioactivity split into a major fraction (*Rf* 0.6, zearalenone) and a minor one (*Rf* 0.3, ZM-I), the former representing 90% of the total radioactivity. As for the labeled com-

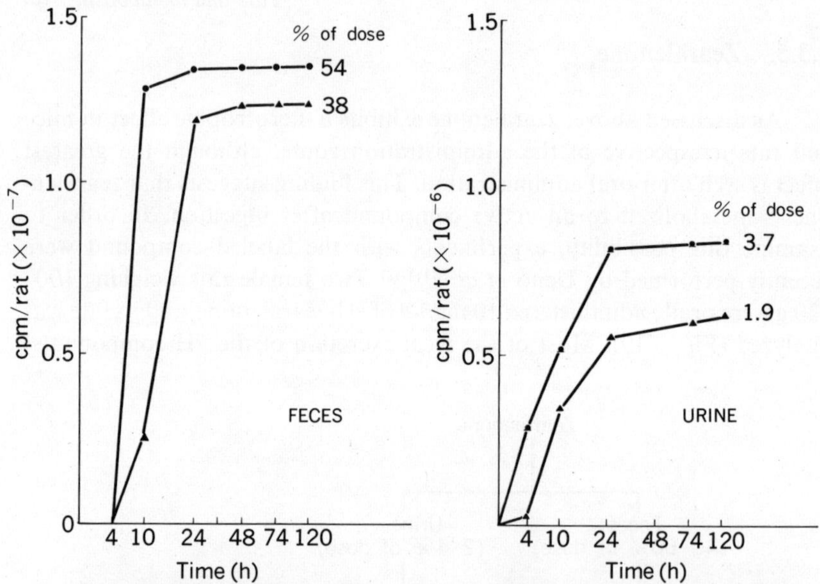

Fig. 3.20. Cumulative excretion of ^{3}H-zearalenone in the feces and urine (cpm/animal) over a 120 h period after p.o. administration of 10 mg/kg of ^{3}H-zearalenone to female rats.

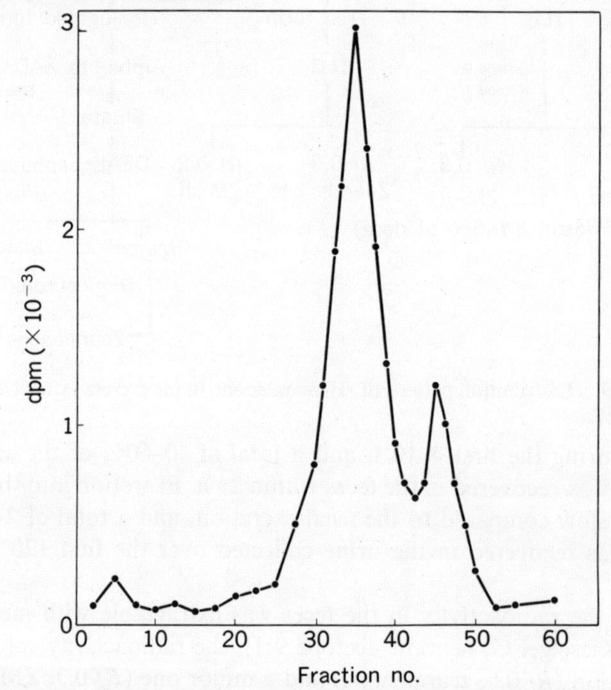

Fig. 3.21. Elution pattern of glucuronide-conjugated ^{3}H-zearalenone in the urine on DEAE-Sephadex chromatography.

pounds in the urine, the free and conjugated forms accounted for 30 and 70% of the total radioactivity, respectively. From the former fraction, zearalenone (90%) and ZM-II (*Rf* 0.2, 10%) were identified. From the latter, two types of glucuronide conjugates of zearalenone were separated by adsorption on Amberlite XAD-2 followed by chromatography on DEAE-Sephadex 25 (Fig. 3.21). These results demonstrate that p.o. administered zearalenone is metabolized into two polar compounds, zearalenone metabolites(ZM)-I and -II, and eliminated via the feces, while a small part of the dose occurs in the urine as glucuronide conjugates.

A further experiment with cell-free preparations of rat liver has demonstrated that the cytoplasmic supernatant is able to convert zearalenone to ZM-I in the presence of an NADPH-regenerating system. From spectral analysis, it appears likely that zearalenone is reduced by the cytoplasmic NADPH-reductase to 6′-hydroxylzearalenone.

3.4. Cytotoxicity of Mycotoxins

For evaluating the toxicity of various chemicals, the mammalian cell culture technique has been introduced with much success. In the field of mycotoxicology, this technique is used for the screening of mycotoxin-producing fungi, fractionation of toxicants from fungal metabolites, chemical and biological characterization of isolated mycotoxins, and evaluation of the comparative toxicity of chemically related analogs. In general, HeLa cells and primary cells from the liver, kidney and lung of animals are used, and the cell toxicity is monitored by

(1) counting the degree of cellular multiplication,

(2) microscopic investigation of any morphological alterations,

(3) chemical determination of cellular constituents such as proteins and DNA,

(4) measurement of the incorporation rate of radioactive precursors of nucleic acids and proteins, such as labeled thymidine, uracil and amino acids,

(5) counting the extent of colony formation, and

(6) analysis of any chromosomal and DNA aberrations.

The actual grade of observed cytotoxicity of mycotoxins depends largely upon the culture cells used. Umeda[133] compared the cytotoxicities of five hepatotoxic mycotoxins and chemicals, as summarized in Table 3.35. Liver parenchymal cells (LPC) exhibited a higher susceptibility to these hepatotoxins than the other cells. In the case of carcinogenic hydrocarbons, primary cells from mice, rats and hamsters are susceptible, while

TABLE 3.35. Comparative cytotoxicity of mycotoxins (after Umeda)[133)]

Mycotoxins and carinogens		Degree of cytotoxicity†1				
		Rat liver†2		Rat kidney	Rat lung	HeLa cells
		LPC	ELC			
(—)Luteoskyrin	1.0 μg/ml	4	3	4	4	4
	0.32	4	2	2	3	3
	0.1	1	0.5	—	0	0.5
Cyclochlorotine	32	4	3	3	—	3
	10	3.5	2	1	—	2
	3.2	3	1.5	1	—	0
Aflatoxin B_1	10	4	3.5	—	3	2
	3.2	4	3	—	2	1
	1.0	3	2	—	—	0
Monocrotaline	10^{-2} M	4	3	—	3	3
	$10^{-2.5}$ M	2	1	—	1	1
DAB	$10^{-3.5}$ M	3	1	0	—	2
	10^{-4} M	1	0	0	—	2

†1The degree of cytotoxicity is expressed as 0 (no cellular damage) through 4 (complete cytolysis or degeneration).
†2LPC, Liver parenchymal cells; ELC, endothelium-like cells.

the transformed cells of these cell lines become resistant to the toxicants. These differences in sensitivity result from the fact that the susceptible cells display different levels of arylhydrocarbon hydroxylase activity. This means that the cytotoxicity of the chemicals is largely dependent on the ability to biotransform the toxicants into toxic metabolites.

Cytotoxicity tests with mammalian cells are very convenient and more economical than tests using whole animals, and provide convenient quantitative information for analysis. Also, when only limited amounts of test material are available, the assay can be performed with a few milligrams. Engelbrecht *et al.*[134)] evaluated the cytotoxicity of 15 analogs related to either sterigmatocystin or aflatoxin B_1, using primary cell cultures. They found that an unsaturated $\Delta^{1,2}$-furobenzofuran ring and the positions of the methoxyl and hydroxyl groups on the xanthone ring of the sterigmatocystin analogs affected the cytotoxicity. Umeda *et al.*[22)] have also compared the cytotoxicity of (−) luteoskyrin and related anthraquinones, as mentioned in section 3.2.1.

The chemical and biological characteristics of mycotoxins are quite diverse. If the broad correlations between morphological alterations in cultured cells and biochemical modes of action could be established, it should be possible to predict the kind(s) of mycotoxins present in particular toxic fungi. Employing well-known metabolic inhibitors such as mitomycin C and actinomycin D, Umeda[133] has attempted to correlate the morphological changes and biochemical modes of action, as summarized in Table 3.36.

Based on such cytomorphological investigations with cultured cells, several mycotoxins have been isolated and characterized. Rubratoxin B,

TABLE 3.36. Cytological alteration of HeLa cells by metabolic inhibitors (after Umeda)[133]

Symbols	Classification of drugs	Cytoplasm	Nucleus	Nucleolus	Mitotic cells	Others
D	Inhibitor of DNA synthesis (FUDR, ara C, (hydroxyurea)	enlarged, polygonal	enlarged: fine granular chromatin	enlarged, irregular contour	scanty, abnormal	
A	Alkylating agent (mitomycin C, nitromine)	enlarged and small	enlarged and small: fine granular chromatin	enlarged, irregular contour	slightly increased, abnormal	polynuclear cells
R	Inhibitor of RNA synthesis (actinomycin D, proflavine)	enlarged, spindle-shaped, faintly stained	enlarged: spotty chromatin (fine nucleoplasm)	very small, round	not decreased	
S	Purine analog (8-azaguanine, 6-mercaptopurine)	small, spindle-shaped	small: clear nucleoplasm	small, round	not decreased, atrophic	polynuclear cells
P	Inhibitor of protein synthesis (cycloheximide, (fusarenon-X)	small, spindle-shaped, scanty	small: thick nuclear membrane	enlarged, irregular or round in shape	decreased, atrophic	
C	Spindle fiber poison (Colcemid, vinblastine)	polymorphic	enlarged and small		increased: clumped chromosomes	polynuclear cells
M	Inhibitor of cytokinesis (cytochalasins, chaetoglobosins)		multiple nuclei of equal size		multipolar division	polynuclear cells
V	Agents inducing cytoplasmic vacuolation	large vacuoles (not stained by fat staining)				

which induced polynucleation and accumulation of mitotic cells, was isolated from a culture filtrate of *Penicillium purpurogenum*,[135] while a new cytochalasin, named chaetoglobosin A, which caused polynucleation and multipolar division in HeLa cells, was obtained from mycelia of *Chaetomiun globosum*.[102] These results indicate the usefulness of the culture technique, although there are several cases in which cultured cells and whole animals have shown different susceptibilities. The oral LD_{50} values of aflatoxin B_1 and sterigmatocystin in rats are 6.2 and 116 mg/kg, respectively, while their lethal concentrations to cultured rat liver cells and HeLa cells are both of the order of 1 μg/ml. Another example is the case of yellowed rice toxins. That is to say, the i.p. LD_{50} values of luteoskyrin and cyclochlorotine are 6.65 and 0.33 mg/kg, respectively, while their lethal concentrations to HeLa cells have been reported as 1 and 30 μg/ml, respectively. These results indicate some disparity between cell culture toxicities and whole animal toxicities. This difference is probably generated by the metabolic activation system or microsomal drug metabolizing system. Actually, as will be discussed below, this system has recently been employed for detecting mutagenicity and carcinogenicity in several mycotoxins.

In addition to the cell culture technique, the organ culture technique should be employed to evaluate cytotoxicity with special reference to tissue or organ specificities. The results can be expected to permit correlations between the *in vitro* and *in vivo* toxicity tests.

3.5. MUTAGENICITY OF MYCOTOXINS

3.5.1. Background

There is little doubt that many diseases, particularly cancer, are caused by environmental pollutants. There is also growing concern and interest over the possible role of chemical carcinogens in activating oncogenic viruses. The World Health Organization has estimated that over 70% of all human cancers are influenced by environmental factors,[136] and other estimates indicate that 60–80% of human cancers are due to chemicals,[137] as reviewed by Epstein.[138]

Toxic fungal metabolites are considered to represent naturally occurring pollutants of many food- and feed-stuffs, and their possible long-term effects such as carcinogenicity as attracting increasing attention among health scientists throughout the world. Numerous studies have confirmed the hepatocarcinogenicity of several mycotoxins such as yellowed rice toxins and aflatoxins in experimental animals, and epidemiological surveys have revealed a significant role in the development of human cancers. However, many fungal metabolites remain whose long-term effects including possible carcinogenicity are unknown. All identified environmental agents should therefore be tested on animals for their potential genetic and carcinogenic effects, although such evaluations would be both time-consuming and expensive. To obtain reliable results that could be extrapolated with confidence to the human population, the number of animals required in each case would be some 30,000 rather than the 50–100 which are currently used.[139] Furthermore, there is the problem of species specificity, i.e. agents affecting one or more species but having no effect on others. For example, 2-naphthylamine is a powerful carcinogen for man but has no marked effect on the experimental animals commonly used in carcinogenicity tests. Clearly, it is extremely difficult to develop a foolproof system of screening.

Microbial systems have recently been introduced for detecting carcinogens. This idea is based on the recognition that a number of agents capable of modifying cellular DNA are endowed with a carcinogenic potential and such substances by virtue of their effect on DNA are also mutagenic. Normal cells exposed to noxious chemicals which alter cellular DNA may overcome this alteration by excising the modified segment of DNA and newly synthesizing the correct sequence. This repair process, both the repair replication step and excision step, is considered to be mediated by DNA polymerase I. Therefore, it can be expected that cells lacking this repair enzyme will tend to be more sensitive to chemicals which react with cellular DNA. Indeed, DNA polymerase-deficient (*pol* A_1^-) strains of *Escherichia coli* are significantly more sensitive than the parent (*pol* A_1^+) strain to a large number of agents known to damage cellular DNA, including known mutagens and carcinogens.[140] DNA polymerase-deficient strains of *Bacillus subtilis* sensitive to UV-light,[141] and a recombination-deficient (*rec*$^-$) mutant of *B. subtilis*,[142] have been used in bioassays for carcinogens similar to that employing *pol* A^-_1 *E. coli*. Since the cell membrane of *B. subtilis* is more permeable than that of the gram-negative *E. coli* to a number of complicated agents such as dyes and water-insoluble chemicals, this technique is expected to have wider application than that using *E. coli*. Also, the polymerase I reaction system of *E. coli* involves several steps for DNA repair, and the spectrum of testable agents in this strain is broader than that in DNA repair-deficient strains.

In general, so-called "carcinogens" should strictly be termed "procarcinogens". In biological systems, procarcinogens are first biotransformed to "proximate carcinogens" and then to "ultimate carcinogens". In parallel to this, "promutagens" are converted to "proximate mutagens" and then "ultimate mutagens" by microbial enzyme systems. Chemicals such as alkylating agents (MNNG, 4NQO, AF-2), which are capable of reacting directly with cellular DNA, are therefore considered to constitute ultimate mutagens and ultimate carcinogens. Several compounds (BP, DAB, MC, fluorenyl acetale, dimethylanthracene) are capable of reacting with DNA after biotransformation by mammalian enzymes, which is beyond the metabolic capacity of microorganisms. To cirumvent this problem experimentally, bacterial systems may be supplemented with a mammalian liver extract and co-factors. For example, Ames *et al.*[143] were able to demonstrate the mutagenicity of many kinds of carcinogens using a combination of His-revertant mutants of *Salmonella typhymurium* and liver extracts (S-9 fraction). McCann *et al.*[144] have developed new strains which contain an R-factor (plasmid carrying antibiotic-resistance genes) and exhibit selectively high sensitivity to mutagenes and carcinogens through an error-prone recombination repair.

Owing to such technical progress in testing mutagens and carcinogens, over 90% of all carcinogens are known to be mutagenic in bacterial systems. Indeed, in the 1960's, carcinogenic HN2 and 4NQO were found to be mutagenic, and in 1966, carcinogenic DMN was shown to be mutagenic, and mutagenic MNNG to be carcinogenic. In 1973, DAB, FAA and BA were found to be both carcinogenic and mutagenic. However, there are still some exceptions. For example, the mutagenic sodium nitrile, bromouracil, NH_2OH, glycidol, 1,2-epoxybutane and anthracene are not carcinogenic to experimental animals, and carcinogenic compounds such as acetamide, ethionine, thioacetamide, thiourea, urethane, succinic anhydride, and bis(*p*-dimethylamino)diphenylmethane are not mutagenic to bacterial systems.

Another approach to the problem of testing *in vitro* carcinogenicity is the tissue culture technique. According to Malling *et al.*,[145] a series of polycyclic hydrocarbons were tested in a tissue culture of Chinese hamster V79 cells for their ability to increase the frequency of azaguanine-resistant cells among sensitive cells, and 3,4-benzpyrene and 9,10-dimethyl-1,2-benzanthracene were found to be positive. Umeda *et al.*[146] also demonstrated DNA damage and mutagenic changes in cultured mouse FM3A cells caused by furylfuramide treatment. As described above, many carcinogens are activated by tissue extracts, and supplementation of the culture technique with a metabolic activation system is thus expected to provide more precise information on the differences between carcinogens and mutagens.

3.5.2. Mutagenicity

Based on the above background knowledge concerning the relationship of mutagenicity and carcinogenicity in chemical carcinogens, Ueno *et al.* carried out microbial screening of carcinogenic mycotoxins. As a first step to screen the ability of mycotoxins to modify cellular DNA, the *Rec* assay method[142)] was employed. This assay is based on the fact that a mutant (M 45) of *B. subtilis* deficient in recombination ability is much more sensitive to agents that alter cellular DNA than is the parent strain (H 17). Inhibition of growth, or the term of the "*Rec*-effect" in the test, reflects this greater sensitivity in the *Rec* (−) mutant to potential mutagens and carcinogens. Among 30 mycotoxins and five chemically modified derivatives tested, six *Penicillium* mycotoxins (citrinin, penicillic acid, patulin, (−)luteoskyrin, (+)ruglosin and PR-toxin), five *Aspergillus* toxins (aflatoxin B_1, aflatoxin G_1, sterigmatocystin, *O*-acetylsterigmatocystin, *O*-acetyldihydrosterigmatocystin) and two *Fusarium* toxins (zearalenone and zearalenol-b) were found to be positive. Among these 13 positive mycotoxins, the following nine compounds have been established as carcinogenic: citrinin, penicillic acid, patulin, (−)luteoskyrin, (+)rugulosin, aflatoxin B_1, aflatoxin G_1, sterigmatocystin, and *O*-acetylsterigmatocystin.[86)] These results, as summarized in Table 3.37, reveal a good agreement between the *Rec*-effect and *in vivo* carcinogenicity of mycotoxins.

One important finding was that the *Rec*-effect is dose-dependent and parallels the carcinogenic activity. Among the aflatoxins, B_1 was the most potent followed by G_1, while B_2 and G_2 were negative. Among the sterigmatocystin derivatives, as illustrated in Fig. 3.22, *O*-acetylsterigma-

TABLE 3.37. Correlation between the *in vivo* carcinogecity and *Rec*-effect of mycotoxins (after Ueno *et al.*)[86)]

In vivo carcinogenicity	*Rec*-effect	Mycotoxins
Positive	positive	aflatoxin B_1, aflatoxin G_1, patulin, penicillic acid, luteoskyrin, rugulosin, sterigmatocystin, *O*-acetylsterigmatocystin, citrinin
Positive	negative	griseofulvin, cyclochlorotine
Unknown	positive	zearalenone, zearalenol-b, PR-toxin
Unknown	negative	T-2 toxin, fusarenon-X, butenolide, moniliformin, aflatoxin B_2, aflatoxin G_2, emodin, erythroskyrine, rubratoxin B, ochratoxin A, citreoviridin, oosponol, cytochalasin B, zearalenol-a, fusaric acid, *O*-methylsterigmatocystin,

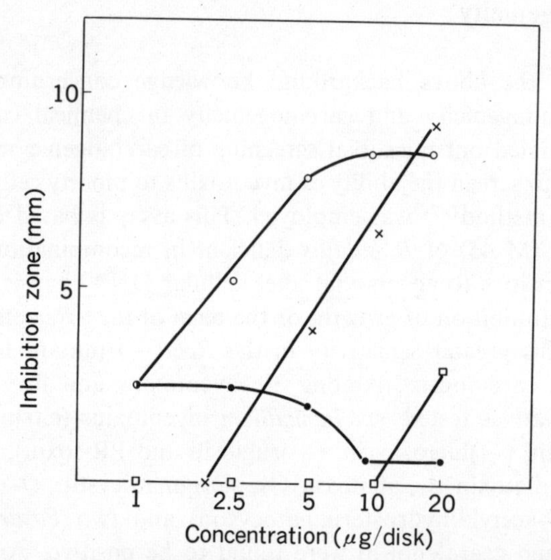

Fig. 3.22. Dose-dependent *Rec*-effect of sterigmatocystin derivatives. ●, Sterigmatocystin; ○, *O*-acetylsterigmatocystin; □, dihydrosterigmatocystin; ×, *O*-acetyldihydrosterigmatocystin. (Source: ref. 86. Reproduced by kind permission of the American Association for Cancer Research, Inc., U.S.A.)

tocystin exhibited a sharp dose-response comparable to that of aflatoxin B_1, while the parent compound sterigmatocystin was positive only at a low dose. The important role of the double bond in the bisfuran nuclei of aflatoxins and sterigmatocystin was thus reconfirmed. As for the anthraquinoid carcinogens, (−)luteoskyrin exhibited a more marked effect than (+)rugulosin.

These observations suggest that the above screening tool provides a reasonable correlation between the *Rec*-effect and *in vivo* carcinogenicity. Tests of the long-term effects of mycotoxins with positive *Rec*-effect but not known carcinogenicity (e.g. zearalenone and PR-toxin) are urgently required. In particular, tests of the long-term toxicity and carcinogenicity of zearalenone and its derivatives are important since these mycotoxins exhibit uterotropic and estrogenic activity in many animals, and zearalienone-producing fungi (mostly *Fusarium roseum* or *F. graminearum*) are widely distributed in cereals and feedstuffs. The secondary metabolites have been responsible for outbreaks of vulvovaginitis in pigs and sterility in cattle. Synthetic zearalanols, hydroxylated products of zearalenone, are already available commercially under the brand name "Ralgo" for use as additives to livestock fodder to promote growth and fattening in young animals.

The following three mycotoxicological problems can be raised in connection with zearalenone: (1) its possible carcinogenicity, (2) contamination of food- and feed-stuffs by zearalenone and other natural estrogens, and (3) secondary pollution in man after ingestion of milk or meat contaminated with zearalenone or its hydroxylated metabolites. As for the first problem, Schoental[84] has emphasized the possible etiologic role of dietary zearalenone in spontaneous tumors developing in laboratory animals. Also, preliminaty data of Ueno *et al.* (*unpublished*) indicate the occurrence of severe changes in mouse tissues after long-term exposure to zearalenone. As for the second problem, Zarrow *et al.*[147] and Drane *et al.*[85] have pointed out that certain commercial rations for laboratory animals possess estrogenic activity. The third problem is a very serious matter, and intentional usage of zearalanols in livestock fodder should be withheld until the safety of this fungal metabolite and its derivatives is clearly established.

A second step to test the mutagenicity of mycotoxins was carried out with a combination of *Salmonella typhimurium* and liver extracts from rats pretreated with phenobarbital, 3-methylcholanthrene or PCB.[148] As summarized in Table 3.38, 0.01–1.0 μg of aflatoxin B_1 and 1.0–10 μg of sterigmatocystin were positive in increasing His-revertants of *S. typhimurium* TA 98 in the presence of the NADPH-fortified S-9 liver fraction from phenobarbital-, 3-methyl-cholanthrene- or PCB-treated rats. This result indicates that these compounds are metabolically activated to "active mutagens". On the other hand, (−)luteoskyrin and (+)ruglosin were negative under the same experimental conditions, although as previously reported by Ueno *et al.*[149] these mycotoxins are actually mutagenic to the yeast, *Saccharomyces cerevisiae*, and the former toxin is several times more

TABLE 3.38. Mutagenicity of mycotoxins to *Salmonella typhimurium* TA 98 (after Ueno and Yanai)[148]

Mycotoxin	S-9†	Histidine revertants/plate					
		Dose (μg/plate)					
		0	0.01	0.1	1.0	10	100
Aflatoxin B_1	+	39	323	533	110	7	10
	−	33	29	30	41	95	827
Aflatoxin G_1	+	23			391	369	0
	−	35			35	34	37
Sterigmatocystin	+	28			133	301	17
	−	17			28	56	107
O-Acetylsterigmatocystin	+	39	48	102	37	9	0
	−	33	30	39	26	53	72

†The S-9 fraction was prepared from the liver of rats administered i.p. with 400 mg/kg PCB (Kanechlor-400).

active than the latter, in parallel to their cytotoxic and carcinogenic potential. Furthermore, trichothecene mycotoxins such as fusarenon-X and nivalenol are capable of inducing RD-mutation in yeast cells.[128)]

The above microbiological approaches have demonstrated that several potent carcinogens of fungal origin are actually mutagenic, and by the use of other microbial systems, as well as activating systems, it should be possible to demonstrate mutagenicity in other toxic mycotoxins. According to Umeda (*personal communication*), aflatoxin B_1, sterigmatocystin, and xanthocillin-X are mutagenic to cultured mammalian cells in the presence of NADPH and a liver extract.

3.5.3. Molecular Basis of Biotransformation

As described above, many chemical compounds are modified by liver extracts to yield active compounds which react with cellular DNA and other biopolymers. This activation system is localized, in most cases, in the endoplasmic reticulum (ER) or microsomes. The ER can be classified into two categories: "rough ER" which contains ribosomes, the site of protein synthesis, and "smooth ER" which is rich in glycogen and pigment-450 (P-450) and carries out many biologically important functions such as lipid metabolism, steroid metabolism, glycogen degradation, ion and molecular transport, and metabolic conversion of foreign chemicals. This means that the last function, which is the subject of this section. is influenced by other related biological phenomena occurring in the same organellae.

Foreign chemicals including carcinogens and toxicologically important pollutants. are metabolized in the smooth ER of liver cells. Biotransformation of lipophilic agents to more hydrophilic compounds represents the basic pattern of metabolism. The reaction is catalyzed by a mixed function oxygenase system requiring P-450, NADPH and molecular oxygen, as summarized in Fig. 3.23. The P-450, protoheme in the b-type, reacts with acceptor substrate (AH) to form a very unstable complex (Fe^{3+}) AH, which is then reduced by an electron from NADPH to give (Fe^{2+}) AH. Molecular oxygen is introduced into this complex in a reaction sensitive to carbon monoxide and ethylcyanide, and the resulting complex $(Fe^{3+}-O^{-}_2)AH$ is rather stable, i.e. the half time of degradation is 6 min at 20°C. An electron from a second molecule of NADPH reacts with this complex to yield $(Fe^{2+}-O_2^{-})AH$, although the position of the electron is undefined. This intermediate then gives rise to a hydroxylated product, AOH, and free P-450. Thus, molecular oxygen is directly introduced into the substrate and the P-450 acts as a catalyst to activate molecular oxygen.

The oxygenase enzyme system is complex and purification of its components is invariably accompanied by denaturation of the enzyme. Recent

$$AH + NADPH + H^+ + O_2 \longrightarrow AOH + NADP + H_2O$$

Fig. 3.23. Mechanism of the cytochrome P-450 enzyme system.

data on the "oxygenase" from the liver of phenobarbital-treated rabbits has revealed that it exists as a hexamer, each monomer consisting of one peptide chain, one protoheme and 5% carbohydrate. The molecular weight of the monomer is of the order of 50,000.

In the electron transfer systems of many biological systems, "dehydrogenase" catalyzes the movement of electrons ($Fe^{3+} \longrightarrow Fe^{2+}$) or movement of hydrogen ion (alcohol⟶aldehyde). This type of electron transfer plays an important role in energy metabolism and is localized in the mitochondria and microsomes. As summarized in Fig. 3.24, the electron transport system of mitochondria essentially requires NAD as hydrogen acceptor, while in microsomes NAD is used for P-450-independent dehydrogenation and NADP for P-450-dependent oxidation. A trans-hydrogenase in mitochondria catalyzes the NADH $\rightleftharpoons$ NADPH reaction. Although microsomes lack this enzyme, cytochrome b_5 in NADH–cyt. b_5 reductase is able to catalyze electron flow in the NADPH-dependent oxygenase system. The P-450-catalyzed oxygenase system is presumably activated in the presence of both NADH and NADPH-regeneration systems, and in this respect, the potential of P-450-independent metabolism such as lipid peroxidation, and desaturation, alcohol oxidation and *N*-oxidation of secondary and tertiary amines, exerts an influence on so-called "drug metabolism" or "oxygenases".

As is well established, microsomal oxygenase systems are inducible by repetitive administration of a wide variety of foreign chemicals such as

[Mitochondria]

substrate → NADH; succinate → Flavin prot.

NADH ⟶ Flavin prot. ⟶ Fe^{2+} ⟶ Quinone ⟶ Cyt. *b*

NADPH → NADH

⟶ Cyt. c_1 ⟶ Cyt. $a + a_3$ ⟶ O_2

[Microsomes]

(a) NADH ⟶ Flavin prot.$_1$ ⟶ Cyt. b_5 ⟶ CSF
(FAD) (NADH-cyt. b_5 reductase)

(b) NADPH ⟶ Flavin prot.$_2$ ⟶ P-450 ⟶ Oxygenase
(FAD, FMN) (NADPH-cyt. *c* reductase

Fig. 3.24. Electron transport systems in mitochondria and microsomes.

polycyclic hydrocarbons, barbiturates, analgesics, steroids and polychlorinated insecticides, and the functional integrity of microsomes is markedly changed by chemical inducers. According to Gurtoo *et al.*,[150] as summarized in Table 3.39, phenobarbital increases aminopyrine demethylase, cytochrome *c* reductase, and P-450-content, while 3-methylcholanthrene is effective in increasing zoxazolamine hydroxylase, arylhydrocarbon hydroxylase and P-448 (which is spectrophotometrically different from P-450). Thus, the P-450 system of hepatic microsomes is multiform. It is therefore highly possible that the enzyme system for activation or biotransformation is variable dependent upon exposure to external toxicants.

TABLE 3.39. Inducers and induced enzymes

Enzyme	Inducers	
	phenobarbital	3-methylcholanthrene
Aminopyrine demethylase	↑	—
Cytochrome *c* reductase	↑	—
Zoxazolamine hydroxylase	—	↑
Arylhydrocarbon hydroxylase	—	↑
P-450	↑	—
P-448	—	↑

Recent progress in drug metabolism has demonstrated that P-450 is distributed not only in microsomes but also in cell membranes, mitochondria and the nuclear envelope. In this respect, metabolic transformation or biotransformation of chemicals will occur in these cell organellae. Although microsomal modification is generally considered to represent the major pathway, the role of other organellae in the metabolic conversion of foreign chemicals cannot be ruled out. In particular, in the case of the biotransformation of carcinogens to obligatory intermediates, it appears unlikely that the "active intermediates" formed by microsomal particles are incorporated into nuclear constituents such as DNA without degradation or trapping on cytoplasmic proteins or nucleic acids. Activating enzyme systems within or around the nuclei of the target cells may thus play some role in the metabolic activation of carcinogenic compounds.

In the toxicological sense, the important enzyme system of microsomes is that involved in the formation and degradation of epoxide derivatives. As shown in Fig. 3.25 and 3.26, the carbon-carbon double bonds of olefins[115] and K-region of aromatic compounds[152] are activated by microsomes in the presence of molecular oxygen and NADPH to yield "epoxides". This reaction is inducible by drug-metabolizing enzyme inducers and inhibited by SKF, an inhibitor of oxygenase. The epoxides are very unstable and reactive with macromolecular tissue constituents. Therefore, chemical detection of "epoxides" *in vitro* or *in vivo* is very difficult without the employment of epoxide analogs which inhibit epoxide hydrolase. This enzyme is heterogenous and hydrolyzes the epoxide to *trans*-diol derivatives. Therefore, the actual concentrations of epoxides or actual potential for chemical carcinogenesis is controlled by "epoxidase" and "epoxide hydrolase" which function in the same organellae of microsomes.

As for the "epoxide hydrolase", its activity can be assayed by (1) radiometric determination of [7-^{3}H]-styrene glycol,[153] (2) gas chromatographic measurement of 3-methylcholanthrene-*trans*-11,12-diol,[154] and (3) photometric determination of safrole glycol,[155] as shown in Fig. 3.27. By these assay methods, it has been established that the hydrolase is localized in the microsomal particles and that the organ highest in this activity is the liver followed by the kidneys and lung.[156] The purified enzyme preparation from guinea-pig liver exhibits maximum activity at pH 8. The K_m and V_{max} values are 5.26×10^{-4}M and 2.7 μmol/mg protein/5 min, respectively, with styrene oxide as the substrate. The enzyme activity was not inhibited by EDTA or NEM. Hepatic epoxide hydrolase activity, known to reach maximal levels during maturation in rats, is inducible by pretreatment of the animals with phenobarbital or 3-methylcholanthrene.[157] Epoxide analogs such as 1,1,1-trichloropropane oxide (TCPO), decane oxide and cyclohexane oxide, inhibit the enzyme competitively.

Based on these properties of "epoxidase" and "epoxide hydrolase", numerous studies on the molecular mechanisms of carcinogenicity of aflatoxins and carcinogenic hydrocarbons have yielded the following interesting results.[158–162] (1) These chemicals are activated *in vitro* by microsomal epoxidase to give obligatory intermediate "epoxides" which react

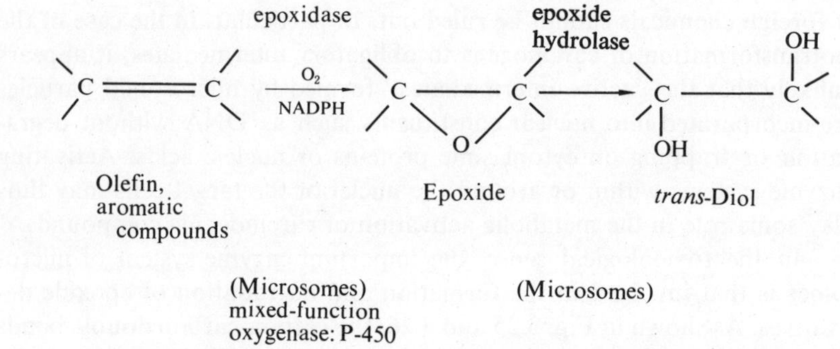

epoxidase

O_2
NADPH

epoxide hydrolase

C=C → C–C (O) → C(OH)–C(OH)

Olefin, aromatic compounds

(Microsomes) mixed-function oxygenase: P-450

Epoxide

(Microsomes)

trans-Diol

Fig. 3.25. Microsomal epoxidation of the carbon-carbon double bond.

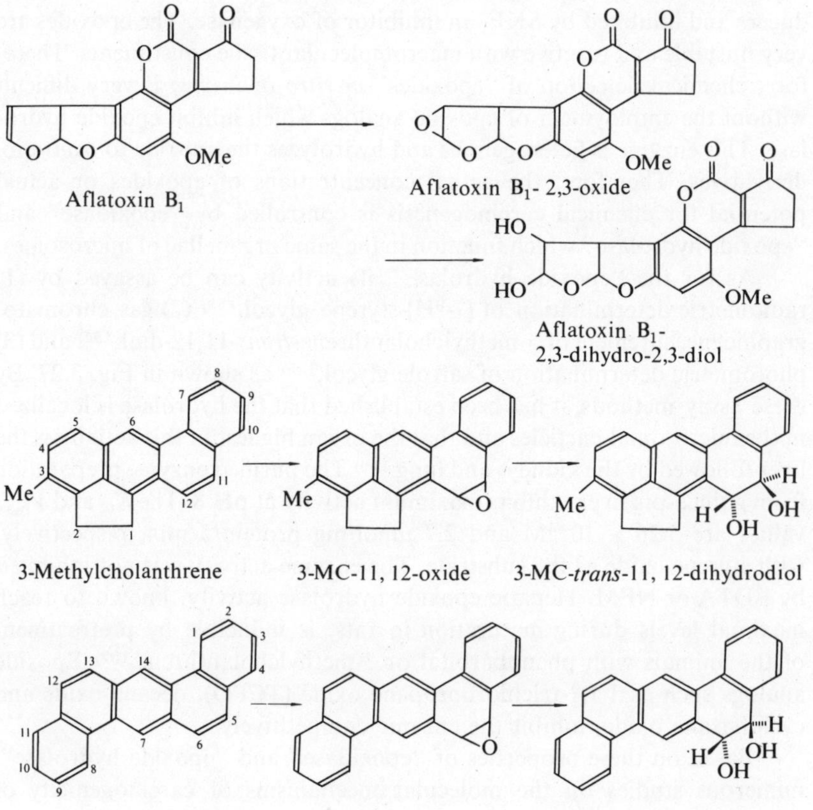

Aflatoxin B_1

Aflatoxin B_1- 2,3-oxide

Aflatoxin B_1- 2,3-dihydro-2,3-diol

3-Methylcholanthrene

3-MC-11, 12-oxide

3-MC-*trans*-11, 12-dihydrodiol

Dibenzanthracene (DBA)

DBA-5,6-oxide

DBA-*trans*-5,6-dihydrodiol

Fig. 3.26. Epoxidation of aflatoxin B_1 and hydrocarbons.

[7-³H]-Styrene oxide → [7-³H]-Styrene glycol → Counting

3-Methylcholanthrene (MC) → 3-MC-*trans*-11,12-diol → GLC

Safrole oxide → Safrole glycol → UV_{288}

Fig. 3.27. Assay of epoxide hydrolase activity.

with DNA, RNA and protein. (2) Microsome-mediated binding of carcinogens to cellular macromolecules is enhanced by pretreatment of the animals with inducers and inhibited by oxygenase inhibitors such as SKF. (3) Microsome-mediated binding of carcinogens to DNA and the skin tumorigenesis induced are elevated in the presence of epoxide hydrolase inhibitors such as TCPO. These observations support the assumption that the "epoxides" of aflatoxins and carcinogenic hydrocarbons are obligatory intermediates responsible for carcinogenicity. The epoxides are cytotoxic as well as transformable, while K-region phenols are cytotoxic but lack transformability.

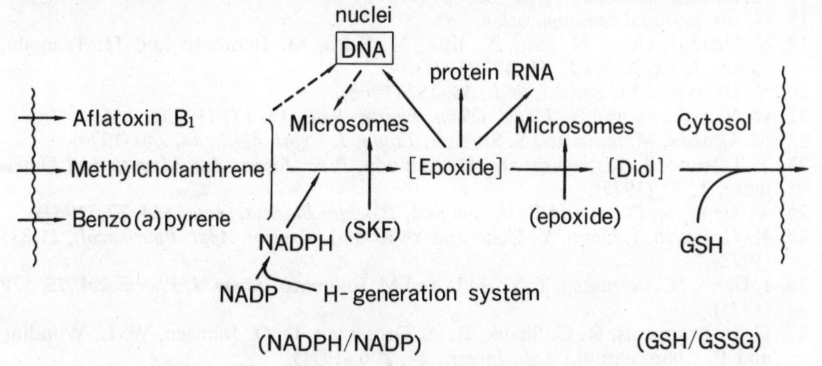

Fig. 3.28. Molecular basis of activation reactions.

Finally, it should be emphasized that the concentrations of cytoplasmic NAD, NADP and their reduced forms, and of glutathione exert a strong influence on the above enzymatic reactions. The former co-factors determine the relative activity of the activation reactions, while the latter cytoplasmic component affects the rate of outflow of activation products. The above processes are summarized diagrammatically in Fig. 3.28.

REFERENCES

1. K. Uraguchi, T. Tatsuno, M. Tsukioka, Y. Sakai, Y. Kobayashi, M. Saito, M. Enomoto and M. Miyake, *Japan. J. Exptl. Med.*, **31**, 1 (1961).
2. R. Allcroft, R. B. A. Carnaghan, K. Sergeant and J. O'Kelly, *Vet. Record*, **73**, 428 (1961).
3. C. J. Mirocha and C. M. Christensen, *Mycotoxins* (ed. I. F. H. Purchase), p. 129, Elsevier (1974).
4. M. Saito and K. Ohtsubo, *ibid.*, p. 271, Elsevier (1974).
5. N. Sato, Y. Ueno and M. Enomoto, *Japan. J. Pharmacol.*, **25**, 263, (1975).
6. K. Uraguchi, *J. Stored Prod. Res.*, **5**, 227 (1969).
7. I. F. H. Purchase, *Mycotoxins* (ed. I. F. H. Purchase), p. 149, Elsevier (1974).
8. C. Takahashi, S. Yoshihira, M. Natori, K. Umeda, K. Ohtsubo and M. Saito, *Experientia*, **30**, 529 (1974).
9. A. M. Guarino, A. B. Mendillo and J. DeFeo, *Biotech. Bioeng.*, **10**, 457 (1968).
10. J. R. Bamburg, N. V. Riggs and F. M. Strong, *Tetrahedron*, **24**, 3329 (1968).
11. M. Saito, M. Enomoto and T. Tatsuno, *Microbial Toxins* (ed. A. Ciegler, S. Kadis and S. J. Ajl), vol. VI, p. 318. Academic Press (1971).
12. G. N. Wogan, G. S. Edwards and P. M. Newberne, *Toxicol. Appl. Pharmacol.*, **19**, 712 (1971).
13. H. Tsukioka, *Folia Pharmacol. Japan.* (Japanese), **55**, 1367 (1959).
14. L. A. Lindenfelser, E. B. Lillehoj and M. S. Milburn, *Dev. Ind. Microbiol.*, **14**, 331 (1973).
15. P. S. Hewlett, *Biometrics*, **25**, 477 (1969).
16. L. A. Lindenfelser, E. B. Lillehoj and H. R. Burmeister, *J. Natl. Can. Inst.*, **52**, 113 (1974).
17. M. Saito, *Proc. Japan. Acad.*, **35**, 501 (1959).
18. N. Ito, *personal communication.*
19. Y. Ueno, I. Ueno, N. Sato, Y. Iitoi, M. Saito, M. Enomoto and H. Tsunoda, *Japan. J. Exptl. Med.*, **41**, 177 (1971).
20. Y. Ueno and M. Saheki, *ibid.*, **38**, 157 (1968).
21. H. Yamakawa and Y. Ueno, *Chem. Pharm. Bull.*, **18**, 177 (1970).
22. M. Umeda, M. Saito and S. Shibata, *Japan. J. Exptl. Med.*, **44**, 240 (1974).
23. T. Tatsuno, T. Kobayashi and H. Tsunoda, *Proc. Japan. Ass. Mycotoxicol.* (Japanese), **1**, 22 (1975).
24. Y. Ueno, A. Platel and P. Fromageot, *Biochim.Biophys. Acta*, **134**, 27, (1967).
25. K. Uraguchi, I. Ueno, Y. Ueno and Y. Komai, *Toxicol. Appl. Pharmacol.*, **21**,335 (1972).
26. I. Ueno, N. Akamatsu, T. Yoshida and M. Enomoto, *Japan. J. Pharmacol.*,**25**, 57P (1975).
27. C. H. Bourgeois, R. C. Shank, R. A. Grossman, D. O. Johnsen, W. L. Wooding and P. Chansavimol., *Lab. Invest.*, **24**, 206 (1971).

28. K. A. V. R. Krishnamachari, R. V. Brat, V. Nagarajan and T. B. G. Tilak, *Lancet*, **May 10**, 1061 (1975).
29. A. S. Salhab and D. P. H. Hsieh, *Res. Commun. Chem. Path. Pharmacol.*, **10**, 419 (1975).
30. I. F. H. Purchase and J. J. van der Watt, *Fd. Cosmet. Toxicol.*, **7**, 135 (1969).
31. H. Kimura, *M. Pharm. Sc. Thesis* (Japanese) (1976).
32. K. Terao and M. Yamazaki, *Ann. Mtg. Japan. Chem. Soc.* (Japanese), **34**, 19 (1975).
33. M. M. R. Kalengayi and V. J. Desmet, *Cancer Res.*, **35**, 2836 (1975).
34. M. M. R. Kalengayi and V. J. Desmet, *ibid.*, **35**, 2845 (1975).
35. D. S. P. Patterson, *Fd. Cosmet. Toxicol.*, **11**, 287 (1973).
36 P. M. Newberne and W. H. Butler, *Cancer Res.*, **29**, 236 (1969).
37. A. Ciegler. *Lloydia*, **38**, 21 (1975).
38. I. F. H. Purchase and J. J. van der Watt, *Fd. Cosmet. Toxicol.*, **6**, 555 (1968).
39. I. F. H. Purchase and J. J. van der Watt, *ibid.*, **8**, 289 (1970).
40. G. M. Zwicker and W. W. Carlton, *ibid.*, **12**, 491 (1974).
41. Y. Ueno, K. Ishii, N. Sato and K. Ohtsubo, *Japan. J. Exptl. Med.*, **44**, 123 (1974).
42. K. Ishii, Y. Ando and Y. Ueno, *Chem. Pharm. Bull.*, **23**, 2162 (1975).
43. S. Morooka, *personal communication* (1975).
44. R. A. Ellison and F. N. Kotsonis, *Appl. Microbiol.*, **26**, 540 (1973).
45. R. F. Vesonder, A. Ciegler and A. H. Jensen, *ibid.*, **26**, 1008 (1973).
46. Y. Ueno, Y. Ishikawa, K. Amakai, M. Nakajima, N. Sato, M. Enomoto and K. Ohtsubo, *Japan. J. Exptl. Med.*, **40**, 33 (1970).
47. H. R. Burmeister and C. W. Hesseltine, *Appl. Microbiol.*, **20**, 437 (1970).
48. R. D. Wyatt, J. R. Harris, P. B. Hamilton and H. R. Burmeister, *Avian Res.*, **16**, 1123 (1972).
49. H. von Stähelin, M. E. Kalberer-Rüsch, E. Signer and S. Lazåry, *Arzneim. Forsch.*, **18**, 989 (1968).
50. M. E. Rüsch and H. von Stähelin, *ibid.*, **15**, 893 (1965).
51. T. Konishi and S. Ichijo, *Res. Bull. Obihiro Univ.* (Japanese), **6**, 242, 258 (1970).
52. J. Forgacs and W. L. Carll, *Advan. Vet. Sci.*, **7**, 273 (1972).
53. R. D. Wyatt, W. M. Colwell, P. B. Hamilton and H. R. Burmeister, *Appl. Microbiol.*, **26**, 757 (1973).
54. M. Saito and K. Ohtsubo, *Mycotoxins* (ed. I. F. H. Purchase), p. 273, Elsevier (1974).
55. R. Schoental and A. Z. Joffe, *J. Path.*, **112**, 37 (1974).
56. W. F. O. Marasas, J. R. Bamburg, E. B. Smalley, F. M. Strong, W. L. Ragland and P. E. Degurse, *Toxicol. Appl. Pharmacol.*, **15**, 471 (1969).
57. B. J. Wilson and C. H. Wilson, *Science*, **144**, 177 (1964).
58. B. J. Wilson, C. H. Wilson and A. W. Hayes, *Nature*, **220**, 77 (1968).
59. B. J. Wilson, T. Hoekman and W. D. Dettbarn, *Brain Res.*, **40**, 540 (1972).
60. P. Stern, *Jugoslav. Physiol. Pharmacol. Acta*, **7**, 187 (1971).
61. B. J. Wilson, *Mycotoxins in Human Health* (ed. I. F. H. Purchase), p. 223, Macmillan (1971).
62. C. W. Holzapfel, *Tetrahedron*, **24**, 2101 (1968).
63. M. Yamazaki, H. Fugimoto and T. Kawasaki, *Tetr. Lett.*, 1241 (1975).
64. M. Yamazaki, H. Fugimoto and T. Kawasaki, *Proc. Japan. Ass. Mycotoxicol.* (Japanese), **2**, 35 (1976).
65. R. J. Cole, J. W. Kirksey, J. H. Moore, B. Blankenship, U. L. Diener and N. D. Davis, *Appl. Microbiol.*, **24**, 248 (1972).
66. R. J. Cole, J. W. Kirksey, R. H. Cox and J. Clardy, *Agr. Food Chem.*, **23**, 1015 (1975).
67. L. Hortujac, P. Stern and R. H. Muftic, VIII Congr. Physiol. Pharmacol. Soc. (Opatia) (1973).
68. F. Dickens, *Potential Carcinogenic Hazards from Drugs* (ed. R. Truhaut), p. 148, Springer-Verlag (1967).

69. F. Dickens and H. F. H. Jones, *Brit. J. Cancer*, **15**, 85 (1961).
70. J. Lovett, *Poultry Sci.*, **51**, 2099 (1972).
71. H. J. Mintzlaff and W. Christ, *Fleischwirtzschaft*, **51**, 1802 (1971).
72. K. Hofmann, H. J. Mintzlaff, I. Alpenden and C. Leistner, *ibid.*, **51**, 1534, 1539 (1971).
73. S. H. Ashoor and F. S. Chu, *Fd. Cosmet. Toxicol.*, **11**, 995 (1973).
74. J. B. Kahn, Jr., *J. Pharmacol. Exptl. Med.*, **121**, 234 (1957).
75. Y. Ueno, H. Matsumoto, K. Ishii and K. Kukita, *Biochem. Pharmacol.*, **25**, 2091 (1976).
76. Y. Ueno, N. Shimada, S. Yagasaki and M. Enomoto, *Chem. Pharm. Bull.*, **22**, 2830 (1974).
77. H. J. Kurtz, M. E. Nairm, G. H. Nelson, C. M. Christensen and C. J. Mirocha, *Am. J. Vet. Res.*, **30**, 551 (1969).
78. R. A. Meronuck, K. H. Garren, C. M. Christensen, G. H. Nelson and F. Bates. *ibid.*, **31**, 551 (1970).
79. R. F. Sherwood and J. F. Peberdy, *Poultry Sci.*, **14**, 127 (1973).
80. C. J. Mirocha, C. M. Christensen, G. Davis and G. H. Nelson, *J. Agr. Food Chem.*, **21**, 135 (1973).
81. T. Sugimoto, M. Minamizawa, K. Takano and Y. Furukawa, *J. Fd. Hyg. Soc. Japan* (Japanese), **17**, 14 (1975).
82. K. Ishii, M. Sawano, Y. Ueno and H. Tsunoda, *Appl. Microbiol.*, **27**, 625 (1974).
83. D. B. R. Johnston, C. A. Sawicki, J. B. Windholz and A. A. Patchett, *J. Med. Chem.* **13**, 941 (1970).
84. R. Schoental, *Cancer Res.*, **34**, 2419 (1974).
85. H. Drane, D. S. P. Patterson, B. A. Roberts and N. Saba, *Fd. Cosmet. Toxicol.*, **13**, 491 (1975).
86. Y. Ueno and K. Kubota, *Cancer Res.*, **36**, 445 (1976).
87. K. Uraguchi, *Pharmacology of Mycotoxins*, I. E. P. T. Section 71, Part V, p. 143, Pergamon Press (1971).
88. K. Uraguchi, *Microbial Toxins* (ed. A. Ciegler, S. Kadis and S. J. Ajl), vol. VI, p. 299, Academic Press (1971).
89. A. M. Roberton, R. B. Beechey, C. T. Holloway and I. G. Knight, *Biochem. J.*, **104**, 54c (1967).
90. R. B. Beechey, D. O. Osselton, H. Baum, P. E. Linnett and A. D. Mitchell, *Membrane Proteins in Transport and Phosphorylation* (ed. G. F. Azzone *et al.*), p. 201, North-Holland (1974).
91. M. O. Moss, *Microbial Toxins* (ed. A. Ciegler, S. Kadis and S. J. Ajl), vol. VI, p. 381, Academic Press (1971).
92. P. M. Newberne, *Mycotoxins* (ed. I. F. H. Purchase), p. 163, Elsevier (1974).
93. A. W. Hayes, R. D. Hood and K. Snowden, *Toxicol. Appl. Pharmacol.*, **29**, 153 (1974).
94. T. Tatsuno, M. Tsukioka, Y. Sakai, Y. Suzuki and Y. Asami, *Chem. Pharm. Bull.*, **3**, 476 (1955).
95. H. Yoshioka, K. Nakatsu, M. Sato and T. Tatsuno, *Chem. Lett.* (*Tokyo*), 1219 (1973).
96. Y. Ueno, M. Kaneko, T. Tatsuno, I. Ueno and K. Uraguchi, *Seikagaku* (Japanese), **35**, 38 (1963).
97. H. Choudhury, C. W. Carlton and G. Semeniuk, *Poultry Sci.*, **50**, 1855 (1971).
98. J. C. Peckham, B. Doupnik Jr. and O. Jone Jr., *Appl. Microbiol.*, **21**, 492 (1971).
99. G. M. Szczech, W. W. Carlton and J. Tuite, *Lab. Invest.*, **26**, 492 (1972).
100. R. C. Doster, R. V. Sinnhuber and N. E. Pawlowski, *Fd. Cosmet. Toxicol.*, **12**, 499 (1974).
101. S. Suzuki, T. Sato and M. Yamazaki *Toxicol. Appl. Pharmacol.*, **32**, 116 (1975).
102. M. Umeda, K. Ohtsubo, M. Saito, S. Sekita, K. Yoshikawa, S. Natori, S. Udagawa, Y. Sakabe and H. Kurata, *Experientia*, **31**, 435 (1975).

103. N. K. Wessells, B. S. Spooner J. F. Ash, M. O. Bradley, M. A. Luduena, E. Y. Taylor, J. T. Wrenn and K. M. Yamada, *Science*, **171**, 135 (1971).
104. T. N. Hayakawa, N. Matsushima, T. Kimura, H. Minato and K. Katagiri, *J. Antibiotics*, **21**, 523 (1968).
105. G. Büchi, Y. Kitaura, S. S. Yuan, H. E. Wright, J. Clardy, A. Demain, T. Glinsukon, N. Hunt and G. N. Wogan, *J. Am. Chem. Soc.*, **95**, 5423 (1973).
106. T. Glinsukon, R. C. Shank, G. N. Wogan and P. M. Newberne, *Toxicol. Appl. Pharmacol.*, **32**, 135 (1975).
107. T. Glinsukon, R. C. Shank and G. N. Wogan, *ibid.*, **32**, 158 (1975).
108. S. A. Patwardhan, R. C. Pandey, S. Dev and G. S. Pendse, *Phytochem.*, **13**, 1985 (1974).
109. I. Ueno, T. Hayashi and Y. Ueno, *Japan. J. Pharmacol.*, **24**, 535 (1974).
110. I. Ueno, *ibid.*, **25**, 171 (1975).
111. N. Sato, *M. Pharm. Sci. Thesis* (Japanese) (1972).
112. G. N. Wogan, G. S. Edwards and R. C. Shank, *Cancer Res.*, **27**, 1729 (1967).
113. J. I. Dalezois, G. N. Wogan and S. M. Weines, *Science*, **171**, 584 (1971).
114. J. I. Dalezois and G. N. Wogan, *Cancer Res.*, **32**, 2297 (1972).
115. G. Büchi, D. Spitzer, S. Poglialungi and G. N. Wogan, *Life Sci.* **13**, 1143 (1973).
116. D. S. P. Patterson and B. A. Roberts, *Fd. Cosmet. Toxicol.*, **9**, 829 (1971).
117. A. Ciegler and R. E. Peterson, *Appl. Microbiol.*, **16**, 665 (1968).
118. D. P. H. Hsieh, J. I. Dalezois, R. I. Krieger, M. S. Masri and U. F. Haddon, *J. Agr. Fd. Chem.*, **22**, 515 (1974).
119. R. I. Krieger, A. S. Salhab, J. I. Dalezois and D. P. H. Hsieh, *Fd. Cosmet. Toxicol.*, **13**, 211 (1975).
120. D. S. P. Patterson and B. A. Roberts, *Experientia*, **28**, 929 (1971).
121. E. Boyland, *Symp. Biochem. Soc.*, **5**, 40 (1950).
122. E. Boyland and P. Sims, *Biochem. J.*, **97**, 7 (1965).
123. R. Schoental, *Nature*, **227**, 401 (1970).
124. G. C. Garner, E. C. Miller and J. A. Miller, *Cancer Res.*, **32**, 2058 (1972).
125. D. H. Swenson, J. A. Miller and E. C. Miller, *Biochem. Biophys. Res. Commun.*, **53**, 1260 (1973).
126. P. G. Thiel and M. Steyn, *Biochem. Pharmacol.*, **22**, 3273 (1973).
127. M. Yamazaki, M. Takano, S. Suzuki and K. Terao, *Proc. Japan. Ass. Mycotoxicol.* (Japanese), **1**, 26 (1975).
128. Y. Ueno, I. Ueno, Y. Iitoi, H. Tsunoda, M. Enomoto and K. Ohtsubo, *Japan. J. Exptl. Med.*, **41**, 521 (1971).
129. Y. Ueno, M. Sawano and K. Ishii, *Appl. Microbiol.*, **30**, 4 (1975).
130. M. Ohta and Y. Ueno, *Proc. Japan. Ass. Mycotoxicol.* (Japanese), **2**, 34 (1976).
131. Y. Ueno and S. Ayaki, *Japan. J. Pharmacol.*, **26** (S), 91 p (1976).
132. Y. Ueno, S. Ayaki, N. Sato and T. Ito, *IUPAC Symposium on Mycotoxins in Foodstuffs* (Paris), p 11 (1976)
133. M. Umeda, *Igaku no Ayumi* (Japanese), **89**, 767 (1974).
134. J. C. Engelbrecht and B. Atenkirk, *J. Natl. Cancer Inst.*, **48**, 1648 (1972).
135. S. Natori, S. Sakaki, H. Kurata, S. Udagawa, M. Ichinoe, M. Saito, M. Umeda and K. Ohtsubo, *Appl. Microbiol.*, **19**, 613 (1970).
136. *WHO Fifth Report of the Joint FAO/SHO Expert Committee on Food Additives*, Evaluation of the Carcinogenic Hazards of Food Additives, *Tech. Rept. Ser.*, p. 220, Geneva (1961).
137. E. Boyland, *Progr. Exptl. Tumor Res.*, **11**, 222 (1969).
138. S. S. Epstein, *Cancer Res.*, **34**, 2425 (1974).
139. S. S. Epstein, *Nature*, **228**, 816 (1970).
140. R. M. D'Alisa, G. A. Carden III, H. S. Carr and H. S. Rosenkranz, *Mol. Gen. Genet.*, **110**, 23 (1971).
141. K. B. Gass, T. C. Hill, M. Goulian, B. S. Strauss and N. R. Corrarelli, *J. Bact.* **108**, 364 (1971).

142. T. Kada, K. Tsutikawa and Y. Sadaie, *Mutat. Res.*, **16**, 165 (1972).
143. B. N. Ames, W. E. Durston, E. Yamasaki and F. D. Lee, *Proc. Natl. Acad. Sci. U.S.A.*, **70**, 2281 (1973).
144. J. McCann, N. E. Springarn, J. Kobori and B. N. Ames, *ibid.*, **72**, 979 (1975).
145. H. V. Malling and E. H. Y. Chu, *Chemical Carcinogenesis* (ed. P. O. P. Ts'o *et al.*), part B, p. 545, Marcel Dekker (1974).
146. M. Umeda, T. Tsutsui, S. Kikyo and M. Saito. *Japan. J. Exptl. Med.*, **45**, 161 (1975).
147. M. X. Zarrow, E. A. Lazo-Wasem and R. L. Shoger, *Science*. **118**, 650 (1953).
148. Y. Ueno and M. Yanai, *unpublished data*.
149. Y. Ueno and M. Nakajima, *Chem. Pharm. Bull.*, **22**, 2258 (1974).
150. H. L. Gurtoo and C. V. Davis, *Cancer Res.*, **35**, 382 (1975).
151. E. W. Maynert, R. L. Foreman and T. Watabe, *J. Biol. Chem.*, **215**, 5234 (1970).
152. I. Y. Wang, R. E. Rasmussen and T. Crocker, *Biochem. Biophys. Res. Commun.*, **49**, 1142 (1972).
153. F. Oesch. D. M. Jerina and J. Daly, *Biochim. Biophys. Acta*, **227**, 685 (1971).
154. T. A. Stroming and E. Bresnick, *Science*, **181**, 951 (1973).
155. T. Watabe and K. Akamatsu, *Biochem. Phramacol.*, **23**, 2839 (1974).
156. T. A. Stroming and F. B. Resnick, *Cancer Res.*, **34**, 2810 (1974).
157. F. Oesch and J. Daly, *Biochim. Biophys. Acta*, **227**, 692 (1971).
158. J. K. Selkirk, E. Huberman and C. Heiderberger, *Biochem. Biophys. Res. Commun.*, **43**, 1010 (1971).
159. E. Huberman, T. Kuroki, H. Marquardt, J. K. Selkirk, C. Heiderberger, P. L. Grove and P. Sims, *Cancer Res.*, **31**, 1391 (1972).
160. H. Marquart, T. Kuroki, F. Huberman, J. K. Selkirk, C. Heiderberger, P. L. Grove and P. Sims, *ibid.*, **32**, 716 (1972).
161. M. M. Coombs, T. S. Bhatt and C. W. Vose, *ibid.*, **35**, 305 (1975).
162. S. H. Blobstein, I. B. Weinstein, P. Dansette, H. Yagi and D. M. Jerina, *ibid.*, **36**, 1293 (1976).

CHAPTER 4

Morphological and Functional Damage to Cells and Tissues

Kiyoshi TERAO
Research Institute for Chemobiodynamics, ~~Narashino~~-shi, Chiba-ken 275, Japan
Yoshio UENO
Tokyo University of Science, Ichigaya, Shinjuku-ku, Tokyo 162, Japan

4.1. CYTOLOGICAL ALTERATIONS

Although early hepatic parenchymal changes induced by aflatoxin B_1 were first demonstrated in 1966 by several investigators,[1-3] the ultrastructural alterations of cellular components have been of relatively little interest to researchers in this field. Apart from some detailed descriptions of the effects of aflatoxin B_1 on liver cells, there is only limited information available on the fine structural changes induced by other mycotoxins. It is the aim of this section to summarize the current information on the ultrastructural and functional events in the nucleus and cytoplasmic organellae and on the chromosomal changes induced by various mycotoxins.

4.1.1. Effects of Mycotoxins on the Nucleus

A. Morphological Changes

a) Nucleolus

One of the most prominent effects of mycotoxins on the nuclear structure is alteration of the hepatocellular nucleolus. This includes a decrease in the nucleolar size, accompanied by a gradual redistribution of nucleolar components and the appearance of dense microspherules (Fig. 4.1). Two types of nucleolar segregation, i.e. the redistribution of nucleolar components, can be discerned, depending on the size of the fibrillar area, its

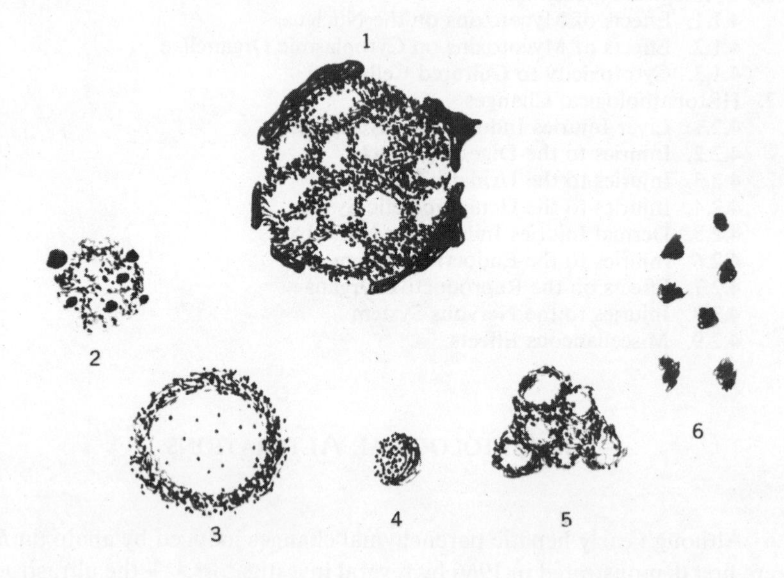

Fig. 4.1. Schematic representation of nucleolar alterations induced by mycotoxins. 1, Normal nucleolus; 2, nucleolus with microspherules; 3, ring-shaped nucleolus; 4, macrosegregation; 5, microsegregation; 6, nucleolar fragmentatation.

purity, and the degree of separation from the granular elements[4] (Fig. 4.2).

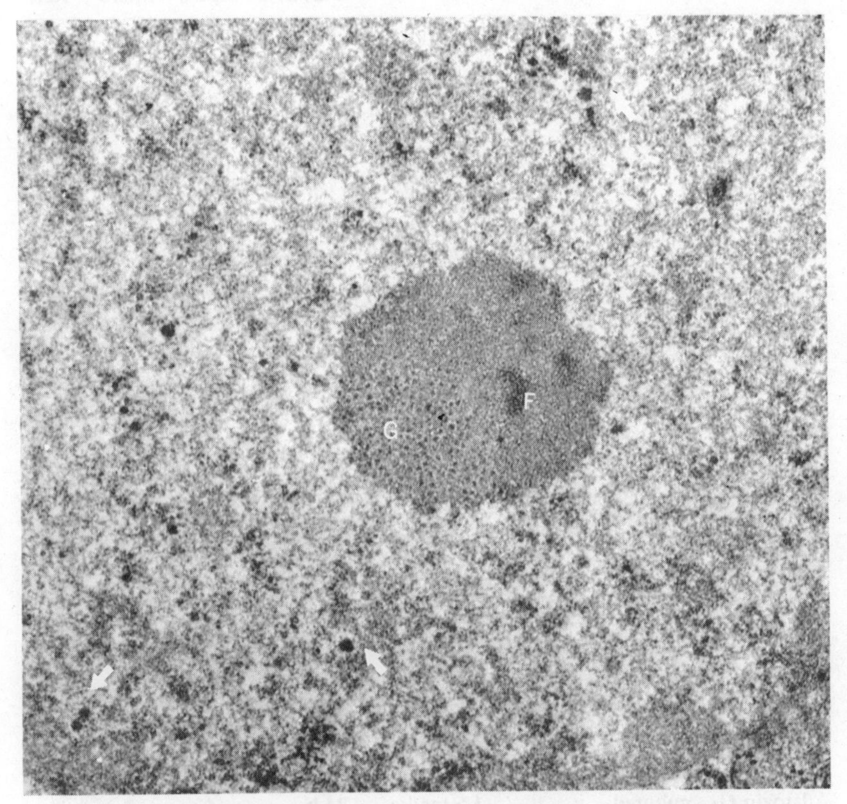

Fig. 4.2. Electron micrograph of a rat hepatocyte treated with aflatoxin B_1. A typical segregation of granular (G) and fibrillar (F) components is seen. Arrows, perichromatin granulae.

In the rat hepatocyte nucleolus, macrosegregation is induced by the administration of sublethal doses of aflatoxin B_1,[1-8] G_1[8] and M_1[9]. Such macrosegregation is also induced by sterigmatocystin in cultured cells,[10,11] while the nucleoli of hepatocytes in *in vivo* systems reveal only slight alterations by the mycotoxin. *O*-Acetylsterigmatocystin,[12] on the other hand, produces severe morphological changes in the nucleoli of both animal and cultured cells. The marked difference between sterigmatocystin and *O*-acetylsterigmatocystin may relate to differing absorbabilities of these mycotoxins in the digestive tract.

Pong and Wogan[9] have investigated the time course of effects in rats

TABLE 4.1. Nucleolar segregation induced by mycotoxins

Mycotoxins	Cells	Dose	Duration	Macro	Micro	Ref.
Aflatoxin B_1	rat hepatocyte	7 mg/kg	9 h	+	+	14
		4 mg/kg	2 h	+	+	15
		3 mg/kg	48 h	+	+	13
		1 mg/kg	15 min	−	+	7
		1 mg/kg	1 h	+	+	7
		1 mg/kg	6 h	+	+	3
		0.75 mg/kg	36 min	+	+	5
		0.5 mg/kg	24 h	+	+	8
		0.5 mg/kg	1 h	+	+	6
		0.45 mg/kg	72 h	+	+	1
		0.3 mg/kg	48 h	−	+	13
		2 ppm		−	+	4
		1 ppm		−	+	4
	monkey hepatocyte	2.6 mg/kg	48 h	+	+	1
	marmoset hepatocyte	0.2 mg/kg	82 days	−	+	16
	cultured primary kidney cells	1 μg/ml	24 h	+	+	10
Synthetic aflatoxin B_1	rat hepatocyte	1 mg/kg	12 h	+	+	9
		0.5 mg/kg	12 h	+	+	9
Synthetic aflatoxin M_1	rat hepatocyte	1 mg/kg	12 h	+	+	9
Aflatoxin G_1	cultured primary kidney cell	1.5 mg/kg	24 h	+	+	8
		2 μg/ml		+	+	10
Aflatoxin B_2	rat hepatocyte	200 mg/kg	24 h	−	+	8
Sterigmatocystin	rat hepatocyte	30 mg/kg	48 h	−	−	13
	cultured primary kidney cell	2 μg/ml	24 h	+	+	11
		1 μg/ml		+	+	10
O-Acetylsterigmatocystin	rat hepatocyte	8 mg/kg	48 h	+	+	12
	cultured primary liver cell	1 μg/ml	1 h	+	+	12
O-Methylsterigmatocystin	cultured primary kidney cell	2 μg/ml	24 h	+	+	11

of 1 mg/kg of aflatoxin B_1 on both RNA polymerase activity and hepatocyte nucleolar ultrastructure. Within 1 h after dosing with aflatoxin B_1, nucleolar macrosegregation was observed. This alteration was still present at 12 h after dosing, but by 36 h, nucleolar macrosegregation was not demonstrable and the fibrillar and granular components were well integrated. Correlating with the morphological changes of the nucleolus, both Mg^{2+} activated and $Mn^{2+}(NH_4)_2SO_4$-activated RNA polymerase activities were maximally inhibited by about 60% 15 min after dosing. This inhibition continued to 12 h. By 36 h, enzyme activity had returned to pretreatment levels. With a dose of 1/10 LD_{50} value or less, aflatoxin B_1 produced not macro- but microsegregation.[4,13] In contrast to the macrosegregation, the fibrillar components with a decreased amount of associated granules appear as coarse strands and not as distinct areas. This microsegregation is probably a less advanced form of macrosegregation, and it has been interpreted as an intermediate stage in the induction of macrosegregation by aflatoxin B_1. Table 4.1 shows the nucleolar segregations induced by various mycotoxins.

A ring-shaped nucleolus is induced by the administration of 1 μg/egg of aflatoxin B_1 in 5-day-old chicken embryonal liver cells.[17] Microscopically, the ring-shaped nucleolus has a doughnut-like configuration with a pale central portion (Fig. 4.3). In the fine structure, the fibrillar and granular elements are found at the periphery of the nucleolus. In the central area, however, there are only a few scattered granular elements in the matrix.

It is now widely accepted[18] that the fibrillar component of the nucleolus is a precursor of the RNA in the granular components, so that the pathway of ribosomal RNA must be as follows: nucleolar DNA⟶ fibrillar component⟶granular component⟶migration into the cytoplasm⟶maturation in polysomes in the cytoplasm. If the first step of this pathway is affected in the central portion of the nucleolus, there may appear an area consisting only of granular ribonucleoprotein (RNP). Such a nucleolus may represent a "ring-shaped nucleolus".

The characteristic nucleolar lesion induced by various mycotoxins with a bisfuran structure in tissue culture cells appears to be nucleolar fragmentation.[12]

b) RNA-containing Granula

An increase in the number of clusters of interchromatin granules as well as perichromatin granules has been reported by several investigators[13,19,20] in rat liver cells after administration of aflatoxin B_1(Fig. 4.4). Interchromatin granules appear as clusters in the interchromatin area of the nucleus, while perichromatin granules always appear around the con-

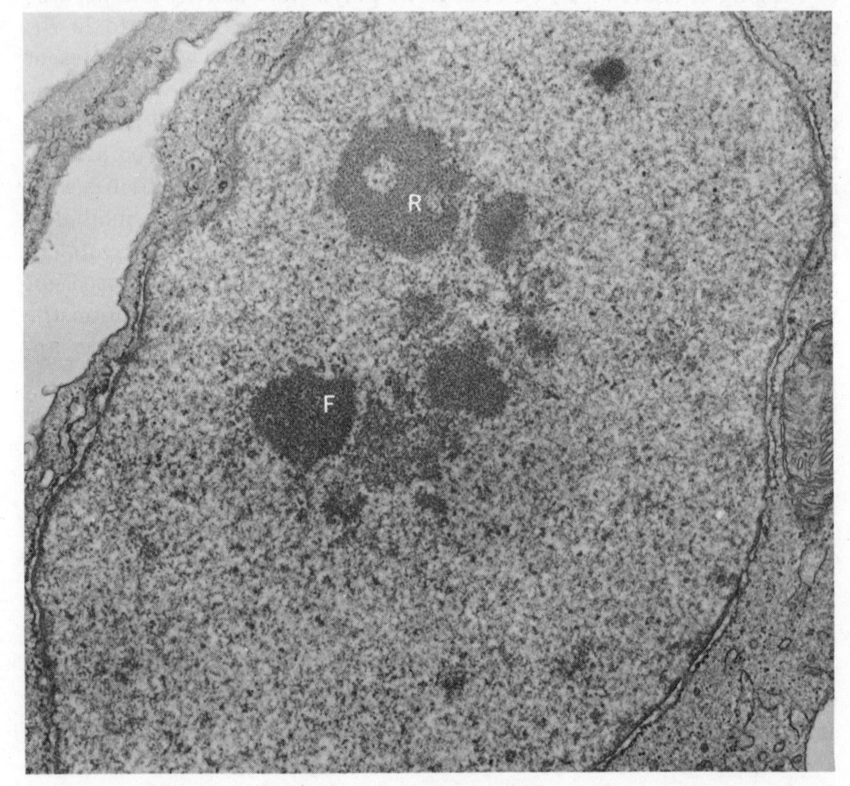

Fig. 4.3. Electron micrograph of cultured chicken embryonal hepatocytes after incubation with aflatoxin B_1 (10 μg/ml for 3 h). Small ring-shaped (R) and fragmented (F) nucleolar remnants are seen.

densed chromatin and these granules are surrounded by halos 25 nm in diameter. No definite conclusion can yet be drawn regarding the function of these RNP-containing-granules.

c) Chromatin

The nucleolus-associated chromatin widens occasionally after the administration of rubratoxin *B* in ddD male mouse hepatocytes.[21] Such pathological changes were also reported[11] after treatment with sterigmatocystin in cultured monkey kidney cells. Yokoyama[21] found that mouse liver cells after s.c. injections of luteoskyrin (2.5 mg/0.2 ml/mouse) revealed edema of the karyoplasm and reduction in both heterochromatin and nucleolus-associated chromatin. Alterations of the dispersed chromatin are

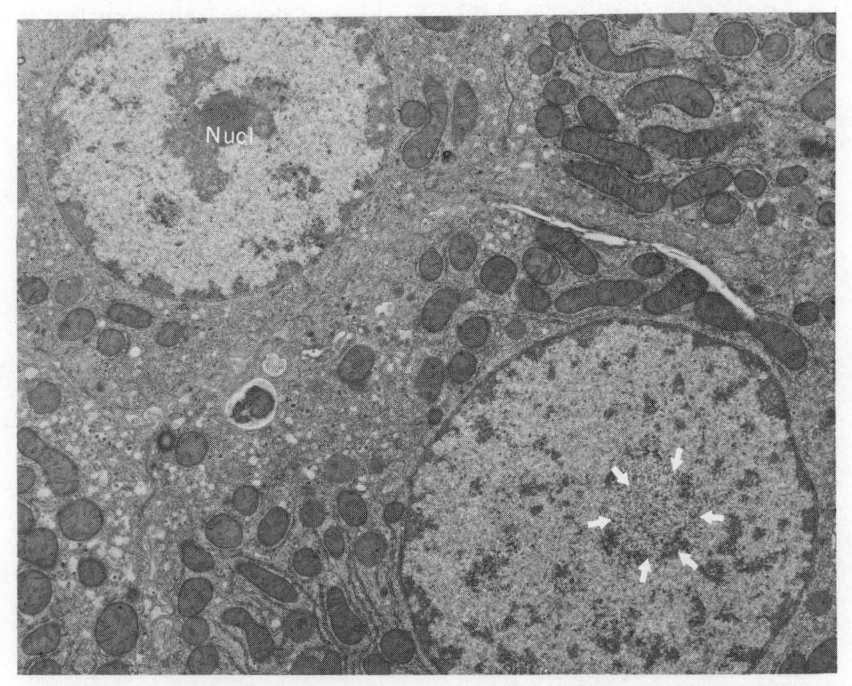

Fig. 4.4. Electron micrograph of rat liver after application of aflatoxin B_1 (3 mg/kg) for 3 days. There are two nuclei with altered nucleoli (Nucl). The nucleolus at the top is a condensed fibrous one. The other at the bottom is pulverized. A cluster of ribosome-like granules (arrow) is surrounded by heterochromatins.

induced by several mycotoxins such as cyclopiazonic acid, fusarenon-X, nivalenol, and penicillic acid.

d) Giant Cells

Multinucleate giant cells in the liver of marmosets[22] and many large hyperchromatic parenchymal cells in rat[23] and pig liver[22] were observed after feeding aflatoxin B_1-containing diets. The slow rate of regeneration is possibly related to the development of such giant cells. The fate of these cells is not known at present. Abnormally high numbers of giant cells are found in human embryonic lung cells after exposure for 8–12 h to 1 ppm of aflatoxin B_1.[24] Ehrlich ascites tumor cells in cultures exposed to luteoskyrin for 3–5 days contained 6–7% of large tri- or multinucleate cells, while control culture cells contained only 1–2% of such cells (mostly binucleate).[25] Such multinuclear cells are also seen in rat liver cells in primary cultures exposed to rubratoxin B.[26]

It is now widely accepted that the shape of the interphase nucleus is closely related to the configuration of the chromosomes in ana- and telophase. For example, a rosette-like configuration of anaphase chromosome clusters results in a ring-shaped interphase nucleus.[27] The "C-mitosis" or dispersed configuration of ana- or telophase chromosomes, causes multiple micronuclei in an interphase cell.[28]

e) Mitotic Aberrations

Relatively few attempts have been made to investigate the action of mycotoxins on chromosomes. Mitotic aberrations caused by mycotoxins have been reported in *in vivo* and *in vitro* systems. Lupinosis toxin causes mitotic stimulation and abnormal mitosis in the liver of domestic and experimental animals.[29] These pathological changes in the animals have been compared with the nuclear effects of colchicine and thioacetamide. Such dispersed "C-mitosis" is also observed in HeLa cells treated with rubratoxin B[30] and aflatoxin B_1,[24] and in kidney cells of the African green monkey in cultures exposed to sterigmatocystin.[11] Diacetoxyscirpenol produces a similar chromosomal abnormality when L cells are exposed to the mycotoxin.[31] Most of the abnormalities observed in the anaphase of mycotoxin-treated cells consisted of chromosome fragments with occasional bridges and stickiness, followed by the formation of an irregularly shaped interphase nucleus.

Chromosome abnormalities have been produced by aflatoxin B_1 in plant,[32,33] animal[34] and human cultured cells.[35] Dolimpio *et al.*[35] reported that chromosomal aberrations of cultured human leucocytes after exposure to aflatoxin B_1 included gaps, breaks, fragments, deletions, and translocations. Each aberration usually affected only one chromatid. Chromosome Nos. 1 and 2 were partially affected, while groups E, F and G exhibited a lower than expected frequency of breaks. The effect of aflatoxin B_1 on human leucocytes is of the delayed type which is characteristic of chromosome breaking compounds such as alkylating agents. Abnormally long and partially "stretched" metaphase chromosomes were observed in luteoskyrin-resistant Ehrlich ascites cells (i.e. those growing in the presence of 1.0 μg/ml and at higher concentrations) after cultivation for 3–5 months in the presence of luteoskyrin (1.0 μg/ml).[25] It is suggested that luteoskyrin produces the described effects primarily by interference with the mechanism of DNA replication.

Patulin also causes chromosome breakage in salamander eggs during mitosis.[36] A high percentage of polyploid cells is found after exposure of human leucocyte cultures to 0.54 μg/ml patulin.[37] Single- and double-strand breaks are induced in HeLa DNA by patulin.[38]

B. Interference with the Functions of Cell Organellae

Recent advances in molecular biology have revealed that the nucleus of living cells has many genetically important functions, and its damage or functional modification greatly influences the cell organization. Nuclear damage is, in general, monitored by (1) DNA binding, (2) biosynthesis of DNA and nuclear RNA, (3) DNA or chromosomal breakage and its repair, and (4) modification of nuclear proteins such as histone.

Ueno *et al.*[39] have established that the hepatocarcinogenic (−)luteoskyrin binds *in vitro* with DNA. An aqueous solution of (−)luteoskyrin exhibits a maximum in its absorption spectrum at 460 nm, at pH 7.4, and on incubation with calf thymus DNA the maximum shifts progressively to 490 nm in the presence of Mg^{2+} ions. Gel-filtration has revealed the formation of a [DNA–Mg^{2+}–luteoskyrin] complex. Data on flow dichroism with the complex indicate that the planar chromophore of (−)luteoskyrin is oriented parallel to the axis of the native DNA molecule. Apurinic acid and pyrimidine homopolymers cause a shift in the visible spectrum of (−) luteoskyrin towards the same region as in the native DNA, while apyrimidinic acid and purine homopolymers shift the spectrum to a shorter region. Since (−)luteoskyrin contains a quinoid group which is capable of chelating Mg^{2+} ion, the pigment presumably binds with pyrimidine bases of DNA through an Mg^{2+}-bridge. Interaction of the pigment with deoxyribonucleohistone brought about a binding of luteoskyrin to the DNA moiety of the nucleoprotein.[40] As for aflatoxin B_1, Sporn *et al.*[41] have reported a shift of the absorption maximum and hypochromism at 362 nm on binding with thymus DNA. In contrast to (−)luteoskyrin, aflatoxin B_1 has a great affinity towards purine bases.[42] King *et al.*[43] conducted an extensive series of studies on the interaction of aflatoxin B_1 and DNA, and concluded that it shows a first-order relationship, with the amino group of adenine required for binding. In contrast to the abundant evidence for the *in vitro* binding of aflatoxin B_1 with DNA, only a few reports have been published concerning such interaction *in vivo*. An autoradiographic study has demonstrated both nuclear and cytoplasmic labeling throughout the liver lobules.[44,45]

In mice poisoned with (−)luteoskyrin, *in vivo* synthesis of RNA is markedly depressed and that of DNA is accelerated (I. Ueno, unpublished) (Table 4.2). In Ehrlich tumor cells, synthesis of *m*RNA synthesis is inhibited.[46] Tashiro *et al.*[47] have purified RNA polymerase (RPase) from rat liver nuclei. The enzymatic activity of the preparation was chromatographically separated into two peaks, RPase I and RPase II. The former enzyme originated from the nucleolus and is responsible for *r*RNA synthesis. The latter enzyme, which is sensitive to α-amanitin, derived from the nucleo-

TABLE 4.2. Synthesis of macromolecules in the liver of mice administered (−)luteoskyrin (after I. Ueno, unpublished)

	Synthetic activity (% of control)		
	RNA	DNA	Protein
Control mice	100	100	100
5 mg/kg luteo. s.c. for 2 days	55	160	99
5 mg/kg luteo. s.c. for 11 days	80	117	75

TABLE 4.3. Localization and characterization of nuclear RNA-polymerases

RNA-polymerase	Localization	RNA synthesized	α-Amanitin sensitivity
I	nucleolus	*r*RNA	insensitive
II	nucleoplasm	DNA-like RNA	sensitive (low concentration)
III	nucleoplasm	*t*RNA and 5S-RNA	sensitive (high concentration)

TABLE 4.4. Inhibitory effect of mycotoxins on RNA synthesis in isolated nuclei and RPases (after Tashiro *et al.*)[47]

Inhibitors and mycotoxins	Concentration (μg/assay)	Inhibition of RNA synthesis (%)			
		Nuclei	RPase I	RPase II	*E. coli*
α-Amanitin	1	56	11	91	0
Actinomycin D	10	83	54	79	52
Rifampicin	50	0	5	14	98†1
Luteoskyrin	50	40	95	50	5
Rugulosin	50	30†1	96	36	58
Aflatoxin B_1	100	3	0	13	0
Sterigmatocystin	100	20	0	27	54
Cytochalasin B	100	0	15	17	25
Citrinin	100	—	69	24	50†2
Patulin	100	38	31	87	0
Penicillic acid	100	20	0	31	15
PR-toxin	100	25	73	15	66
Cyclochlorotine	100	4	0	0	0
Fusarenon-X	100	12	0	0	0
Zearalenone	50	—	0	3	0

†1 100 μg/assay.
†2 50 μg/assay.

plasm and isconcerned with the synthesis of DNA-like RNA (Table 4.3). The inhibitory effects of several carcinogens and cytotoxic mycotoxins on these RPases have been compared with data obtained for the whole nuclei. As summarized in Table 4.4, α-amanitin inhibits nuclear RNA synthesis by 60%, indicating that 40% of the total synthesis in the isolated nuclei is carried out by RPase I and the remaining 60% by RPase II. No contamination of each RPase preparation with mitochondrial RPase was proved by the fact that rifampicin induced no enzyme inhibition.

Actinomycin D, (−)luteoskyrin and (+)regulosin are inhibitory to nuclear RNA synthesis to the same extent, as well as to each of the two RPases, while sterigmatocystin, penicillic acid and patulin exhibit a higher affinity to RPase II, similarly to α-amanitin, and PR-toxin and citrinin inhibit RPase I to a greater extent than RPase II. Based on these preliminary findings, these mycotoxins are characterized as specific inhibitors of the RPase reaction in animal cells. The functional damage to the nuclei will be clarified on this molecular basis.

Employing purified RPase of *E. coli*, Ueno *et al.*[48] demonstrated the preferential binding of ^{3}H-luteoskyrin to the [DNA–polymerase] complex and ensuing inactivation of the enzyme activity. Ruet *et al.*[49] suggested that (—)luteoskyrin first binds loosely to the DNA-bound enzyme, and upon initiation of RNA synthesis, the mycotoxin binds to an exposed single-strand segment of DNA in front of the enzyme, thus preventing chain elongation.

As for aflatoxin B_1, which impairs RNA synthesis and inhibits the template activity of chromatin *in vivo*,[50] no profound effects are noted *in vitro*[47] (Table 4.4), suggesting metabolic alteration of aflatoxin *B in vivo.*

Several lines of research on RPase have suggested that RNA hybridase (RNase H) may be a key enzyme which connects the two reaction processes of DNA synthesis and RNA synthesis occurring in the nucleus. Tashiro and Ueno[51] have attempted to characterize mycotoxins from the viewpoint of affinity to RNase H. Purified enzyme from the nuclei of rat liver required Mg^{2+} for activity and was inhibited by rifampicin derivatives and SH-blockers. Among the mycotoxins tested, patulin, PR-toxin, penicillic acid, (−)luteoskyrin and (+)rugulosin affect the activity, as summarized in Table 4.5. Further studies will resolve whether these mycotoxins cause inhibition through binding to the hybrid or inactivating the enzyme protein. It seems highly possible that mycotoxins which preferentially inhibit RPase and RNase H from animal cells exist among toxic fungal metabolites, and that these mycotoxins will serve as biological tools for clarifying the molecular mechanisms of control in the nuclei of animal cells.

By means of sucrose or CsCl density gradient sedimenation analysis,

TABLE 4.5. Inhibition of RNase H of rat liver nuclei (after Tashiro and Ueno)[51]

Inhibitors and mycotoxins	Concentration (μg/assay)	Inhibition (%)
Rifampicin	50	0
	25	0
AF/013	30	99
	10	90
AF/ABDPcis	30	0
	10	0
Actinomycin D	30	9
	10	3
α-Amanitin	2	5
	1	14
N-Ethylmaleimide	50	72
	10	0
Cycloheximide	30	0
	10	0
Patulin	50	62
PR toxin	100	83
Penicillic acid	100	30
Sterigmatocystin	100	4
Aflatoxin B_1	100	9
Rugulosin	100	83
Luteoskyrin	100	93
Cytochalasin B	100	0

radiation and several carcinogenic substances have been found to induce reduction of sedimentation profiles of DNA molecules from mammalian cells *in vivo*, and this reduction is reversible on further incubation. From tracer experiments and radioautography, these phenomena were judged to represent breakage and repair of DNA or chromosomal strands. During the repair process, errors in base sequence or rearrangement of chromosomes occur. This may lead to malformation or tumorous changes. (−) Luteoskyrin is reported to induce chromosomal aberrations in Ehrlich tumor cells[25] and to inhibit DNA repair in *Tetrahymena*,[52] while aflatoxin B_1 as well as sterigmatocystin cause incomplete DNA repair in epithelial cells from monkey kidney.[53] Umeda *et al.*[38] have compared the potencies of various mycotoxins to induce DNA breakage in HeLa cells and demonstrated that aflatoxin B_1, penicillic acid, and patulin induce breakage, whereas (−)luteoskyrin, rubratoxin *B* and fusarenon-X induce little DNA strand damage.

In microbial cells, the mechanisms of induction and repression of enzymes are well established and the "operon-repression" theory has been proposed. In animal cells, the processes are more complex than in bacterial cells. Numerous studies have demonstrated that several enzyme systems are inducible by pretreatment of animals with foreign chemicals which modify the nuclear DNA or its regulatory system, and the induction may be suppressed by several chemicals which impair RNA function or the translation processes of *m*RNA. Carcinogens such as 3-methylcholanthrene and 3,4-benzo(*a*)pyrene are capable of inducing several microsomal enzymes through modification of DNA template activity, and actinomycin *D* inhibits the enzyme induction through depression of *m*RNA synthesis in nuclei. Several mycotoxins have been tested for their effects on enzyme induction and repression. Ueno *et al.* (*unpublished*) examined the effect of (−)luteoskyrin on enzyme induction and, as shown in Table 4.6, consecutive daily administration of 5 mg/kg (−)luteoskyrin for 3 days depressed the cortisone- or trypthophan-induced activity of trypthophan pyrrolase in the liver of mice. Aflatoxin B_1 has also been reported to impair the 3,4-benzo(*a*)pyrene induction of trypthophan pyrrolase and hydroxylase activities.[54)]

TABLE 4.6. Inhibitory effect of (−)luteoskyrin on the induction of tryptophan pyrrolase in mouse liver (after I. Ueno unpublished)

Inducer	Kynurenine formed (μmol/h)	
	control mice	poisoned mice†
—	3.42	3.53
Cortisone	8.97	6.46
Trypthophan	9.17	5.17

†5 mg/kg of (−)luteoskyrin was daily administered s.c. for 3 days.

The DNA molecule, acidic in nature, binds with basic proteins such as histone and the resulting deoxyribonucleohistone complex is the functional form in many nuclear processes. The rates of association and dissociation of the complex are influenced by chemical modifications of histone. Masking of amino groups of basic proteins by methylation, acetylation and formylation, and phosphorylation of the hydroxyl groups of hydroxyl amino acids result in changes in affinity of the protein molecule towards DNA, followed by conformational changes and alteration of the replication and transcription capacity of the DNA for the polymerase systems.

In vitro addition of aflatoxin B_1 causes a loss of histone and some acidic protein relative to DNA, suggesting a possible combination with histones. *In vivo* experiments have also revealed that short-term exposure to aflatoxin B_1 induces acceleration of deacetylation of histone prior to the dramatic fall in RNA-polymerase activity.[55)] Such approaches are very useful for studying the nuclear functions as a whole, and may lead to an understanding of the mechanisms of the earliest changes occurring in poisoned nuclei.

4.1.2. Effects of Mycotoxins on Cytoplasmic Organellae

A. Mitochondria

Mitochondrial alterations are encountered after the administration of several mycotoxins. These alterations are characterized irregular profiles, swelling, and bleb-like evaginations. The blebs and infoldings of membranes are composed of either the outer or the inner membrane, or include a double membrane. Aflatoxin B_1,[1,3] luteoskyrin,[56,57] ochratoxin A[58] and cyclochlorotine[59] cause severe mitochondrial injuries.

Among these mycotoxins, (−)luteoskyrin is fully characterized as a potent inhibitor of mitochondrial function in animal cells. According to the data of I. Ueno *et al.*, respiration of rat liver, kidney and heart is inhibited by (−)luteoskyrin and the coupled phosphorylation of mitochondria is markedly affected. As for the latter phenomenon, detailed experiments demonstrated that (−)luteoskyrin acts as an inhibitor of oxidative phosphorylation and at a certain concentration also behaves as an uncoupler like 2,4-dinitrophenol (DNP).[60] *In vitro* incubation of mitochondria with (−)luteoskyrin caused swelling, and inorganic phosphate- or calcium-induced swelling was inhibited.[61] In rat liver poisoned intravenously, the mitochondria appeared swollen by electron microscopy and the p/o ratio was decreased. These observations suggest that the mycotoxin may bind with some mitochondrial components to induce a fixation of the flexible integrity of the mitochondria and that this biophysical loss of flexibility may be connected with the morphological damage to the organellae. This high affinity of (−)luteoskyrin to mitochondria explains the development of liver injuries in animals administered with luteoskyrin. As described in section 3.3.1, (−)luteoskyrin accumulates preferentially in the mitochondrial fraction of the liver *in vivo*, and the rate of this accumulation depends on the age, sex and strain of the animals in parallel to the toxicological susceptibility. Such *in vitro* and *in vivo* investigations have contributed much to our understanding that the preferential accumulation of (−) luteoskyrin and ensuing damage to mitochondria are responsible for the development of necrogenic changes in liver cells in the acute hepatotoxicity of this mycotoxin.

Mitochondrial damage was put forward early as one of the initial causes of liver damage by CCl_4 intoxication. Two possible mechanisms are conceivable: (1) damage of the mitochondrial membrane, and (2) calcium accumulation. As for the first possibility, Christie *et al.*[62] proposed that CCl_4 or its metabolite(s) may attack the membrane structure directly or through peroxidation reactions. In the case of luteoskyrin hepatotoxicity, I. Ueno *et al.* (*unpublished*) have found that labeled luteoskyrin preferentially binds with the mitochondrial membrane. As for the second possibility, there is much evidence to indicate that mitochondria *in vitro* take up an enormous amount of calcium from media, and the accumulated calcium which exists as calcium phosphate eventually destroys the mitochondria through (1) destruction of the energy-conservation mechanism of oxidative phosphorylation, (2) loss of intramitochondrial components such as nucleotides and metal ions, and (3) irreversible changes in the membrane.

In the case of (−)luteoskyrin, it is highly possible that reduction of ATP formation is associated with the necrogenic as well as fatty changes of liver cells.

Citreoviridin, one of the yellowed rice mycotoxins, is produced by *Penicillium citreo-viride* Biourge, and is believed to be responsible for acute cardiac beriberi. In acutely poisoned animals, ascending paralysis is followed by a decrease in blood pressure and respiratory arrest. In an attempt to disclose the mode of action of citreoviridin, Ueno *et al.* (unpublished) examined its effect on energy metabolism in brain and other tissues. As a result, citreoviridin was found to inhibit Mg^{2+} -ATPase at low concentrations.

Doherty *et al.*[63] have examined the effect of aflatoxin B_1 on rat mitochondria. When the mycotoxin was added to an actively respiring liver mitochondrial preparation, electron transport was inhibited by 25–44% at $2.5\text{–}4.8 \times 10^{-4}$ M of aflatoxin B_1. The major site of inhibition of electron transport appears to be between cytochromes *b* and *c* (c_1). The significance of this inhibitory effect in aflatoxin hepatotoxicity remains to be clarified.

In the field of mycotoxicology, no direct evidence for functional impairment of mitochondrial DNA has been reported. From the fact that (−)luteoskyrin binds with DNA, accumulates in mitochondria and induces RD-mutants in yeast cells, it is considered that this mycotoxin interferes with the function of mitochondrial DNA.

B. Endoplasmic Reticulum

The endoplasmic reticulum is usually affected in the early stage of treatment with exogenous chemical substances. At the earliest phase, the ultrastructural change in this microorganella is dilatation of its cisternae. Generally, at this time, other organellae in the cytoplasm such as the mitochondria, Golgi complexes, microbodies and lysosomes, appear normal. In some instances, ribosomal detachment from cytoplasmic membranes follows thereafter. This phenomenon appears to be accompanied by diminished protein synthesis. It is well known that nucleolar morphology and function relate closely to protein synthesis. However, Butler[3,14] found no correlation between the two cytoplasmic changes, disruption of granular endoplasmic reticulum (RER) and nucleolar segregation, in aflatoxin B_1-poisoned rats.

Proliferation of agranular endoplasmic reticulum (SER) is reported not infrequently. Butler[3] found a rapid increase of SER after the administration of aflatoxin B_1 in male rat hepatocytes. In most instances, continuity between the proliferated SER and RER is demonstrable. The ultrastructural changes consist of proliferation of masses of smooth membranes within the liver cell cytoplasm referred to as "fingerprints" or "membrane whorls". Such membrane whorls in hepatocytes are induced by aflatoxin B_1[1], *O*-acetylsterigmatocystin,[64] and ochratoxin A.[58] In several instances the membranes of the whorl associate with many isolated glycogen particles, so that they form a "glycogen-membrane array". Whorls of agranular membranes with or without associated glycogen, are a common response to several types of pathologic stimuli. The phenomenon of proliferated SER is usually considered to be an adaptive response to a toxic agent.[4] A proliferation of SER is seen not only in the cytoplasm of hepatocytes but also in the cytoplasm of proximal convoluted tubules of the kidney (Fig. 4.5).

In general, there is a negative correlation between the proliferation of SER and glycogen. After the application of certain exogenous chemical substances, there is always an observed increase in SER associated with a loss of glycogen. When the proliferation of SER ceases, glycogen particles reappear between the membranes of the SER.

The fundamental profile of the SER is recognized as the site of drug metabolism. The precise molecular mechanism has been discussed in section 3.5. Functional interference of this organella may be ascribed to (1) direct and indirect inhibition of the enzyme system, (2) competitive requirement for NADPH, UDP-glucuronic acid and glutathione, and (3) inactivation of P-540 by reaction with the protoheme. Actually, during the course of toxicological analysis of (−)luteoskyrin hepatotoxicity in mice, Ueno *et al.* (unpublished) observed that prior administration of (−)luteoskyrin prolonged sleeping time in mice which had been injected with pentobarbital and reduced the P-450 content in the liver. Therefore, it is highly possible that fungal metabolites are able to interfere with and modify the fundamental pattern of drug metabolizing activity through damage of SER. There is interesting evidence that the SER activity is altered by foreign compounds. As summarized in Table 4.7, I. Ueno *et al.*[65] demonstrated that pretreatment of mice with 3-methylcholanthrene, promethazine, dibenamine or phenobarbital affords protection against (−)luteoskyrin toxicity which correlates with elevation of s-GPT and the content of (−)luteoskyrin in the liver.

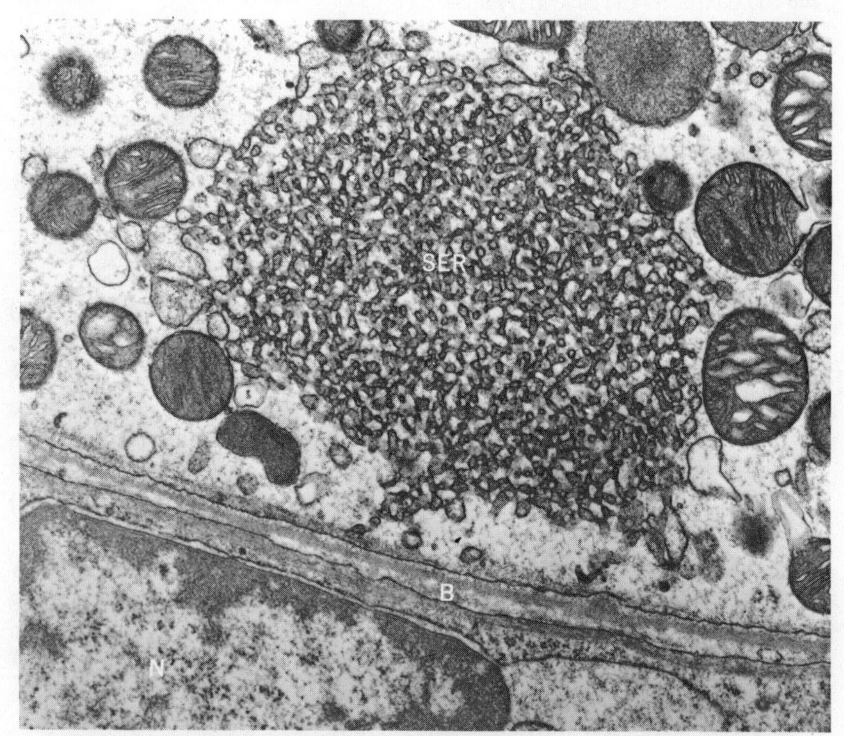

Fig. 4.5. Electron micrograph of proximal convoluted tubules of a rat kidney after injection of ochratoxin A (5 mg/kg for 3 days). There is marked proliferation of agranular endoplasmic reticulum (SER) at the bottom of degenerated tubular cells. Degenerated mitochondria and dilated vesicules are also seen. B, Basement membrane; N, nucleus of capillary endothelial cell.

TABLE 4.7. Protection against luteoskyrin hepatotoxicity (after I. Ueno *et al.*)[65]

Pretreatment	(μg/g/day)	(day)	Dose of luteoskyrin (μg/g)	s-GPT activity†1 (units/ml)	Luteoskyrin in liver†1 (μg/g)
Phenobarbital	80	3	50	45 ± 15 (5)†2	0.7
	0		50	884 ± 331 (5)	11.6
Promethazine	25	3	50	64 ± 43 (4)	0.6 ± 0.4 (5)
	0		50	958 ± 355 (4)	4.8 ± 2.2 (4)
Methylcholanthrene	100	1	50	130 (4)	3.4 ± 1.7 (4)
	0		50	195 ± 102 (4)	6.4 ± 1.0 (4)
Dibenamine	25	2	50	4 ± 2 (4)	1.0 ± 0.1 (4)
	0		50	371 ± 211 (5)	9.1 ± 3.2 (5)

†1 s-GPT activity and luteoskyrin content were measured 2 days after oral administration of luteoskyrin (50 μg/g).
†2 No. of mice used.

According to Aust,[66] the salivation effect of slaframine is markedly altered by pretreatment of animals with various inducers and inhibitors. Pretreating mice with phenobarbital and DDT greatly enhances the activity of slaframine, as measured by the time for which salivation occurs after slaframine administration. Inhibition of the oxygenase system of SER delays onset of the salivation response. These results suggest that slaframine is biotransformed to an active form by the drug metabolizing enzyme of liver SER.

Investigations of the effects of hepatotoxic mycotoxins such as (−) luteoskyrin and aflatoxin have revealed a close similarity with the effects of chemical hepatotoxins such as CCl_4 and ethionine insofar as the RER is concerned. Numerous studies on the acute hepatotoxicity of these toxicants suggest the following common sequence of events: (1) disaggregation of polysomes, (2) failure in protein synthesis, especially of lipoproteins, (3) accumulation of triglycerides and lipid peroxide, and (4) necrogenic changes in the liver cells.

Disaggregation or breakage of the polyribosomal structure is an early event in the structural and functional modification of the RER. According to Sato and Ueno,[67] polysomes of the liver of mice given (+)rugulosin s.c. begin to disaggregate before any elevation of s-GPT or damage of other liver functions, and this breakage persists for a long time even if the s-GPT falls below the normal level.

Roy[68] reported a gradual disaggregation of polysomes after i.p. administration of aflatoxin B_1 at a dose of 1.5 mg/kg. Within 6 h after treatment, an extensive breakdown of polysomes into di- and monomers occurs. Sarasin *et al.*[69] found that in rats administered i.p. 1 mg/kg of aflatoxin B_1, the disaggregation of liver polysomes starts 3–4 h after administration of aflatoxin, and progressively increases to reach a maximum of 50–60% at 15–24 h after the administration.

Such disaggregation of the polysomal structure would result from many functional changes such as (1) breakage of the *m*RNA which holds the polysomes in spiral form, (2) cessation of *m*RNA synthesis, (3) inhibition of release of nascent peptides from the polysomes, (4) interference with the ribosomal cycle, and (5) direct damage of the ER membrane on which the polysomes are attached. If these possibilities are correct, it may be expected that other well-defined inhibitors cause similar phenomena.

Interesting findings on a quite different property of aflatoxin have been reported by Williams *et al.*[70] They found that aflatoxin B_1 exerted an effect on the binding of polysomes to microsomal membranes. On *in vitro* incubation of the mycotoxin and hepatic microsomes, displacement of the polysomes from the membrane took place, and furthermore, membranes which had been exposed to the toxin lost their ability to bind the polysomes. One significant finding is that corticosterones antagonize this effect. These results lead to the hypothetical conclusion that aflatoxin B_1 and other carcinogens may occupy or destroy a site specific for steroid hormones on the membranes of the ER.

Robert *et al.*[71] demonstrated in ethionine-intoxicated rats that RER damage would involve alteration of (1) membranes, (2) ribosomes, and (3) the association mechanism of the ribosomes and membranes. The peroxidation process of microsomal membranes exerts a significant effect on drug metabolism as well as on the toxic effects. This peroxidation reaction is controlled *in vivo* by several soluble factors and enzymes. The soluble fraction of the liver possesses glutathione peroxidase which catalyzes the reaction between intracellular peroxides and nucleophiles such as glutathione. In the absence of glutathione, microsomes and mitochondria are capable of decomposing hydroperoxides by several systems that involve free radicals and cytochromes. Because the ER lacks glutathione peroxidase and abounds in enzymes that catalyze peroxidation, the ER membranes should be easily attacked by several peroxidation reactions *in vivo*. This susceptibility of the ER structure to peroxidation may lead to damage of the organellae, particularly of the association of ribosomes and membranes. On the other hand, mitochondria are rich in antioxidant content.

The protection from polysomal disturbance observed in the liver of rats administered glutathione prior to CCl_4, and the prevention of (−) luteoskyrin liver injuries by prior administration of promethazine, may involve such peroxidation and antioxidant reactions.

Protein synthesis is one of the most important functions of the RER, and its disturbance results in impairment of cellular growth and tissue differentiation. In the field of antibiotics, numerous studies have clarified the molecular mechanisms of protein synthesis in microbial cells and the sites of action of the antibiotics. However, the mechanism of protein synthesis in eukaryotes is so complex that precise information is not yet available on the molecular mechanisms of agents which influence protein synthesis in them. In the field of mycotoxins, the trichothecene compounds are known to be inhibitors of protein synthesis in animal cells. In the early stage of biological examinations of toxic principles from *Fusarium*-molded rice, crude toxin from *F. nivale* exhibited karyorrhexis and caused cellular damage in actively dividing tissues of poisoned animals. The histopathological findings led Ueno *et al.* to investigate the inhibitory mode of action of the toxic principles. Employing rabbit reticulocytes as an assay tool, it was demonstrated that the most effective fraction of the crude toxin contained nivalenol, a toxic principle of *F. nivale*, and that all the trichothecene-type mycotoxins are powerful inhibitors of protein synthesis in animal cells.[72,73] Furthermore, it was established that the high affinity of trichothecenes to animal and plant cells is a result of their high affinity to 80-S ribosomes. In this respect, the trichothecenes represent "real" mycotoxins.

Concerning other effects of mycotoxins on protein synthesis, trichodermin has been reported to block termination by preventing peptidyl *t*RNA hydrolysis,[74] and T-2 toxin and verrucarin have been reported to block polypeptide chain initiation in eukaryotes.[75,76] According to Ueno *et al.*,[77] all the trichothecenes inhibit protein synthesis in Ehrlich ascites tumor cells, rabbit reticulocytes and rat liver, both *in vitro* and *in vivo*, and the degree of inhibition differs markedly depending on the chemical nature of the particular trichothecene. At the whole cell level, T-2 toxin is the most effective followed by diacetyoxyscirpenol, HT-2 toxin, fusarenon-X and nivalenol, and at the cell-free level, no marked differences are observed among the different compounds. In general, macrocyclic trichothecenes and lipophilic agents exhibit a potent activity at the whole cell level, and the hydroxyl group at C-3 exerts a great influence on the activity. Ueno *et al.* have demonstrated that one early effect of trichothecenes on animal cells is to disaggregate the polysomal structure, as mentioned above (see Fig. 4.6). The binding of trichothecenes to polysomes and microsomes interferes with the ribosomal cycle of eukaryotes.[78] Thus, insofar as inhibitory effects on protein synthesis are concerned, antibiotics and mycotoxins can be to classified into two groups according to their affinities to 70-S and 80-S ribosomes, as summarized in Table 4.8.

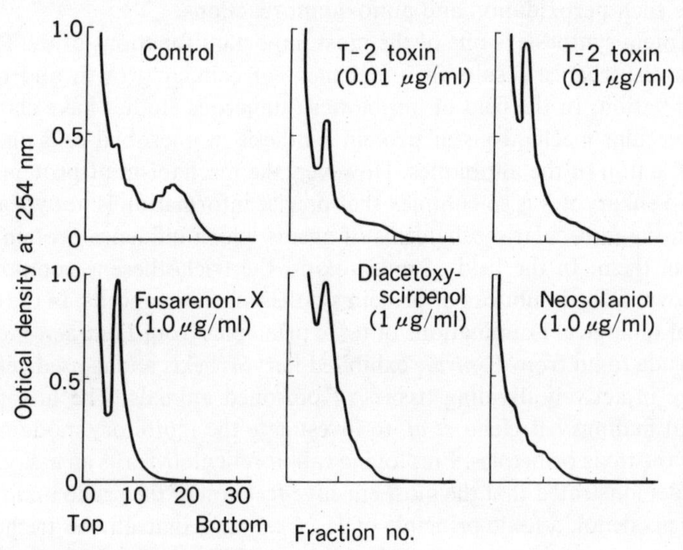

Fig. 4.6. Polyribosomal disaggregation in rabbit reticulocytes by trichothecenes (after Ueno *et al.*)[78]

TABLE 4.8. Inhibitors of protein synthesis acting on eukaryotic ribosomes

	Acting on ribosomes of the 80-S type		Acting on ribosomes of the 70-S and 80-S types
Antibiotics	Anisomycin		Puromycin
	Tylophorine		Gougerotin
	Enomycin		Actinobolin
	Phenomycin		Sparsomycin
	Emetine		Edeine
	Tenuaxonic acid		Pactamycin
	Cycloheximide group:	cycloheximide	Blasticidin S
		actiphenol	Fusidic acid
		streptovitacin A	Aurin tricarboxylic acid
		streptimidone	Tetracycline group
	Amicetin		
Mycotoxins	Trichothecene group:	nivalenol	
		fusarenon-X	
		T-2 toxin	
		verrucarins	

C. Lysosomes

In mice administered 5 mg/kg of (−) luteoskyrin s.c. for 3 consecutive days, serine dehydrogenase, acid phosphatase and β-glucuronidase in the liver were elevated by 30–50% of the control level (I. Ueno, unpublished). Lysosomal damage in aflatoxin intoxication is also known in chickens.[79] *In vitro* experiments[80] have revealed that upon incubation for 3 h, the release of five lysosomal enzymes was accelerated by aflatoxin B_1, indicating a direct interaction between the mycotoxin and the lysosomal membranes. Schabort *et al.*[81] and Pitout *et al.*[82] have demonstrated the *in vitro* binding of aflatoxin B_1 to bovine pancreatic DNase and studied the *in vivo* effect of the mycotoxin on DNase from rat and mouse liver. The bovine pancreatic enzyme was strongly activated by aflatoxin and the *in vivo* injection elevated only the activity in the liver of rats, not that in mice.

The role of lysosomal damage in the development of liver cell necrosis remains unclear. Several experiments including data on (−)luteoskyrin have indicated damage of ER followed by lysosomal alteration.

D. Membranes

All cells and cell organellae are surrounded by biomembranes, and recent interest has mostly focussed on the functions of membranes and the molecular mechanisms by which agents specifically affect the membrane permeability. In general, the transport mechanisms are classified into several groups according to the requirements of energy and a carrier model, as summarized in Table 4.9. Various naturally occurring as well as synthetic substances, primarily of peptide nature, have been found to be capable of selectively increasing the alkaline ion permeability of artificial and biological membranes. (−)Luteoskyrin, a hepatotoxic pigment from *Pencillium islandicum,* inhibits the activity of Na-K-ATPase prepared from ghosts of red cells and microsomes of the brain. Reticulocytes have been shown to be capable of concentrating glycine and alanine in the presence of extracellular Na^+. This transport system is considered to be mediated by an energy-independent carrier system which is active in the presence of a favorable Na^+-gradient across the membranes. Taking this into consideration, Ueno attempted to select mycotoxins which affect the glycine transport system of reticulocytes. Among the 23 mycotoxins examined, patulin and cyclochlorotine caused inhibition of glycine uptake at 100 μg/ml, and the ID_{50} of patulin was calculated to be 30 μg/ml (2×10^{-4}M). Experiments with SH-compounds and kinetic analysis revealed that patulin may inactivate SH-groups of the glycine binding center of the reticulocyte membranes.[83)]

TABLE 4.9. Transport mechanisms in biomembranes

Mechanism	Examples
(A) Passive transport (or diffusion)	membrane pores
	membrane lipids
(B) Active transport	
(1) Energy-independent passive mediated transport	
	facilitated diffusion
	exchange diffusion
(2) Energy-dependent transport	
i. carrier transport	ATPase system
	cotransport system
	binding protein system
ii. group translocation	
iii. energy production-coupled ion transport	
iv. bulk transport	exocytosis
	endocytosis

As described earlier, the trichothecene mycotoxins exhibit biochemical and cytological effects on animal cells, and their chemical structure, especially the positions of the hydroxyl and alkyl groups, has a strong influence on their relative toxicity. One interesting point is that T-2 toxin-type trichothcenes affect the cell function more effectively at the whole cell level than do nivalenol-type ones, whereas at the cell-free level no marked differences are observed between the two types. From these results, Ueno *et al.* have concluded that the former toxins have a higher affinity to the biomembrane structure than the latter. In fact, tracer experiments with ^3H-labeled T-2 toxin and fusarenon-X have clearly demonstrated the higher affinity of T-2 toxin to reticulocyte cell membranes than fusarenon-X.[78]

Several species of fungi produce a number of chemically related metabolites which show an unusual biological activity. The so-called cytochalasins were isolated from *Helminthosporium dematioideum, Metarrhizium antisopliae, Rosellinia necatrix, Aspergillus clavatus,* and *Zygosporium masonii.* The chaetoglobosins, in which the benzyl group of the cytochalasins is replaced by a 3-indole group, were isolated from *Chaetomium globosum.* The cytochalasins as well as chaetoglobosins inhibit cytoplasmic division (cytokinesis) without inhibiting nuclear division (karyokinesis) and also inhibit cell movement.[84] Subsequent reports have indicated a wide range of biological effects, including inhibition of phagocytosis, pinocytosis, secretion of thyroid and growth hormones, and inhibition of morphogenesis. One interesting finding is that cytochalasin B competitively impairs the transport of glucose, deoxyglucose and glucosamine by Novikoff hepatoma cells and fibloblasts, without affecting their intracellular phosphorylation and metabolism. No inhibition of macromolecular synthesis is observed. Thus, the binding of the mycotoxins to the transport site on the plasma membranes is presumed to cause an interference with the biological function of the cell membranes.[85]

E. Miscellaneous

Large autophagic vacuoles are frequently observed in rat liver adjacent to a necrotic site induced by aflatoxin B_1.[14] These autophagic vacuoles show a gradual reduction in size as the central vein is approached. Butler[14] suggested that large autophagic vacuoles in themselves are not lethal to the cell. It appears that they are a response to focal cytoplasmic degradation.

There have been only a few detailed descriptions of ultrastructural alterations in organellae other than those mentioned above. A marked dilatation of Golgi cisternae in hepatocytes has been reported following administration of aflatoxin B_1 in rats.[1]

Specific inhibition of enzyme systems other than those mentioned above is briefly summarized next. Mycophenolic acid, a product of several *Pencillium* species, contains a five-membered lactone ring fused to a benzene moiety. It exhibits toxic properties towards a limited number of bacteria, fungi and cultured cells, and possesses an antiviral effect. The development of virus-induced Rous sarcomas and Friend leukemia is also inhibited by mycophenolic acid. Acute and subacute toxicities to mice, rats and dogs demonstrated weight loss, diarrhea, inflammatory changes of the small and large intestine, and hematopoietic changes.[86] Cline *et al.*[87] reported that inhibition of viral growth by this agent was prevented when either guanine, guanosine or GMP was supplemented. Sweeney *et al.*[88] demonstrated that IMP dehydrogenase which converts IMP → XMP, and GMP dehydrogenase which converts XMP → GMP, are inhibited *in vitro* by 10^{-4} M of mycophenolic acid.

The experiments of Ueno *et al.* with purified enzyme preparations have revealed that the trichothecenes, fusarenon-X, neosolaniol and T-2 toxin, cause inactivation of SH-enzymes such as creatine phosphokinase, lactate dehydrogenase and alcohol dehydrogenase when the enzymes were preincubated with the mycotoxins in the absence of substrates. A supplement of dithiothreitol prevents this inactivation. Gel-filtration revealed the formation of a complex [trichothecene–enzyme] with a molar ratio of 4:1. These data indicate that the epoxytrichothecenes bind with the reactive thiol residues of SH-enzyme proteins.[89]

The above findings suggest an interaction between specific proteins (or enzymes) and specific mycotoxins. This approach may lead to technical progress in the radioimmunoassay of mycotoxins. Chu[90] has investigated the interaction of ochratoxin A with bovine serum albumin, and by employing such techniques as spectro-fluorometry, equilibrium dialysis and Sephadex gel filtration, it was found that 1 mole of albumin binds with 1.87, 2.23 and 2.47 moles of ochratoxin A with binding constants of 3.17×10^5, 1.86×10^6 and 3.17×10^6 M^{-1} at 25, 12 and 6°C, respectively. No binding between ochratoxin A and γ-globulin has been observed.

4.1.3. Cytotoxicity to Cultured Cells

The available quantities of mycotoxins are usually limited, so that it is difficult to carry out the requisite extensive tests. Tissue culture systems have therefore been widely used in the bioassay of mycotoxins. The parameters of toxicity include inhibitions of growth rate, the mitotic index, morphological abnormalities both in the mitotic phase and interphase, and inhibitory effects on the synthesis of biopolymers such as DNA, RNA, proteins, lipids and glycogen.

The following mycotoxins have been tested with various cultured cells: aflatoxin B_1,[24,45,91–97] sterigmatocystin[10,11] and its derivatives,[12] luteoskyrin[25,98] and various anthraquinoids,[99] cyclochlorotine,[98,100] rubratoxin B,[30] fusarenon-X[101] and toxic 12, 13-epoxytrichothecenes,[102] ochratoxin A,[103] penicillic acid,[103,104] and patulin.[104] Established cell lines as well as primary cultures are used for estimation of the toxicities

In vivo systems

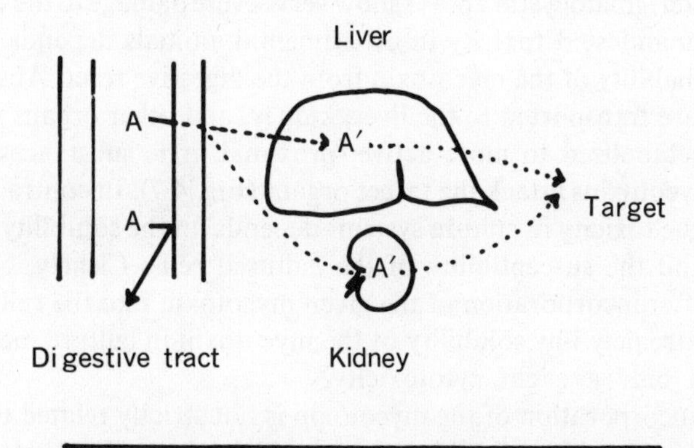

In vitro systems

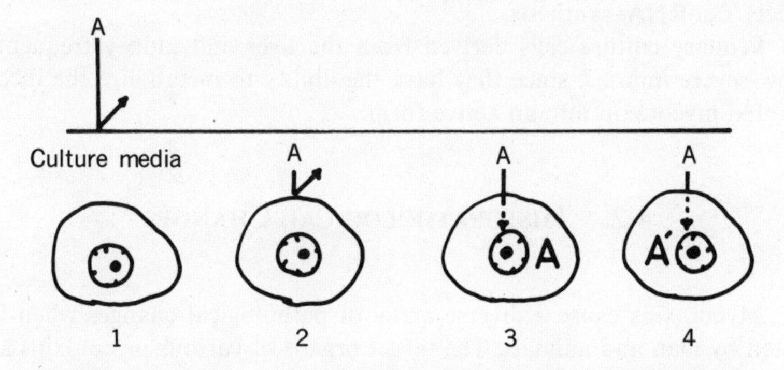

Fig. 4.7. Schematic representation of the incorporation of activated mycotoxins into target cells in *in vivo* and *in vitro* systems. A, Mycotoxin; A′, activated mycotoxin. *In vivo* systems: mycotoxins are absorbed from the digestive tract and transported to the liver and/or kidney where they may be activated and further affect the target. Some mycotoxins cannot be absorbed from the digestive tract. *In vitro* systems: 1, some mycotoxins which are insoluble in culture media reveal no cytotoxicity; 2, cytotoxicity depends on the permeability of the cultured cell to the mycotoxin; 3, although mycotoxins are incorporated into the cell bodies, no cytotoxicity is displayed, and the mycotoxin is either very stable or not activated by the cell; 4, the cell is capable of metabolizing the mycotoxin, and the activated mycotoxin injures the cell body.

of mycotoxins. A dose-response relationship in cytotoxicity is usually demonstrable in the mycotoxin-treated cultures.

Most mycotoxins studied show a good correlation between toxicity in *in vivo* systems and that in *in vitro* systems. However, occasionally a difference exists between them. Sterigmatocystin, a related compound of aflatoxin B_1, produces only slight injuries in rats; however, tissue cultures exposed to sterigmatocystin always show very severe damage to the cells.[10,11]

The manifested toxicity in experimental animals depends mainly on the absorbability of the mycotoxin from the digestive tract. Absorbed mycotoxins are transported to the liver, kidney, and other organs where they can be metabolized to more active, proximal toxic substances. Such activated mycotoxins attack the target organs (Fig. 4.7). In contrast to *in vivo* systems, the toxicity in culture systems depends on the solubility of the mycotoxin and the susceptibility of the cultured cells. Clearly, cytotoxicity appears after incorporation of the given mycotoxin into the cell bodies, so that an extremely low solubility of the mycotoxin in culture media always results in only a slight cytotoxicity.

The incorporation of the mycotoxin is not strictly related to the cytotoxicity. Using tritium-labeled aflatoxin derivatives, Terao *et al.*[45] demonstrated that there were no significant differences in the patterns of incorporation of ^{3}H-aflatoxin B_1, B_2 and tetrahydrodesoxoaflatoxin B_1 into liver cells *in vitro,* although they show a wide variation in their inhibitory effects on RNA synthesis.

Primary culture cells derived from the liver and kidney frequently show severe injuries, since they have the ability to metabolize the incorporated mycotoxin into an active form.

4.2. HISTOPATHOLOGICAL CHANGES

Mycotoxins cause a diverse array of pathological changes when ingested by man and animals. The target organs of various mycotoxins are shown in Table 4.10. Organs affected by the mycotoxins are those in the digestive, urinary, hematopoietic, dermal, endocrine, reproductive, nervous, circulatory, and respiratory systems. There are no clear relationships between chemical structure and organ-specific injuries. Usually one mycotoxin affects not one organ, but many organs. For example, aflatoxin B_1 produces pathological changes in organs such as the liver and kidney, and in the circulatory and central nervous systems. However, it is the liver which is injured most frequently and severely, so that aflatoxin B_1 is called a hepatotoxic mycotoxin.

TABLE 4.10. Target organs of mycotoxins

Organs affected	Myoctoxins	Animal species
Digestive organs:		
Liver	aflatoxin B_1, B_2, G_1 and M_1	turkey,[105] duckling,[105] chicken[105] cattle,[105] pig,[105] guinea-pig,[105] dog,[105] cat,[105] rabbit,[105] monkey,[105] rat[105]
	sterigmatocystin	rat,[106,107] monkey[108]
	luteoskyrin	mouse,[59] rat,[59] rabbit,[59] monkey[59]
	erythroskyrine	mouse,[109] rat[109]
	cyclochlorotine	mouse,[59,109] rat,[59,109] dog[59]
	rugulosin	mouse,[59,109] rat[59,109]
	rubratoxin B	rat,[110] mouse,[110] guinea-pig[110] chicken,[110] cat,[110] dog[110]
	ochratoxin A	duckling,[58] rat,[58,111] chicken[112]
	patulin	mouse[113]
	lupinosis toxin	sheep,[114,115] cattle,[116] horse[117] pig,[114] rabbit,[118] mouse[29,119]
	12, 13-epoxytrichothecenes	rat[120–2]
Bile duct and gall bladder		
	cyclopiazonic acid	rat[123]
	sporidesmin	sheep[124,125]
	aflatoxin B_1	pig,[105] dog[105]
	rubratoxin B	cat[110]
	lupinosis toxin	sheep,[114] horse[117]
Digestive tract: mouth, esophagus, stomach, intestine		
	ochratoxin A	chicken[112]
	sporidesmin	sheep[125]
	lupinosis toxin	sheep[114]
	fusarenon-X	mouse,[126,127] rabbit[128]
	aflatoxin B_1	guinea-pig[105]
	12, 13-epoxy-trichothecenes	man,[128] mouse,[127,128] rabbit[128] dog,[128]
Urinary organs:		
Kidney		
	aflatoxin B_1	monkey[105,129]
	aflatoxin G_1	rat[105,130]
	sterigmatocystin	rat,[106,107] monkey[108]

TABLE 4.10—*Continued*

Organs affected	Mycotoxins	Animal species
	cyclopiazonic acid	rat[123]
	penitrem A	chicken,[131] rabbit,[131] guineapig,[131] rat,[131] mouse[131]
	rubratoxin B	dog,[110] chicken,[110] rat,[110] mouse,[110] guinea-pig,[110] cat[110]
	ochratoxin A	rat,[58,111,132] dog[133]
	sporidesmin	sheep,[125]
	lupinosis toxin	sheep,[114] horse[117]
	patulin	mouse[113]
	citrinin	mouse,[134] rat,[135–137] rabbit,[138] guinea-pig[138]
	12, 13-epoxy-trichothecenes	dog[128]
Ureters, bladder		
	sporidesmin	sheep[125]
Hematopoietic organs:		
Bone marrow		
	12, 13-epoxy-trichothecene	mouse,[127] rat[127]
	ochratoxin A	chicken[112]
Lymph nodes		
	aflatoxin B_1	dog[139]
	rubratoxin B	cat[110]
	12, 13-epoxy-trichoothecenes	mouse,[127] rat[127]
	lupinosis toxin	sheep[140]
Spleen		
	aflatoxin B_1	guinea-pig[105]
	cyclopiazonic acid	rat[123]
	rubratoxin B	mouse,[110] rat,[110] cat[110]
	ochratoxin A	chicken[112]
	lupinosis toxin	sheep[140]
Bursa of Fabricius		
	ochratoxin A	chicken[112]
	zearalenone	chicken[141]
Thymus		
	fusarenon-X	mouse[126]

TABLE 4.10—*Continued*

Organs affected	Mycotoxins	Animal species
Skin:		
	12, 13-epoxy-trichothecenes	man,[142] mouse,[142] rat[142]
	patulin	man,[143] rabbit,[144] mouse[113]
Endocrine Organs:		
Adrenal body		
	aflatoxin B_1	guinea-pig[105]
	aflatoxin G_1	rat[130]
	sporidesmin	sheep[125]
Islets of Langerhans		
	cyclopiazonic acid	rat[123]
Reproductive Organs:		
Female (ovary, uterus vulva and mammary glands)	zearalenone	swine,[145,146] poultry,[141] cattle[147]
Male (testes)		
	fusarenon-X	mouse[127]
Nervous system:		
	aflatoxin B_1	monkey,[129] man[129]
	rubratoxin B	horse[148]
	citreoviridin	mouse,[134] rat,[134] guinea-pig,[134] dog,[134] monkey,[134] rabbit[134]
	penitrem A	mouse,[131,149] rat[131,149]
	fumitremorgen A	mouse,[149,150] rabbit[150]
	patulin	mouse[151]
Circulatory organs:		
Heart		
	aflatoxin B_1	human,[129] monkey[129]
	cyclopiazonic acid	rat[123]
	rubratoxin B	rat,[110] mouse,[110] cat[110]
	zearalenone	swine[145]
	xanthoascin	mouse[152]
Respiratory Organs:		
Lung		
	aflatoxin B_1	guinea-pig[105]
	patulin	mouse[113,151]
	sporidesmin	mouse,[124] rat[153,154]
	12, 13-epoxy-trichothecenes	mouse[121,122,141]

Mycotoxins can be divided into several groups, i.e. they are categorized as hepatotoxic, nephrotoxic, hemotoxic, dermatotoxic, endocrine-toxic, reproductive organ-toxic, and neurotoxic mycotoxins.

4.2.1. Liver Injuries Induced by Mycotoxins

Many mycotoxins affect the liver since it is the central organ of chemical metabolism and is the organ which receives topographically concentrated mycotoxins after ingestion.[155] The most consistent histopathological findings of liver injuries after administration of various mycotoxins are hepatocellular necrosis, hemorrhage and fat infiltration. Levels representing 1/4–1/2 of the LD_{50} dose usually result in massive cell necrosis, whereas 1/10 or less of the LD_{50} dose results in single cell necrosis.[13]

The zonal distribution of the massive necrosis in the liver lobules induced by various mycotoxins is shown in Fig. 4.8 and Table 4.11. The distribution of hepatic lesions is not consistent from species to species, or from mycotoxin to mycotoxin. A periportal zone of parenchymal cell necrosis develops in rat and duckling livers during a 3-day period after dosing with aflatoxin B_1.[105,156] Occasionally, a variation in response to a given dose is observed among individual animals. Fig. 4.9 shows the typical periportal necrosis in a rat liver after administration of aflatoxin B_1. One rat in the same experimental group, however, revealed centrilobular necrosis (Fig. 4.10). Cyclochlorotine is another typical periportal toxic agent.[59,109] Animals intoxicated with cyclochlorotine developed vacuolation and degeneration of the liver cells, almost exclusively in the periportal regions of the liver (Fig. 4.11). Necrosis of periportal cells in the rat liver is also observed after dosing with the LD_{50} value of ochratoxin A.[58]

The liver lesions induced by rubratoxin B[110] and *O*-acetylsterigmatocystin[12,156] usually originate in the midzonal area and spread peripherally in the lobule to the periportal area.

Necrosis in the centrilobular region of the liver has been reported in guinea-pigs, pigs, dogs, and sheep after the application of aflatoxin B_1.[105] Centrilobular necrosis also develops after ingestion of luteoskyrin[60] (Fig. 4.12), erythroskyrine,[59] and lupinosis toxin.[114] It is of interest to note that centrilobular necrosis is induced following oral administration of sterigmatocystin, but periportal necrosis is encountered when the mycotoxin is injected intraperitoneally.[108]

The mechanism of the zonal distribution of massive necrosis is not yet clear. However, topographical differences in concentrations of the mycotoxins in the liver lobules may be significant, and the structural and functional differences of the hepatocytes in the periportal, midzonal and centrilobular regions may play an important role.

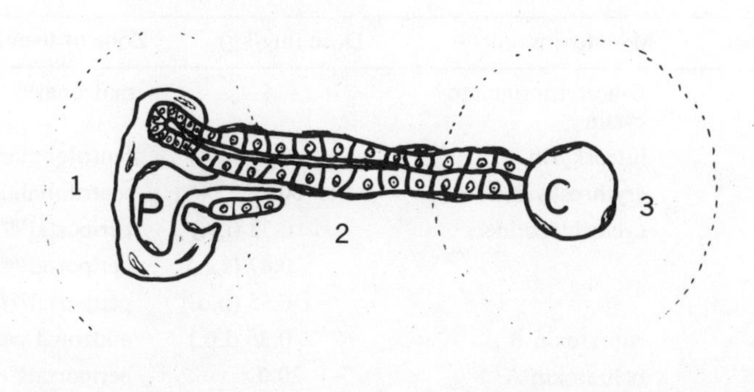

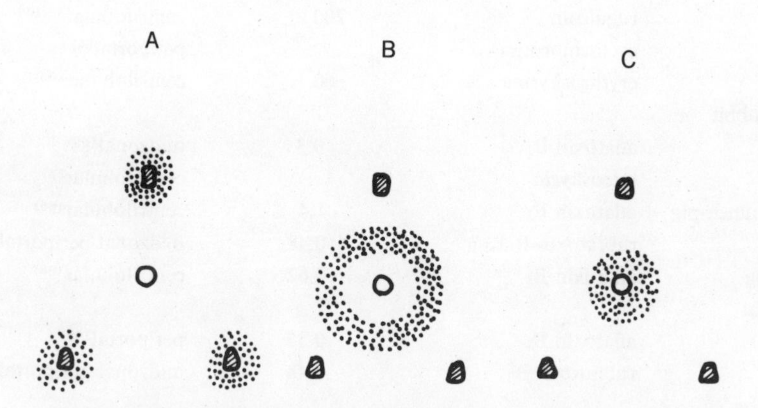

Fig. 4.8. Schematic representation of the distribution of massive necrosis in liver lobules. 1, Periportal zone; 2, midzonal zone; 3, centrilobular zone. P, Portal vein; C, central vein; A, periportal necrosis; B, midzonal necrosis; C, centrilobular necrosis; dotted area, massive necrosis.

TABLE 4.11. Zonal distribution of massive necrosis in the liver induced by mycotoxins

Animal	Mycotoxins	Dose (mg/kg)	Zone of liver lesions
Rat	aflatoxin B_1		
neonate		*ca.* 0.56	diffuse[105]
weanling		5.5	periportal[105]
weanling		7.4	periportal[105]
100g		7.2	periportal
150g		17.9	periportal
	sterigmatocystin	166 (p.o.)	centrilobular[155]
		60 (i.p.)	periportal[155]

TABLE 4.11—*Continued*

Animal	Mycotoxins	Dose (mg/kg)	Zone of liver lesions
	O-acetylsterigmato-cystin	8	midzonal[12]
	luteoskyrin		centrilobular[59]
	erythroskyrine	60	centrilobular[109]
	cyclochlorotine	0.33 (i.v.)	periportal[109]
		0.47 (s.c.)	periportal[109]
		6.55 (p.o.)	periportal[109]
	rubratoxin B	0.36 (i.p.)	midzonal-periportal[110]
	ochratoxin A	20.0	periportal[58]
Mouse			
	luteoskyrin	7.4	centrilobular[59]
	rugulosin	200	centrilobular[59,109]
	cyclochlorotine		periportal[59]
	erythroskyrine	60	centrilobular[109]
Rabbit			
	aflatoxin B_1	0.3	midzonal[105]
	luteoskyrin		centrilobular[58]
Guinea-pig	aflatoxin B_1	1.4	centrilobular[105]
	rubratoxin B	0.48	midzonal-periportal[110]
Pig	aflatoxin B_1	0.62	centrilobular[105]
Cat			
	aflatoxin B_1	0.55	periportal[105]
	rubratoxin B	0.48	midzonal-periportal[110]
Dog			
	aflatoxin B_1	1.0	centrilobular[105]
	cyclochlorotine	1.0	periportal[59]
	rubratoxin B	5.0	midzonal-periportal[110]
Sheep			
	aflatoxin B_1	1.0	centrilobular[105]
	lupinosis toxin		centrilobular[114]
	sporidesmin	1.0	periportal[123]
Chick			
	rubratoxin B	4.0	midzonal-periportal[110]
	ochratoxin A	1.0	diffuse[58]
Duckling			
	aflatoxin B_1	0.335	periportal[105]
Monkey			
	sterigmatocystin	32	centrilobular[108] midzonal
	luteoskyrin		centrilobular[59]

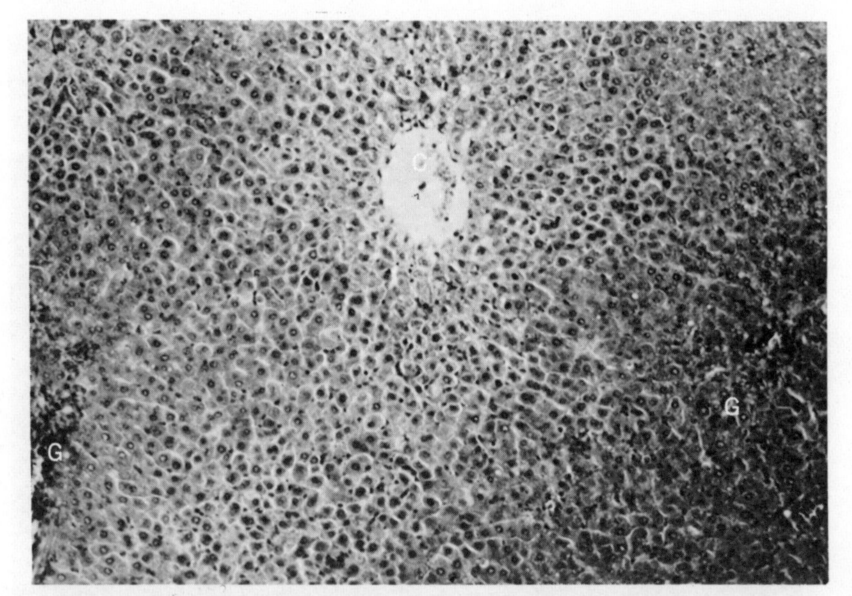

Fig. 4.9. Section of liver lobules of a rat dying after a single (3.5 mg/kg) dose of aflatoxin B_1, showing typical periportal necrosis (hematoxylin and eosin). C, Central vein; G, Glisson's sheath.

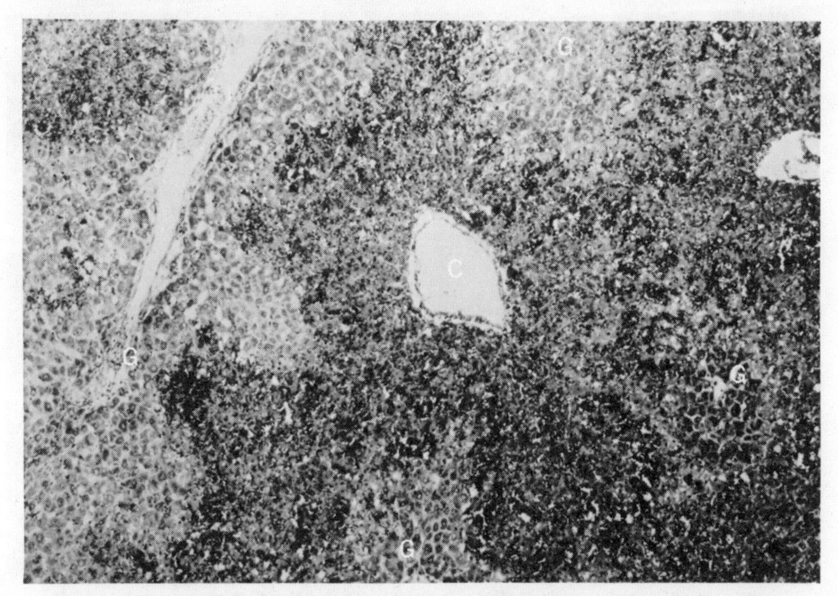

Fig. 4.10. Section of liver lobules of a rat dying after a single (3.5 mg/kg) dose of aflatoxin B_1, showing severe centrilobular necrosis. The hepatocytes survive only around periportal areas (hematoxylin and eosin). C, Central vein; G, Glisson's sheath.

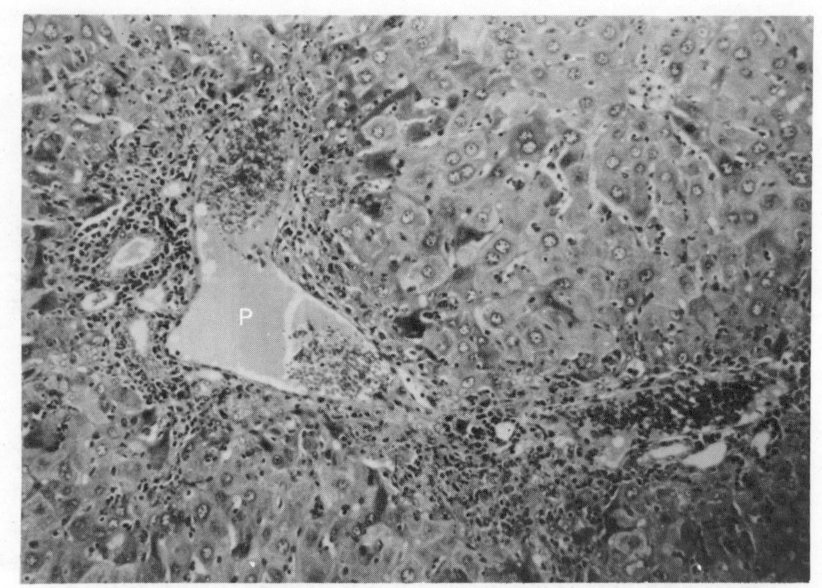

Fig. 4.11. Section of liver lobules of a mouse dying after repeated doses of cyclochlorotine, showing periportal round cell infiltration (hematoxylin and eosin). P, Portal vein.

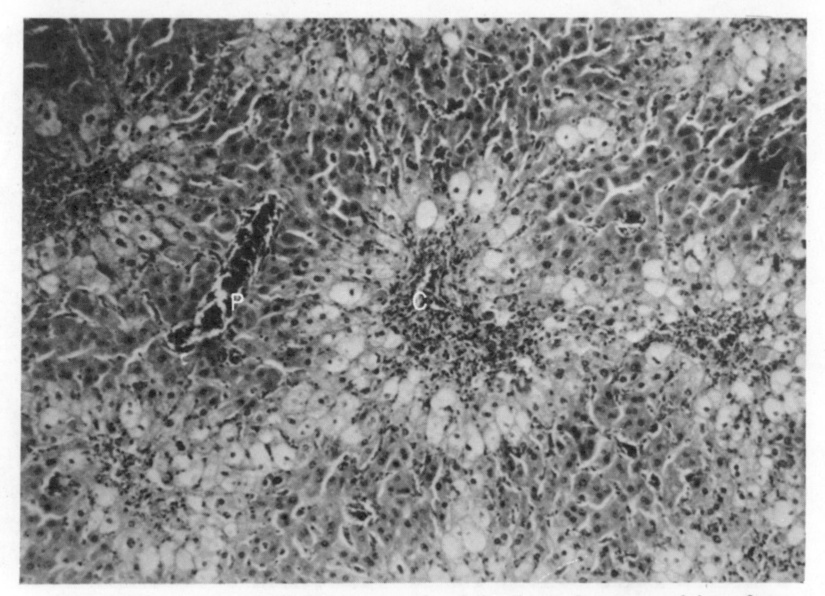

Fig. 4.12. Acute centrilobular necrosis of the liver of a mouse dying after administration of 500 μg luteoskyrin/day for 6 days (hematoxylin and eosin). There are marked vacuolated hepatocytes around the centrilobular necrosis. P, Portal vein; C, central vein. (By courtesy of Dr. M. Enomoto.)

In most instances, the susceptibility of hepatocytes to various mycotoxins is greater than that of Kupffer's cells. In necrotic lesions induced by aflatoxin B_1,[13] rubratoxin B,[110] luteoskyrin,[59] erythroskyrine,[59] ochratoxin A,[58] and sporidesmin,[125] the Kupffer's cells always survived. In contrast, both parenchymal and Kupffer's cells are affected by lupinosis toxin,[114] *O*-acetylsterigmatocystin,[12] and cyclochlorotine.[59]

Generally, there is little or no inflammatory reaction in hepatic injuries after ingestion of the mycotoxins. The necrotic debris is removed by macrophages, but rapid regeneration of the lesions is not seen in acute aflatoxicosis[13,105] or rubratoxicosis.[110] Only a small increase in collagen is seen after the administration of aflatoxin B_1.[4] In contrast, after the application of cyclochlorotine the mesenchymal reaction proceeds around the portal triads with piecemeal necrosis and cellular infiltration[59] (Fig. 4.13). The pathological changes that occur in the mouse liver after administration of sporidesmin are mild but appear to repair rapidly.[124]

In animals treated at a lower dose rate (1/10 LD_{50} or less), aflatoxin B_1 and/or sterigmatocystin induces in rat[13] and monkey liver[1] a single cell necrosis resembling the acidophilic body in viral hepatitis. The distribution pattern of single cell necrosis in the liver lobules is not consistent. In electron micrographs, the single cell necrosis is not a cytolytic process

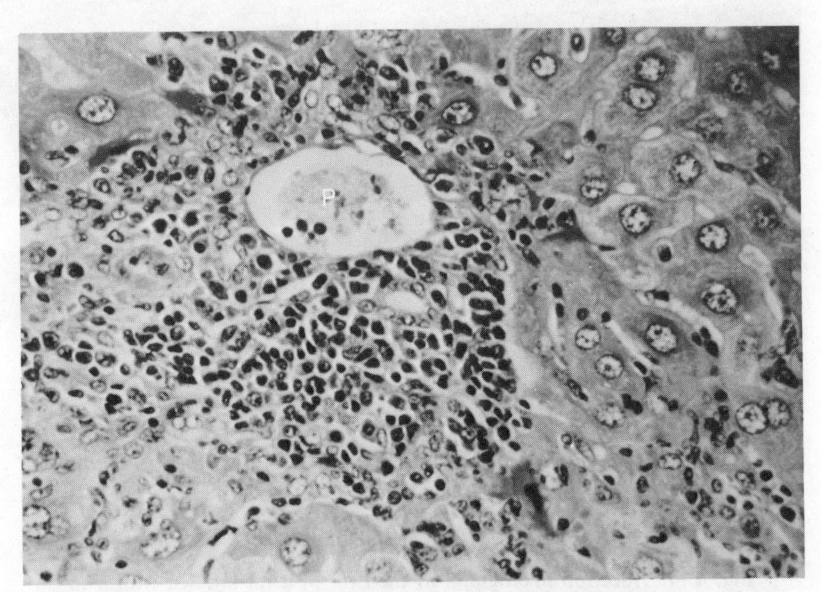

Fig. 4.13. Section of the liver of a mouse dying after repeated doses of cyclochlorotine, showing a mesenchymal reaction around portal triads with piecemeal necrosis (hematoxylin and eosin). P, Portal vein.

such as can be observed in the lesions of massive necrosis, but a dehydration of isolated hepatocytes. Therefore, most microorganellae such as the mitochondria, endoplasmic reticulum, free ribosomes, glycogen granules, nucleus, and nucleolus are well preserved, although these organellae are markedly increased in electron density. The dehydrated, degenerated hepatocyte is released from the liver trabecula and consumed by Kupffer's cells (Fig. 4.14). Single cell necrosis is also observed in the liver after treatment with luteoskyrin,[59,109] cyclopiazonic acid,[123] sterigmatocystin and its derivatives.[12,13,157]

Occasionally, increased numbers of enlarged Kupffer's cells are seen in the liver of rats fed with luteoskyrin[59] and/or in sporidesmin intoxication,[125] and in lupinosis in sheep.[114] Sometimes the enlarged Kupffer's

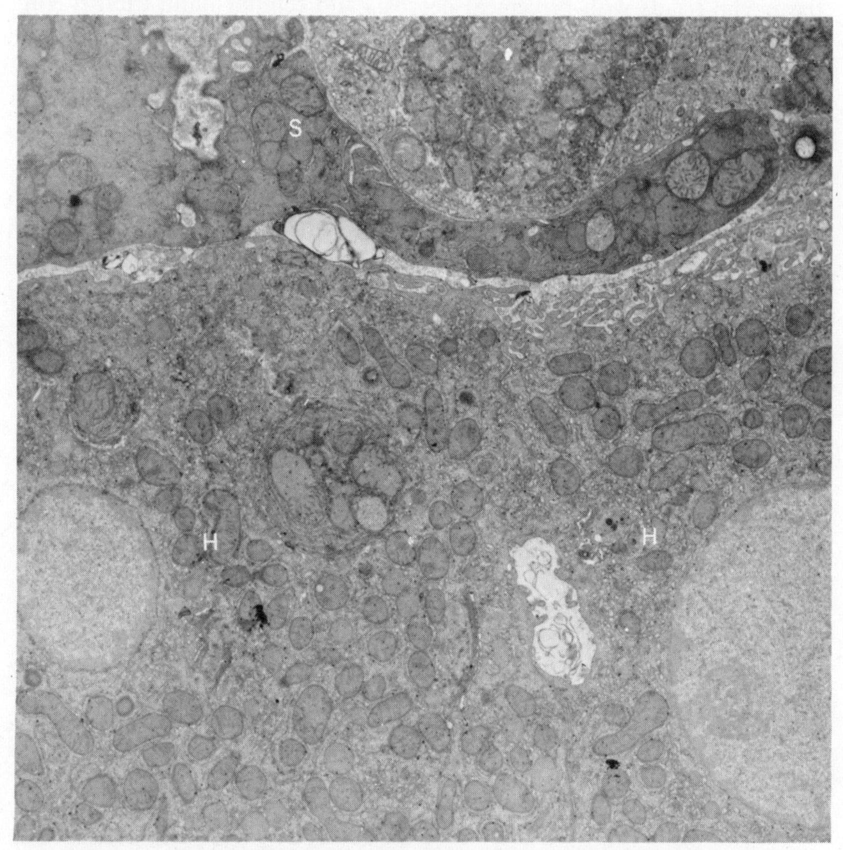

Fig. 4.14. Electron micrograph of the liver of a rat injected with a single (30 mg/kg) dose of sterigmatocystin. Single cell necrosis (S) is seen at the top of the picture. Microorganellae in the necrotic cell are well preserved. H, Hepatocyte.

cells contain bile pigment or ceroid-like pigment. Pigment has also been reported in the cytoplasm of hepatocytes in sheep lupinosis[114,140,158] and luteoskyrin intoxication.[159]

Development of biliary proliferation is always found after the administration of aflatoxins to various animal species. It was therefore at first thought to be specific for aflatoxicosis. However, similar lesions are also associated with lupinosis and/or with the administration of luteoskyrin, dimethylnitrosamine, and cycasin.[4] Usually a biliary proliferation occurs when the degenerative process spreads to the periportal region (Fig. 4.15). There are no definite relationships between the degree of biliary proliferation and necrotic areas. The bile ductule cells are also highly resistant to aflatoxins[13] and luteoskyrin,[59,109] and in contrast to hepatocytes, those in the portal triads remain unaffected.

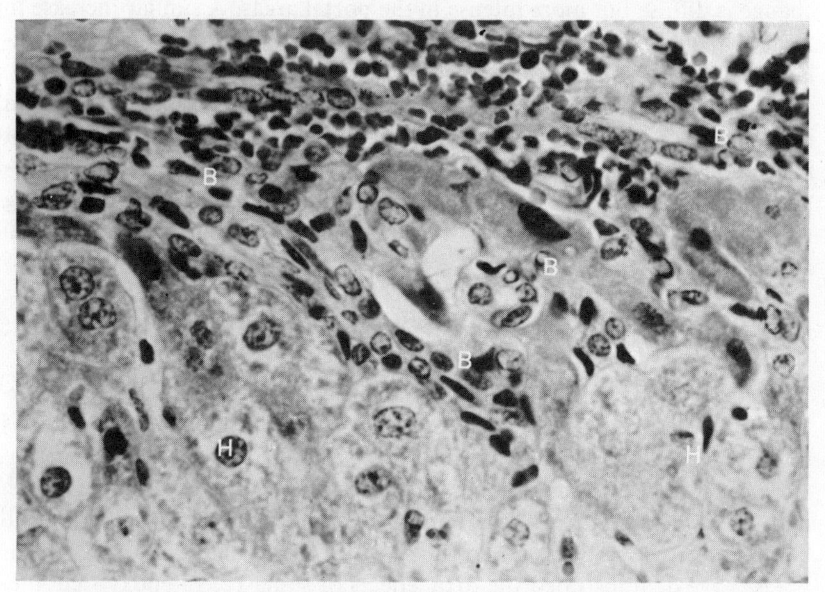

Fig. 4.15. Section of the liver of a rat dying after a single (3.5 mg/kg) dose of aflatoxin B_1, showing a proliferation of bile ductules (B) in the periportal area (hematoxylin and eosin). H, Hepatocyte.

In contrast with aflatoxicosis, sporidesmin-[125] and cyclopiazonic acid-treated[123] animals have revealed severe injuries in bile duct cells. In the early stages of sporidcsmin intoxication in sheep, edematous, inflammatory changes center around the bile ducts. The severity of the bile duct lesions is proportional to the size of the bile duct. Histopathological examinations of the liver in cyclopiazonic acid-treated rats have revealed swollen bile ductule cells.[123] Such swollen bile ductule cells occur at higher doses. In the sporidesmin intoxication of sheep,[125] some of the larger intrahepatic bile ducts are edematous. Many intrahepatic ducts are completely obstructed with plugs, and the liver is stained with bile.

Fatty infiltration in the liver is the most prominent and consistent pathological feature when toxic exogenous chemical substances are ingested by animals. The predominant hepatic lesion in acute lupinosis in sheep,[115] cattle and horses is extremely severe fat degeneration of the liver.[114] Severe fatty infiltration has also been observed in rats treated with rubratoxin B.[111] The principal change in the parenchymal cells of the liver induced by aflatoxin B_1[23,105] or sporidesmin[125] is the appearance of large quantities of lipid in the cytoplasm of the liver cells surrounding necrotic sites in the periportal areas. The main damage found in the monkey liver during aflatoxicosis is fatty degeneration.[129] In severely affected animals, the change is diffuse but more intense in the portal areas. A similar increase in fat is also demonstrated in the heart and kidney. There is a striking similarity between this reaction in macagues and human Rey's syndrome.[129]

The gall bladder is the primary organ affected by sporidesmin in sheep.[125] Marked edemas and multiple petechial hemorrhages are not infrequently seen on the serosal surface and in the submucosa of the organ. Similar pathological changes are induced by aflatoxin B_1 in the dog and pig.[105] Rubratoxin B also produces marked edemas of the serosal surface of the gall bladder.[110]

4.2.2. Injuries to the Digestive Tract

The digestive tract is one of the main target organs of 12, 13-epoxytrichothecenes, which may be the causative agents of stachybotryotoxicosis in domestic animals and human ATA.[128] The first sign of stachybotryotoxicosis in various animals is a marked inflammatory process in the mouth and gastrointestinal tract.[160] Hemorrhage and necrosis are observed all along the digestive tract and in the mucous membrane. Saito and Ohtsubo[161] have demonstrated the toxic effects of some 12, 13-epoxytrichothecenes such as nivalenol, fusarenon and fusarenon-X in the mouse, and reported that marked injuries occurred in the digestive organs. The grade of sensitivity may be summarized as follows: ileum $=$ jejunum $>$ colon $\gg$ duodenum $>$ stomach. Histopathological examinations revealed massive necrosis of the crypt cells of the intestine.[127] Similar lesions are also observed in experimental animals after administration of T-2 toxin, aflatoxin B_1[105,139] and rubratoxin B,[110] and in birds fed with ochratoxin A.[112] The direct contact of these mycotoxins with the mucosal cells may play an important role in the toxicity.

Severe ulcerative enterogastritis has also been observed in sheep after dosing with sporidesmin.[125] In contrast to the trichothecenes, the effect of sporidesmin is not a direct one. The histopathological changes induced by this mycotoxin in the small intestine are characterized by a fibrinoid necrosis of the arterioles and veins associated with the lesions.[125] Such necrotizing vessels are accompanied by marked necrosis or ulceration of the overlaying tissues and hemorrhage. Mice dying 3 days after receiving fusarenon-X at the LD_{50} level showed a marked exudation between the villous epithelial cells of the duodenum[127] (Fig. 4.16). As seen with the electron microscope, the exudation occurs within the cytoplasm, and the cytoplasmic membranes are ruptured in some places. The interdigitations between neighboring cells persist in the exudation (Fig. 4.17).

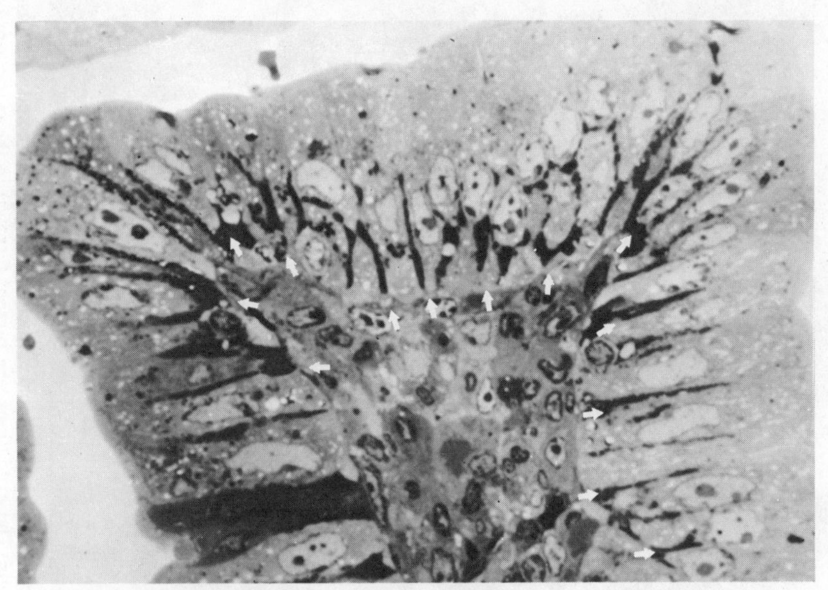

Fig. 4.16. Section of the duodenal epithelium of a mouse treated with fusarenon-X (Epon embedded semithin section; toluidine blue staining). Exudation is prominent in the infranuclear region of the epithelial cells (arrows).

4.2.3. Injuries to the Urinary System

The kidney is not infrequently affected by various mycotoxins. Macroscopically the affected kidneys are always enlarged and pale in color. Nephrotoxic mycotoxins can be divided into two groups depending on the target tissues. The first group includes mycotoxins which act preferentially on the nephron, and the second comprises those which act on the collect-

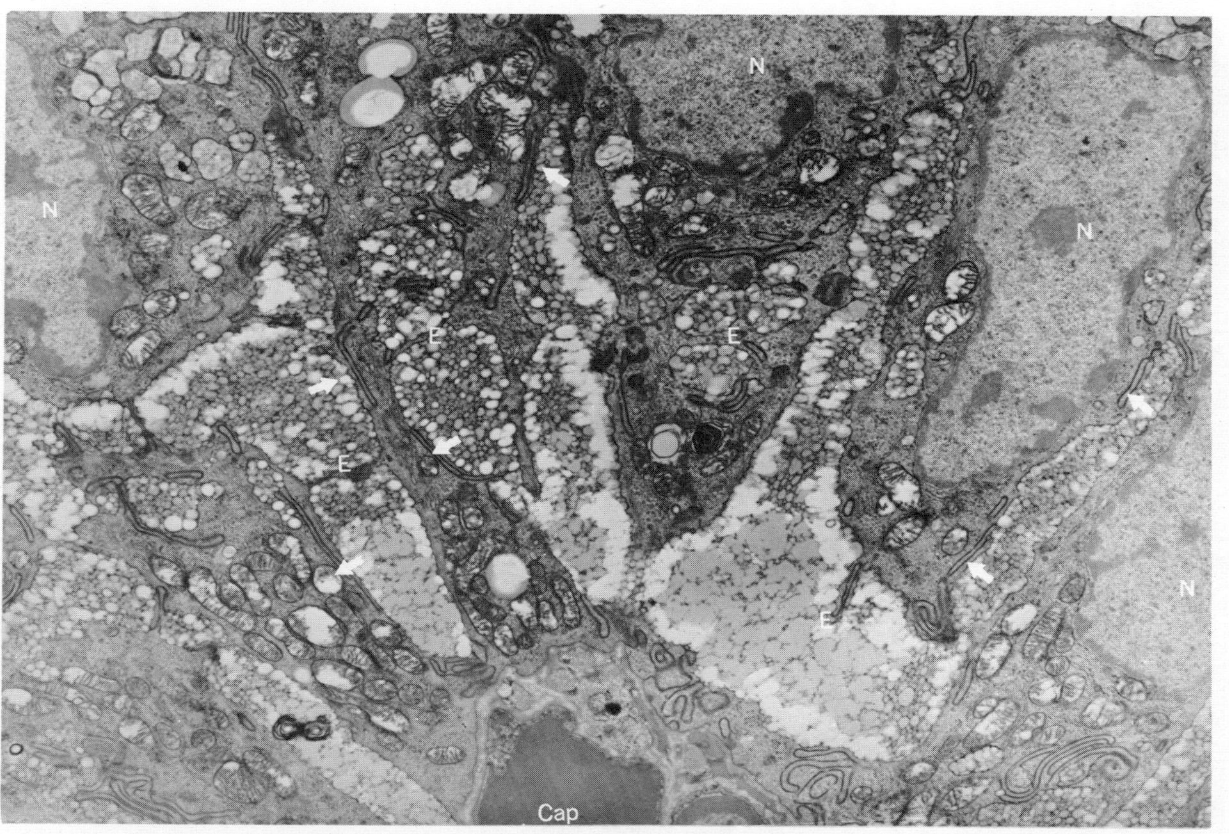

Fig. 4.17. Electron micrograph of the duodenal epithelium of a mouse treated with fusarenon-X. Marked exudations (E) are present between the cells. The exudations are present (arrows) in the cytoplasm of the epithelial cells, although the interdigitations persist. Mitochondria are severe degenerated. Cap, Capillary.

ing tubules. The clinical signs and histopathological changes of the lesions induced by the former mycotoxins are those of nephrosis. The clinical signs of the intoxication induced by the latter are not prominent and consistent. Usually the histopathological changes induced by the latter mycotoxins spread to the mucous membrane and submucous layers of the ureters and urinary bladder.

Animals receiving ochratoxin A,[58,111,132] citrinin,[134] aflatoxin B_1,[105] aflatoxin G_1,[130] sterigmatocystin,[108] and cyclopiazonic acid[123] show typical pathological changes of the nephron (Fig 4.18,4.19). Marked by enlarged nuclei containing marginal chromatin, and necrosis are often seen in the proximal and/or distal tubules in the rat and sheep. In contrast with the urinary tubules, the glomerulus is usually affected only slightly. Dogs seem to be more susceptible to ochratoxin A than rats.[133] Pigs intoxicated with citrinin display degeneration of the proximal tubules, especially in the last two thirds of the convoluted portions.[162] Similar but milder renal damage is observed in rubratoxicosis in various animals, but the finding is not consistent within a species or even within any one group.[111]

A degenerative process in the epithelial cells of the collecting tubules and papillary ducts is occasionally seen in sheep treated with sporidesmin.[125] Such degenerative changes sometimes spread to the pelvic transi-

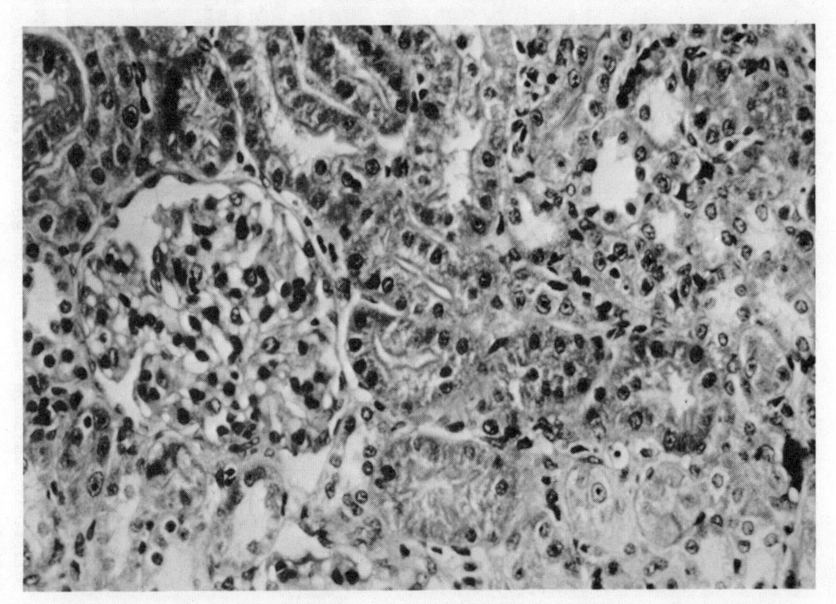

Fig. 4.18. Section of the kidney of a rat dying after application of ochratoxin A. Degeneration of proximal convoluted tubules are prominent (hematoxylin and eosin). (By courtesy of Dr M. Kanisawa).

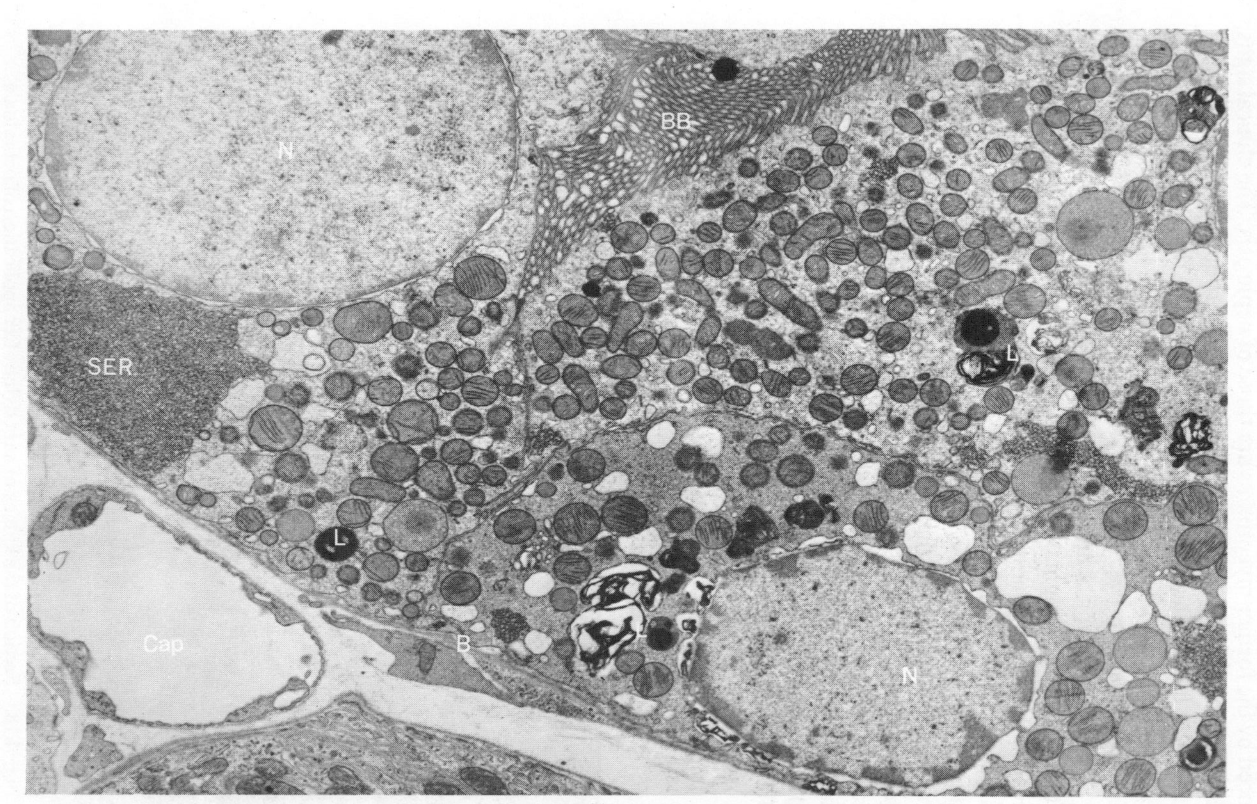

Fig. 4.19. Electron micrograph of the proximal convoluted tubules in the kidney of a rat treated with ochratoxin A. Marked degeneration with multiple lipids (L) and proliferation of SER is noted. BB, Brush border; N, nucleus; Cap, capillary.

tional epithelium, ureters and urinary bladder. In severe instances, muscular layers are separated by edemas and some of the subepithelial capillary loops are ruptured.

Diuresis is produced by the administration of penitrem A in rats and mice, since this mycotoxin prevents kidney tubular reabsorption.[131] In contrast, patulin shows a marked antidiuretic effect in rats.[113]

4.2.4. Injuries to the Hematopoietic System

It is now widely accepted that toxic 12, 13-epoxytrichothecenes are the causative agents of human ATA.[128] The most severe effects are on the hematopoietic system and result in depression of leucopoiesis, erythropoiesis, and thrombopoiesis. However, this depression is temporary and the disturbance is reversible; there is no destruction of the bone marrow.[163] It is also well known that mouse hematopoietic organs are very susceptible to toxic trichothecenes. Ueno *et al.*[126] reported that fusarenon-X causes necrosis of the thymus, and of cells in the immune system, such as undifferentiated lymphoid cells of the follicles in the spleen, lymph nodes and other parts of the lymph apparatus. Depression and lymphocytic destruction of lymph nodes are characteristic of acute and chronic aflatoxin B_1 toxicosis in dogs.[139] Cats treated with rubratoxin B have been observed to develop hemorrhagic and congestive lesions of the lymph nodes.[110] Suppression of lymphoid elements from the spleen and bursae of Fabricius has also been described in chickens and hens after the application of ochratoxin A.[112] An enlargement of the bursae of Fabricius[141] was also noted in birds treated with zearalenone. The increase in bursal weight appeared to be related to the presence of cyst development. Hematopoietic cells of the bone marrow are injured by 12,13-epoxytrichothecenes[126] and ochratoxin A.[112] Depletion of lymphocytes and necrosis are commonly found in the spleen of rubratoxin B-treated animals.[110] Cyclopiazonic acid also causes necrosis of lymphocytes in the white pulp of sheep spleen.[123]

4.2.5. Dermal Injuries Induced by Mycotoxins

All trichothecenes so far studied are capable of inducing dermal reactions consisting of severe local irritation, inflammation, and desquamation when applied to the skin of experimental animals. This irritating effect on the skin has been employed for bioassay of these toxic mycotoxins. Most of the trichothecenes studied are cytotoxic[164–7] when tested against various plant and animal tissue culture systems. It is interesting to note that there is a good correlation between dermal toxicity and cytotoxicity among the various trichothecenes. Therefore, their dermal toxicity can be attributed to direct effects on the epidermal cells. Intense edemas were observed within 48 h in patch tests carried out with patulin on epidermal as well as dermal tissue of man[143] and the rabbit.[144] A keroid later formed at the site of reaction.

4.2.6. Injuries to the Endocrine System

There are relatively few reports concerning the effects of mycotoxins on the endocrine system. Hemorrhage and necrosis of the zona reticularis and zona fasciculata of the adrenal bodies are frequently observed after administration of aflatoxin G_1 in rats[130] and guinea-pigs.[168] Aside from hemorrhage, hypertrophy is noted following the feeding of sporidesmin to sheep.[125] In 75% of the animals, degenerative cells were found in the adrenal zona reticularis, and polymorphonuclear phagocytes were active in most.

The islets of Langerhans in rats treated with cyclopiazonic acid[123] showed distinct degenerative changes and necrosis, particularly of cells in the periphery of the islets. There was an absence of granules in the islets, suggesting that the coma preceding death in orally dosed animals was a result of hypoglycemia.[123]

A marked reduction in aflatoxin B_1 carcinogenesis has been reported in hypophysectomized rats.[169] It is possible therefore that the hypophysis plays an important role in the metabolism of aflatoxin to some substance which is then an effective carcinogen for the liver.[169]

4.2.7. Effects on the Reproductive Organs

Estrogenism in swine has been associated with the ingestion of moldy corn. The fungus, *Fusarium graminearum,* and other related organisms growing on corn produce the potent estrogenic substance, zearalenone. This substance causes precocious sexual development in young female swine and inhibits the normal development of the testes in males.[145] According to Nelson *et al.,*[146] enlarged vulvas, mammae and nipples, and prolapse of the vagina are noted in swine within 5–7 days after feeding zearalenone at a rate of 5 mg/day. Microscopically, distinct edemas and cellular proliferation appear in all layers of the uterus. In the mammary glands, interstitial edemas and ductular proliferation are observed. Metaplasia of the mucosal epithelium to a stratified squamous epithelium is always seen in the cervix and vaginal region.[145] Suppression of the development of corpora lutea in the ovary may contribute to infertility. Atrophy of the testes is encountered in male swine after treatment with zealarenone.[170] Partial atrophy of the seminiferous tubules[126] and destruction of granulosa cells of the ovary are occasionally noted in fusarenon X-treated mice.

4.2.8. Injuries to the Nervous System

Several neurogenic disorders or symptoms in domestic animals are known or suspected to be of mycotoxic origin. The clinical signs of the intoxications are similar: tremors, hyperexcitability, and ataxia consistently appear in all intoxications, but gross visible lesions are usually not associated with them.

Aflatoxin B_1[129] causes marked cerebral edemas in the macaque monkey. Light microscopy has revealed that the perivascular space is dilated and that swelling and separation of the myelin sheaths occurs. Shrinkage of cytoplasm, piknosis and distortion of ganglion nuclei are seen. Aside from encephalopathy, generalized fatty degeneration of the viscera is characteristic of aflatoxicosis in the macaque. The similarity between such aflatoxin intoxication in the macaque and Reye's syndrome in young children from tropic areas has been well documented by Bourgeois *et al.*[129]

Citreoviridin is one of the classical neurotoxic mycotoxins.[134] After administration of citreoviridin, several mammalian species show similar clinical symptoms. These include progressive paralysis in the extremities, ataxic posture, cardiovascular disturbance, and dyspnea. Animals treated with penitrem A,[131] ergots[171] and fumitremorgen A[149,150] also show typical neurological signs such as generalized irritability, ataxia and sustained tremors. However, no reports of histopathological findings are yet available.

4.2.9. Miscellaneous Effects

Fatty infiltration of heart muscle and sub-endocardial hemorrhage are frequently observed in monkeys, cats, and dogs after application of aflatoxin B_1,[105] rubratoxin B[110] and lupinosis toxin. Recently, Ohtsubo *et al.* reported that xanthoascin (toxin B) of *Aspergillus candidus* produces severe injury to the cardiac muscle, especially in the nuclei. This injury leads to severe dilatation of the right and left ventricles.

Small petechial hemorrhages in the lung are observed after the administration of aflatoxin B_1 to guinea-pigs.[110] Pulmonary congestion is occasionally seen in sporidesmin-,[153] patulin-,[113,151] and trichothecene-intoxicated animals.

Some mycotoxins cause increased permeability in the capillaries of the serosal surface. As a result of such increased capillary permeability, ascites and pleural effusions were found after sporidesmin intoxication in mice and rats.[153)] Aflatoxin B_1-intoxicated dogs,[139)] rubratoxin B poisoned cats[110)] and guinea-pigs,[110)] and sheep with lupinosis[140)] have also been reported to show ascites and pleural effusions.

REFERENCES

1. D. Svoboda, H. J. Grady and J. Higginson, *Am. J. Path.*, **49**, 1023 (1966).
2. W. Bernhard, *National Cancer Inst. Monograph,* **23**, 13 (1966).
3. W. H. Butler, *Am. J. Path.*, **49**, 113 (1966).
4. D. Svoboda and J. Higginson, *Cancer Res.*, **28**, 1703 (1968).
5. T. Unuma, H. P. Morris and H. Busch, *ibid.*, **27** (part 1), 2121 (1967).
6. L. R. Floyd, T. Unuma and H. Busch, *Exptl. Cell Res.*, **51**, 423 (1968).
7. R. S. Pong and G. N. Wogan, *Cancer Res.*, **30**, 294 (1970).
8. G. S. Edwards, G. N. Wogan, B. Sporn and R. S. Pong, *ibid.*, **31**, 1943 (1971).
9. R. S. Pong and G. N. Wogan, *J. Natl. Cancer Inst.*, **47**, 585 (1971).
10. J. C. Engelbrecht, *S. African Med. J.*, **44**, 153 (1970).
11. J. C. Engelbrecht and B. Altenkirk, *J. Natl. Cancer Inst.*, **48**, 1647 (1972).
12. K. Terao, M. Takano and M. Yamazaki, *Chem. Biol. Interactions,* **11**, 507 (1975).
13. K. Terao, *Trans. Soc. Path. Japan,* **60**, 19 (1971).
14. W. H. Butler, *Chem. Biol. Interactions,* **4**, 49 (1971).
15. W. H. Butler and G. E. Neal, *Cancer Res.*, **33**, 2878 (1973).
16. J. J. Lin, C. Liu and D. Svoboda, *Lab. Invest.*, **30**, 267 (1974).
17. K. Terao, Y. Sakakibara, M. Yamazaki and K. Miyaki, *Exptl. Cell Res.*, **66**, 81 (1971).
18. H. Busch and K. Smetana, *The Nucleolus*, Academic Press (1970).
19. R. Simard, *Int. Rev. Cytol.*, **28**, 169 (1970).
20. A. Monneron, C. Lafarge and M. C. Frayssinet, *Compt. Rend.*, **267**, 2053 (1968).
21. T. Yokoyama, *Virchows Arch. Abt. B. Zellpath.*, **11**, 133 (1972).
22. D. J. Svoboda, J. K. Reddy and C. Liu, *Arch. Path.*, **91**, 452 (1971).
23. W. H. Butler, *Brit. J. Cancer,* **18**, 756 (1964).
24. M. Legator, *Bact. Rev.*, **30**, 471 (1966).
25. D. O. Schachtschabel, F. Zilliken, M. Saito and G. E. Foley, *Exptl. Cell Res.*, **57**, 19 (1969).
26. M. Umeda, *Japan. J. Exptl. Med.*, **41**, 195 (1971).
27. K. Terao and A. Gropp, *Exptl. Cell Res.*, **40**, 686 (1966).
28. D. M. Phillips and S. G. Phillips, *J. Cell Biol.*, **49**, 803 (1971).
29. M. R. Gardiner, *J. Path. Bact.*, **94**, 452 (1967).
30. M. Umeda, A. Saito and M. Saito, *Japan. J. Exptl. Med.*, **40**, 400 (1970).
31. H. von Stähelin, M. E. Kalberer-Püsch, E. Signer and S. Lazáry, *Arzneim. Forsch.*, **18**, 265 (1968).
32. L. J. Lilly, *Nature,* **207**, 434 (1965).
33. J. Reiss, *Experientia*, **27**, 971 (1971).
34. S. Green, *Mammalian Chromosomes News Lett.*, **8**, 36 (1967).
35. D. A. Dolimpio, C. Jacobsen and M. Legator, *Proc. Soc. Exptl. Biol. Med.*, **127**, 559 (1968).
36. R. F. J. Withers, cited from ref. 113.
37. P. Sentein, *Compth. Rend.*, **149**, 1621 (1955).
38. M. Umeda, T. Yamamoto and M. Saito, *Japan. J. Exptl. Med.*, **42**, 527 (1972).

39. Y. Ueno, A. Platel and P. Fromageot, *Biochim. Biophys. Acta,* **134**, 27 (1967).
40. Y. Ueno, I. Ueno, K. Mizumoto and T. Tatsuno, *J. Biochem.*, **63**, 395 (1968).
41. M. B. Sporn, C. W. Dingman, H. L. Phelps and G. N. Wogan, *Science,* **151**, 1539 (1966).
42. J. I. Clifford and K. R. Rees, *Biochem. Pharmacol.*, **18**, 2783 (1969).
43. A. M. Q. King and B. H. Nicholson, *Biochem J.*, **114**, 679 (1969).
44. W. H. Butler, *Progress in Liver Disease* (ed. H. Popper and F. Schaffner), vol. 3, p. 408, Grune and Stratton (1970).
45. K. Terao, M. Yamazaki and K. Miyaki, *Z. Krebsforsch.*, **78**, 303 (1972).
46. Y. Ueno, I. Ueno, K. Ito and T. Tatsuno, *Experientia,* **23**, 1001 (1967).
47. F. Tashiro, N. Hirai and Y. Ueno, 96th Ann. Mtg. Japan. Pharm. Soc. (Japanese), **5S**, 10-2 (1976)
48. Y. Ueno, I. Ueno and T. Tatsuno, *Seikagaku* (Japanese), **38**, 687 (1966).
49. A. Ruet, A. Sentenac, E. J. Simon, J. C. Bouhet and P. Fromageot, *Biochemistry*, **12**, 2318 (1973).
50. R. S. Pons and G. N. Wogan, *Cancer Res.*, **30**, 294 (1970).
51. F. Tashiro and Y. Ueno, *unpublished data.*
52. R. E. Mouton and P. Fromageot, *FEBS Lett.*, **15**, 45 (1971).
53. J. C. Seegers and M. J. Pitout, *Suid-Afr. Med. Tydshrif,* **23**, 961 (1973).
54. H. L. Gurtoo and T. C. Campbell, *Biochem. Pharmacol.,* **19**, 1729 (1970).
55. G. S. Edwards and V. G. Allfrey, *Biochem .Biophys. Acta,* **299**, 354 (1973).
56. M. Saito, *Acta Path. Japon.*, **9**, 785 (1959).
57. B. Takagi, *Electron Microscop.*, **8**, 154 (1959).
58. J. J. Theron, K. J. van der Merwe, N. Liebenberg, H. J. B. Joubert and W. J. Nel, *J. Path. Bect.*, **91**, 521 (1966).
59. M. Enomoto and I. Ueno, *Mycotoxins* (ed. I. F. H. Purchase), p. 303, Elsevier (1974).
60. I. Ueno, Y. Ueno, T. Tatsuno and K. Uraguchi, *Japan. J. Exptl. Med.,* **34**, 135 (1964).
61. Y. Ueno, I Ueno, T. Tatsuno and K. Uraguchi, *ibid.*, **34**, 197 (1964).
62. G. S. Christie and J. D. Judah, *Proc. Roy. Soc.* (B), **142**, 241 (1954).
63. W. P. Doherty and T. C. Campbell, *Chem. Biol. Interactions,* **7**, 63 (1973).
64. K. Terao, *unpublished data.*
65. I. Ueno, *unpublished data.*
66. S. D. Aust, *Mycotoxins* (ed. I. F. H. Purchase), p. 106, Elsevier (1974).
67. N. Sato and Y. Ueno, *unpublished data.*
68. A. K. Roy, *Biochim. Biophys. Acta,* **169**, 206 (1968).
69. A. Sarasin and Y. Moule, *FEBS Lett.*, **29**, 329 (1973).
70. D. J. Williams, R. P. Clark and B. R. Rubin, *Brit. J. Cancer,* **27**, 283 (1973).
71. K. Robert and W. Larry, *Cancer Res.*, **34**, 3421 (1974).
72. Y. Ueno, M. Hosoya, Y. Morita, I. Ueno and T. Tatsuno, *J. Biochem.*, **64**, 479 (1968)
73. Y. Ueno, M. Hosoya and Y. Ishikawa, *ibid.*, **66**, 419 (1969).
74. M. E. Stafford and C. S. McLaughlin, *J. Cell. Physiol.*, **82**, 121 (1973).
75. E. Cundriffe, M. Cannon and J. Davies, *Proc. Natl. Acad. Sci. U.S.A.*, **71**, 30 (1974).
76. C. M. Wei and C. S. McLalghlin, *Biochem. Biophys. Res. Commun.*, **57**, 838 (1974).
77. Y. Ueno, M. Nakajima, K. Sakai, K. Ishii, N. Sato and N. Shimada, *J. Biochem.*, **74**, 285 (1973).
78. Y. Ueno and H. Matsumoto, Proc. 1st Intersect. Congr. IAMS (Tokyo), **4**, 314 (1975).
79. H. T. Tung, W. E., Donaldson and P. B. Hamilton, *Biochim. Biophys. Acta,* **222**, 665 (1970).
80. A. A. Pokrovsky, L. V. Kravchenko and V. A. Tutelyan, *Toxicon.*, **10**, 25 (1972).
81. J. C. Schabort and M. J. Pitout, *Enzymologia*, **41**, 201 (1971).
82. M. J. Pitout, H. A. McGee and J. C. Schabort, *Chem. Biol. Interactions*, **3**, 353 (1971)
83. Y. Ueno, H. Matsumoto, K. Ishii and K. Kukita, *Biochem. Pharmacol.,* **25**, 2091 (1976).

84. S. B. Carter, *Nature,* **213**, 261 (1967).
85. R. D. Estensen and P. G. W. Plageman, *Proc. Natl. Acad. Sci. U.S.A.,* **69**, 1430 (1972).
86. M. J. Sweeney, K. Gerzon, P. N. Harris, R. E. Holmes, G. A. Poore and R. H. Williams, *Cancer Res.,* **32**, 1795 (1972).
87. J. C. Cline, J. D. Nelson, K. Gerzon, R. H. Williams and D. C. DeLong, *Appl. Microbiol.,* **18**, 14 (1969).
88. M. J. Sweeney, D. H. Hoffman and M. A. Esterman, *Cancer Res.,* **32**, 1803 (1972).
89. Y. Ueno and H. Matsumoto, *Chem. Pharm. Bull.,* **23**, 2439 (1975).
90. F. S. Chu, *Arch. Biochem. Biophys.,* **147**, 359 (1971).
91. S. Juhász and E. Gréczi, *Nature,* **203**, 861 (1964).
92. M. R. Daniel, *Brit. J. Exptl. Path.,* **46**, 183 (1965).
93. J. Gablichs, W. Schaeffer, L. Friedman and G. N. Wogan, *J. Bact.,* **90**, 720 (1965).
94. *K. Terao, Exptl. Cell Res.,* **48**, 151 (1967).
95. A. J. Zucherman, K. R. Rees, D. R. Inman and I. A. Robb, *Brit. J. Exptl. Path.,* **49**, 33 (1968).
96. K. Terao and K. Miyaki, *Z. Krebsforsch.,* **71**, 199 (1968).
97. E. H. Harley and K. R. Rees, *Biochem. J.,* **114**, 289 (1969).
98. M. Umeda, *Acta Path. Japon.,* **14**, 373 (1964).
99. M. Umeda, M. Saito and S. Shibata, *Japan. J. Exptl. Med.,* **44**, 249 (1974).
100. M. Umeda, *ibid.,* **41**, 195 (1971).
101. K. Ohtsubo and M. Saito, *Japan. J. Med. Sci. Biol.,* **23**, 217 (1970).
102. K. Ohtsubo, M. Yamamoto and M. Saito, *ibid.,* **21**, 185 (1968).
103. S. Natori, S. Sakaki, H. Kurata, S. Udagawa, M. Ichinoe, M. Saito and M. Umeda, *Chem. Pharm. Bull.* **18**, 2259 (1970).
104. I. Kawasaki, T. Oki, M. Umeda and M. Saito, *Japan J. Exptl. Med.,* **42**, 327 (1972).
105. P. M. Newberne and W. H. Butler, *Cancer Res.,* **29**, 236 (1969).
106. I. F. H. Purchase and J. J. van der Watt, *Fd. Cosmet. Toxicol.,* **7**, 135 (1965).
107. J. J. van der Watt and I. F. H. Purchase, *S. African Med. J.,* **44**, 159 (1970).
108. J. J. van der Watt and I. F. H. Purchase, *Brit. J. Exptl. Path.,* **51**, 183 (1970).
109. M. Saito, M. Enomoto and T. Tatsuno, *Microbial Toxins* (ed. A. Ciegler, S. Kadis and S. J. Ajl), vol. VI, p. 299, Academic Press (1971).
110. G. N. Wogan, G. S. Edwards and P. M. Newberne, *Appl. Pharmacol.,* **19**, 712 (1971)
111. I. C. Murno, P. M. Scott, C. A. Moodie and R. F. Willes, *J. Am. Vet. Med. Ass.,* **163**, 1269 (1973).
112. J. C. Peckham, B. Doupnik, Jr. and O. H. Jones, Jr., *Appl. Microbiol.,* **21**, 492 (1971).
113. W. A. Broom, E. Büllring, C. J. Chapman, J. W. F. Hampton, A. M. Thomson, J. Unger, R. Wien and G. Woolfe, *Brit. J. Exptl. Path.,* **25**, 195 (1944).
114. M. R. Gardiner, *Advan. Vet. Sci.,* **11**, 85 (1967).
115. K. T. van Warmelo, W. F. O. Marasas, T. F. Adelaar, T. S. Kellerman, I. B. J. van Rensburg and J. A. Minne, *Mycotoxins in Human Health* (ed. I. F. H. Purchase), p. 185, Macmillan (1971).
116. M. R. Gardiner, *J. Comp. Path. Therap.,* **77**, 63 (1967).
117. M. R. Gardiner and H. D. Seddon, *Australian Vet. J.,* **42**, 242 (1966).
118. J. Dobberstein and W. Walkiewicz, *Arch. Path. Anat.,* **291**, 695 (1933).
119. M. R. Gardiner and D. S. Petterson, *J. Comp. Path. Therap.,* **82**, 5 (1972).
120. E. B. Smalley, W. F. O. Marasas, F. M. Strong, J. R. Bamburg, R. E. Nichols and N. R. Kosuri, Proc. 1st US-Japan Conf. Toxic Microorganisns (ed. M. Herzberg), *Unnumb. Pub. U.S. Dept. of the Interior and UJNR Panels on Toxic Microoganisms, Washington, D. C.*, p. 163 (1970).
121. N. R. Kosuri, E. B. Smalley and R. E. Nichols, *Am. J. Vet. Res.,* **32**, 1843 (1971).
122. P. H. Mortimer, J. Campbell, M. E. Di Menna and E. P. White, *Res. Vet. Sci.,* **12**, 508 (1971).

123. I. F. H. Purchase, *Toxicol. Appl. Pharmacol.*, **18**, 114 (1971).
124. P. H. Mortimer, *New Zealand J. Agr. Res.*, **13**, 437 (1970).
125. P. H. Montimer, *Res. Vet. Sci.*, **4**, 166 (1963).
126. Y. Ueno, I Ueno, Y. Iitoi, H. Tsunoda, M. Enomoto and K. Ohtsubo, *Japan J. Exptl. Med.*, **41**, 521 (1971).
127. M. Saito, M. Enomoto and T. Tatsuno, *Gann*, **60**, 599 (1969).
128. Y. Ueno, *J. Fd. Hyg. Soc. Japan* (Japanese), **14**, 403 (1973).
129. C. H. Bourgeois, R. C. Shank, R. A. Grossman, D. O. Johnsen, W.L. Woodling and P. Chandavimol, *Lab. Invest.*, **24**, 206 (1971).
130. W. H. Butler and W. Lijinsky, *J. Path.*, **102**, 290 (1970).
131. I. F. H. Purchase, *Mycotoxins* (ed. I. F. H. Purchase), p. 158, Elsevier (1974).
132. I. F. H. Purchase and J. J. Theron, *Fd. Cosmet. Toxicol.*, **6**, 479 (1968).
133. G. M. Szezech, W. W. Carlton and J. Tuite, *Lab. Invest.*, **26**, 492 (1972).
134. K. Uraguchi, *Microbial Toxins* (ed. A. Ciegler, S. Kadis and S. J. Ajl), vol. VI, p. 367, Academic Press (1971).
135. P. Krogh, E. Hasselager and P. Friis, *Acta Path. Microbiol. Scand.* (B), **78**, 401 (1970).
136. F. Sakai, *Nihon Yakurigaku Zasshi* (Japanese), **51**, 431 (1955).
137. J. Nagai, M. Hayashi and K. Mizobe, *Fukuoka Igaku Zasshi* (Japanese), **48**, 311 (1957).
138. A. M. Ambrose and F. De Eds, *Proc. Soc. Exptl. Biol. Med.*, **59**, 289 (1945).
139. P. M. Newberne, *J. Am. Vet. Med. Ass.*, **163**, 1262 (1973).
140. M. R. Gardiner, *Path. Vet.*, **2**, 417 (1965).
141. G. M. Speers, R. A. Meronuck, D. M. Barnes and C. J. Mirocha, *Poultry Sci.*, **50**, 627 (1971).
142. M. Saito and K. Ohtsubo, *Mycotoxins* (ed. I. F. H. Purchase), p. 263, Elsevier (1974).
143. J. E. Dalton, *Arch. Derm. Syph.*, **65**, 53 (1952).
144. K. Hofmann, H. J. Mintzlaff, I. Alperden and L. Leistner, *Fleischwirtschaft*, **51**, 1534 (1971).
145. H. J. Kuntz, M. E. Nairn, G. H. Nelson, C. M. Christensen and C. J. Mirocha, *Am. J. Vet. Res.*, **30**, 551 (1969).
146. G. H. Nelson, D. M. Barnes, C. M. Christensen and C. J. Mirocha, Symp. Proc. 70–0 Extension Service, Univ. of Nebraska, p. 90 (1970).
147. C. J. Mirocha, J. Harrison, A. A. Nichols and M. McClintock, *Appl. Microbiol.*, **16**, 797 (1968).
148. J. E. Burnside, W. L. Forgacs, W. T. Carll, W. T. Atwood and E. R. Doll, *Am.J. Vet. Res.*, **18**, 817 (1957).
149. A. Ciegler, *Lloydia*, **38**, 21 (1975).
150. M. Yamazaki, S. Suzuki and K. Miyaki, *Chem. Pharm. Bull.*, **19**, 1739 (1971).
151. W. A. Hopkins, *Lancet*, **245**, 631 (1943).
152. K. Ohtsubo, M. Enomoto, T. Ishiko, M. Saito, F. Sakabe, S. Udagawa and H. Kurata, *Japan. J. Exptl. Med.*, **44**, 477 (1974).
153. T. F. Slater, U. D. Sträuli and B. Sawyer, *Res. Vet. Sci.*, **5**, 450 (1964).
154. C. Pilmington, T. F. Slater, W. G. Spector, U. D. Sträuli and D. A. Willoughby, *Nature*, **194**, 1152 (1962).
155. M. Enomoto and M. Saito, *Ann. Rev. Microbiol.*, **26**, 279 (1972).
156. K. Terao, *Chiba Med. J.*, **50**, 159 (1974).
157. J. J. van der Watt and I. F. H. Purchase, *Mycotoxins in Human Health* (ed. I. F. H. Purchase), p. 209, Macmillan (1971).
158. M. R. Gardiner and W. H. Parr, *J. Comp. Path. Therap.*, **77**, 51 (1972).
159. N. Uchida and Y. Egashira, *Trans. Soc. Path. Japan*, **7**, 463 (1957).
160. J. V. Rodricks and R. M. Eppley, *Mycotoxins* (ed. I. F. H. Purchase), p. 181, Elsevier (1974).
161. M. Saito and K. Okubo, Proc. 1st US-Japan Conf. Toxic Microorganisms (ed. M.

Herzberg), *Unnumb. Publ. U.S. Dept. of the Interior and UJNR Panels on Toxic Microorganisms, Washington, D. C.*, p. 82 (1970).
162. P. Friis, E. Hasselager and P. Krogh, *Acta Path. Microbiol. Scand.*, **77**, 559 (1969).
163. A. Z. Toffe, *Mycotoxins* (ed. I. F. H. Purchase), p. 229, Elsevier (1974).
164. D. Perlman, W. L. Lummis and H. J. Geiersbach, *J. Pharm. Sci.*, **58**, 633 (1969).
165. J. P. Helgeson and G. T. Haberlach, *Pl. Physiol.*, **52**, 660 (1973).

CHAPTER 5

Carcinogenicity of Mycotoxins

Makoto Enomoto

Sagamihara Kyodo Hospital,
Sagamihara-shi, Kanagawa-ken 229, Japan

5.1. Carcinogens Produced by Fungi

Carcinogenic mycotoxins display three fundamental properties, and may accordingly be characterized under the following three categories: (1) carcinogens of microbial origin, (2) chemical carcinogens, and (3) naturally occurring carcinogens.

The recent development of applied microbilogy for the industrial production of fermented foods, amino acids, enzymes and antibiotics is based largely on increased knowledge of bacterial metabolism and heredity. However, microorganisms synthesize biologically active compounds which exhibit not only antimicrobial activity but also toxic and even carcinogenic effects. Toxin-producing fungi were occasionally encountered during exploration for antibiotics but were given little attention. Carcinogenicity to rats and mice was experimentally demonstrated with some antibiotics including actinomycin D, L and S, azaserine, daunomycin, elaiomycin, mitomycin C, streptozotocin, penicillin G, and griseofulvin. However, the production of cancer is usually seen in animals only with parenteral administration of these drugs, except for griseofulvin. The hepatocarcinogenicity of griseofulvin was demonstrated by the oral administration of a daily dosage of 5000–10,000 ppm in mice,[1,2] but no tumors were produced in rats. Considering the high dose needed to produce cancer in mice, there should be no risk of cancer with the regular levels of griseofulvin used for the treatment of dermatophytoses. The important point is that contamination by carcinogenic mycotoxins—biologically active metabolites of molds—commonly occurs in a variety of human foods, especially such staple materials as rice in S.E. Asia and Japan, and peanuts, beans and corn in South Africa and India.

Mycotoxins vary widely in their chemical structures and it is almost impossible to speculate about their biological effects, including carcinogenicity, on the basis of chemical structure alone. However, it is interesting that almost all known mycotoxins showing carcinogenic activity belong to phenols biosynthesized via the acetate-malonate pathway. These are luteoskyrin, rugulosin, aflatoxins, sterigmatocystin and griseofulvin. Extensive research on the animal species or strains susceptible to them, the tissues which serve as targets, and their metabolic fate and biological effects has yielded many interesting results, similar to those observed in experiments with synthetic chemical carcinogens. Table 5.1 shows the animals susceptible to both the carcinogenic and toxic effects, the regular carcinogenic dose in representative animals, and the animal species sensitive only to the toxic effects.[3–9]

Proved carcinogenic mycotoxins such as aflatoxin B_1 and luteoskyrin, are known to show dose-response relationships in their carcinogenic effects. Feeding of aflatoxin B_1 at 40 μg/day to inbred Fischer strain rats for 10 days (total dose, 0.4 mg) induced hepatocellular carcinomas around 80 weeks later (incidence, 16%). A total dose of 3.6–10 mg of aflatoxin B_1 was sufficient to induce hepatocellular carcinomas with a frequency of 100%. Random-bred Porton-Wistar rats fed 5 ppm (around 0.1 mg/day) for 6 weeks, followed by 47 weeks of normal diet, produced hepatocellular carcinomas (total dose, 4.2 mg; incidence, 60%). The frequency of hepatocellular tumors was 100% in rats of the same strain fed a total dose of 6.3 mg.

Administration of sterigmatocystin via a stomach tube or in the diet at doses of 0.15–2.25 mg per rat per day for 52 weeks produced hepatocellular carcinomas in 78% of the animals in 42 weeks. Luteoskyrin is hepa-

TABLE 5.1. Mycotoxins known to produce cancer in animals by oral administration[3–8)]

Carcinogenic mycotoxin	Animal species susceptible to both the carcinogenic and toxic effects (*target organs*)	Regular[†1] dosage (ppm in diet)	Animal species sensitive only to the toxic effects
Aflatoxin B_1	Rat (*liver, kidney, colon*); trout, duck, guppy, ferret, mouse, monkey (*liver*); sheep (*liver, nose*)	0.5—1.5 (rat)	Turkey, mink, cattle, guinea-pig, swine, dog, hamster, rabbit, pheasant, salmon, quail, chicken, cat, frog
Aflatoxin G_1	Rat (*liver, kidney*); duck (*liver*)	1–3 (rat)	
Aflatoxin B_2	Rat, duck (*liver*)		
Aflatoxin M_1	Rat, trout (*liver*)		Duckling
Sterigmatocystin[†2]	Rat, velvet monkey (*liver*)	10–100 (rat)	Mouse
Luteoskyrin	Mouse, rat (*liver*)	30–100 (mouse)	Rabbit, chicken, monkey
Rugulosin	Mouse (*liver*)	200 (mouse)	Rat
Griseofulvin	Mouse (*liver*)	5000–10,000 (mouse)	Rat

[†1]Oral carcinogenic dosage per day in the representative animal shown in parentheses.

[†2]*O*-Acetylsterigmatocystin is more carcinogenic than sterigmatocystin in rats. Hepatocellular carcinomas were produced in 53% of rats fed 10 ppm *O*-acetylsterigmatocystin for 52 weeks.[9)]

tocarcinogenic in mice following oral administration in the diet[10,11)] (Table 5.2).

Adult mice appear to be resistant to the toxic and carcinogenic effects of aflatoxins and sterigmatocystin. However, limited exposure of 4- or 7-day-old mice to intraperitoneally injected aflatoxin B_1 resulted in a high incidence (40–75 and 85–100%, respectively) of liver tumors in male mice by 52 and 82 weeks. Controls of the same strain, $C_{57}BL \times C_3H\ F_1$, developed no tumors at 52 weeks, but 3 of 100 mice bore hepatomas at 82 weeks.[12)] A limited number (1, 3 or 5) of subcutaneous injections of sterigmatocystin within 24 h after birth (0.5, 1 and 5 μg/animal in doses, respectively) to newborn mice resulted in the development of hepatocellular tumors in 15, 27 and 37% of male mice and 0, 0 and 9% of female mice, respectively. Six percent of the control male mice developed hepatocellular tumors; no female mice developed tumors during the 1-yr experimental period.[13)]

TABLE 5.2. Incidence of hepatic lesions in mice fed luteoskyrin (LS) or cyclochlorotine (CC) for up to 2 yr[10,11]

Group code†1	Intake of toxin (μg/day)	No. of mice autopsied/no. in group	Hepatic lesions: Necrosis		Fibrosis	Cirrhosis	Cholangiofibrosis	Nodular hyperplasia	Hepatoma†3		
			Zonal	Scattered					I	II	III
LS											
ddNi/M/R–B	0	10/13	0	0	0	0	0	0	0	0	0
	50	18/19	1	2	0	0	0	0	0	0	0
	150	24/29	0	5	0	0	0	0	5	1	0
	500	29/30	1	19	0	1	3	1	6	2	0
DDD/M	0	0/18	0	0	0	0	0	0	0	0	0
	160†2	26/27	2	8	0	0	1	5	12	4	1
ddNi/M/R	0	9/10	0	0	0	0	0	0	0	0	0
	150	29/29	0	3	0	3	1	1	8	0	0
ddNi/M/B	0	10/10	0	0			0	0	0	0	0
	150	29/30	1	2	0	0	1	1	1	0	0
ddNi/F/R–B	0	7/ 8	0	0	0	0	0	0	0	0	0
	150†2	26/30	1	2	0	0	0	0	2	1	0
CC											
ddNi/M/R–B	0	11/19	0	0	0	0	0	0	0	0	0
	40	18/20	0	0	11	2	0	0	0	0	0
	60	19/20	0	0	10	2	0	0	2	1	0

†1ddNi or DDD strain of mouse; M = male; F = female; R = rice as basal diet; B = barley as basal diet; R–B = mixed (50:50) diet of rice and barley.
†2One spindle cell sarcoma of the liver was present in each of these groups.
†3Hepatoma I, liver cell adenoma; hepatoma II, well-differentiated liver cell carcinoma; hepatoma III, undifferentiated liver cell carcinoma.

The direct and delayed effects of aflatoxin B_1 on rat fetuses have been investigated extensively by Tanaka.[14] No abnormalities in fetus development or number of animals born were observed when less than 1 mg/kg aflatoxin B_1 was injected i.p. during the 38 days of pregnancy. However, administration of 2, 4 and 8 mg/kg aflatoxin B_1 on the 17th and 19th days of gestation resulted in an increase in the incidence of tumor formation in the newborn rats. The observed tumors were hepatocellular tumors (7.2 and 8.3% incidence, respectively; 0% in controls), lung adenomas (11.5 and 12.5%, respectively; 0% in controls), leukemia (5.7 and 4.1%, respectively; 1.2% in controls), and others. One interesting finding was the early death of exposed progeny within 3 months after birth, when aflatoxin B_1 was injected during the 38 days of gestation. The baby rats were born with malformations, including exencephaly and skeletal abnormalities, when the mycotoxin was administered during days 9–14 of gestation (Fig. 5.1).

Among the factors which influence susceptibility to carcinogenic fungal toxins, nutrition and the combined effects of other mycotoxins or carcinogens may play significant roles in the carcinogenesis of fungal toxins. However, with regard to nutritional factors, the results obtained so far are not consistent. Table 5.3 shows enhancing and inhibiting factors that in-

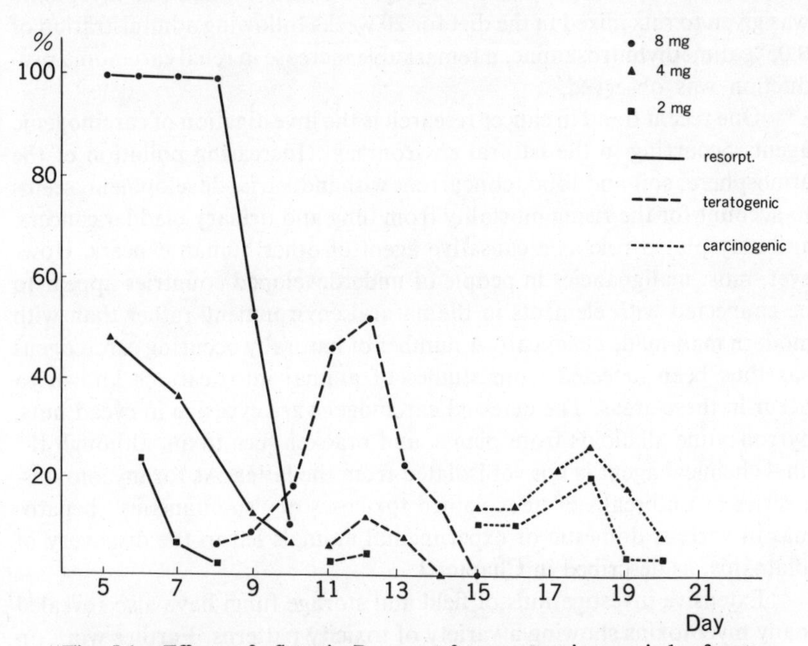

Fig. 5.1. Effects of aflatoxin B_1 on rat fetuses at various periods of gestation. (By courtesy of Dr. T. Tanaka, Laboratory of Cell Biology, Aichi Cancer Center, Nagoya, Japan.)

TABLE 5.3. Factors influencing susceptibility to carcinogenic fungal toxins[15]

Carcinogenic mycotoxins (Animal)	Enhancing factors	Inhibiting factors
Aflatoxin B_1 (Rat)	Low protein diet (9%) Lipotrope deficiency 0.2% Methionine	Low protein diet (5%) Low dietary vitamin A† Phenobarbitone Hypophysectomy Urethane (0.1–0.6%) Diethylstibestrol
Aflatoxins (Trout)	Cyclopropenoid fatty acid	

†Production of colon cancer in rats was observed, although the incidence of hepatocellular carcinomas was less than that in a group of rats fed aflatoxin B_1 in their regular diet.[17]

fluence susceptibility to carcinogenic mycotoxins.[15,17] Lithocholic acid, β-naphthoflavone[16] and cyclopropenoid fatty acids have little, if any, modifying effect on aflatoxin carcinogenesis in rats.

Citrinin, a nephrotoxic mycotoxin, seems to have no carcinogenicity itself. However, as reported recently by Ito *et al.*,[18] when this mycotoxin was given to rats mixed in the diet for 20 weeks following administration of 0.05% dimethylnitrosamine, a remarkable increase in renal carcinoma production was observed.

One recent trend in cancer research is the investigation of carcinogenic agents occurring in the natural environment. Increasing pollution of the atmosphere, soil and food, concurrent with industrial development, seems to account for the rising mortality from lung and urinary bladder cancers, and may play a role as a causative agent of other human cancers. However, most malignancies in people of underdeveloped countries appear to be connected with elements in the natural environment rather than with modern man-made chemicals. A number of naturally occuring carcinogens has thus been detected from studies of animal intoxications known to occur in these areas. The detected carcinogens are cycasine in cycad nuts, pyrrolizidine alkaloids from plants, and bracken fern toxin, although the final chemical agent is not yet isolated from the latter. As for mycotoxins, a series of outbreaks of unexplained toxicoses or "spontaneous" hepatomas in various domestic or experimental animals led to the discovery of aflatoxins, as described in Chapter 1.

Extensive investigations of field and storage fungi have also revealed many mycotoxins showing a variety of toxicity patterns. Further work on their chronic effects will disclose more precise information about the new carcinogenic fungal toxins. In particular, important contributions are expected from research on fungal toxins with specific biological effects, such as the trichothecene compounds from *Fusaria* species which show "radiomimetic" effects on actively dividing cells of animals,[19–21] and any toxins which give a positive reaction in mutagenic and tetratogenic screening tests.

5.2. MYCOTOXICOSIS AND CANCER DEVELOPMENT IN DOMESTIC ANIMALS and MAN

5.2.1. Tumors of Mycotoxic Origin: Pathogenic and Etiologic Considerations

There has been much speculation about the etiology of hepatocellular carcinomas. It has been suggested that the disease was more common in populations consuming foods contaminated by fungal growths. Of major importance in this regard is a group of aflatoxins produced by *Aspergillus flavus*. Field investigations have demonstrated a statistical correspondence between levels of aflatoxin intake and hepatocellular cancer rate in various areas including Thailand, Uganda, Kenya, New Guinea, Malaysia, Japan, Mozambique, and West India.

A Japanese team[22,23] has carried out repeated field studies on the dietary habits of Japanese people in several areas, including the Goto Islands and Saku (Nagano Pref.), during the past 10 yr. The Goto Islands are known to show the highest mortality rate from hepatocellular carcinomas in Japan, but the mortality from hepatocellular carcinomas is not high at Saku. The latter is, however, one of the areas showing a high incidence of gastric cancer in Japan. Toyokawa,[24] an etiologist, has pointed out that the recent disappearance of a specific pattern of dietary habits in Japan renders it difficult to disclose the mycotoxin-cancer association. At the same time, he suggested a possible mosaic pattern of mycotoxin contamination in foods. Diagnostic uncertainties inherent in the medical certification of death also make it difficult to establish causal relationships between human hepatocellular carcinomas and fungal contamination of foodstuffs.

The following three reports provide incriminating evidence that the carcinogenic mycotoxin, aflatoxin B_1, plays a significant role in the etiology of human malignancies.

(1) A joint American-Thai team detected aflatoxin in the tissues of 22

out of 23 Thai children who died of encephalopathy and fatty degeneration of the viscera (EFDV) or Rey's syndrome. The levels of aflatoxin reached 93 μg/kg in the liver, 123 μg/kg in stool, and 127 μg/kg in the stomach of these children.[25] The close resemblance of human EFDV to clinical and pathological findings in acutely intoxicated macaques after ingestion of more than 4.5 mg/kg aflatoxin B_1, has also been emphasized by Bourgeois and his colleagues.[26]

(2) A joint African-Portuguese team found through a program of large-scale sampling of prepared food immediately before consumption, and a concurrent cancer registration program, that aflatoxin-containing foods (9.3% of all samples) resulted in a mean daily per capita consumption of 222.4 ng/kg body wt (Table 5.4). These data represent both the highest known primary liver cancer rate and the highest known aflatoxin intake.[27]

TABLE 5.4. Primary liver cancer incidence rate and aflatoxin intake

Locality	Cancer rate (10^5/yr)	Aflatoxin intake (ng/kg body wt/day)
Kenya-high altitude	0.7	3.5
Thailand-Songkhla	2.0	5.0
Kenya-middle altitude	2.9	5.8
Kenya-low altitude	4.2	10.0
Thailand-Ratburi	6.0	45.0
Mozambique-Inhambane	25.4	222.4†

†This level represents a daily intake of 15 μg per adult (from ref. 27).

(3) Outbreaks of hepatitis affecting both humans and dogs occurred in October, 1974 and January, 1975 in West India. A total of 397 patients in two states (Gujarat and Rajasthan) were affected, and 106 died. A team from the National Institute of Nutrition, Indian Council of Medical Research, observed that the outbreaks began with the consumption of spoiled maize that had been stored under adverse conditions during unseasonal rains, and lasted only until the stock had been consumed. A large number sharing the food of affected households also developed liver injuries and died within 2–3 weeks of onset. The maize was heavily contaminated with *Aspergillus flavus*. Analysis of contaminated samples showed that the affected people could have consumed between 2 and 6 mg aflatoxin daily over a period of one month. Histology of the liver injuries showed bile duct proliferation, cholestatic changes and occasional giant-cell transformation of hepatocytes. Detection of aflatoxin B_1 in two of seven serum samples suggested a causal role for aflatoxin.[28]

5.2.2. Toxic Liver Damage, Liver Cirrhosis, and Primary Hepatocellular Tumors Due to Mycotoxins

In addition to research on the experimental production of cancer with mycotoxins and on the etiological significance of their presence in the human environment, as described in the preceding section, other important work has concerned the histogenesis of hepatocellular carcinomas and a demonstration of the pathognomonic role of mycotoxins in the process of tumor development. The first significant work was done by Japanese researchers on hepatotoxic mold metabolites, and it was shown that changes in the liver, varying from acute damage to cirrhosis and tumors, were produced in mice and rats depending on the amount of metabolites produced by the mold, *Pencillium islandicum.*

It is well known that most cases (60–75%) of human hepatocellular carcinoma observed in areas showing a high incidence of primary liver cancer, are accompanied by cirrhosis, especially of the postnecrotic or posthepatitic type with a multilobular pattern of pseudoacinus. Postnecrotic cirrhosis was produced in rats fed moldy rice at high levels in the diet for a few months (Fig. 5.2). Large scar areas were occasionally observed in the cirrhotic liver. On the other hand, posthepatitic cirrhosis developed after

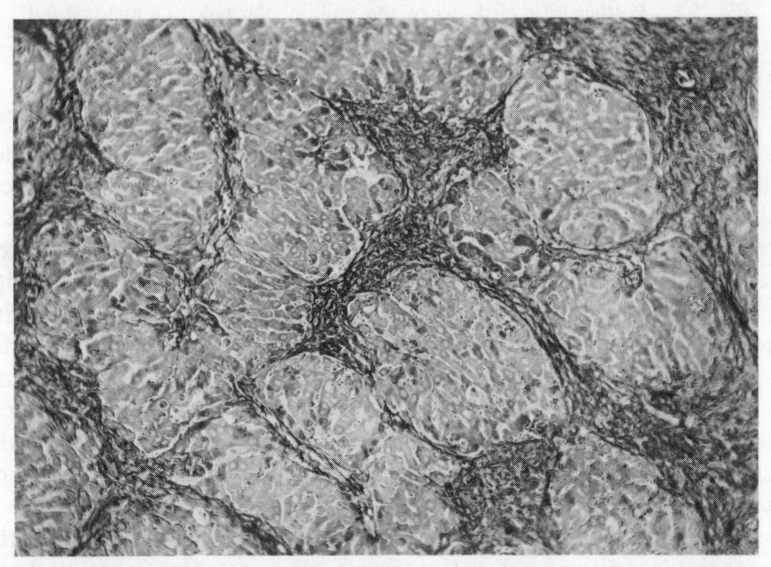

Fig. 5.2. Postnecrotic cirrhosis of the liver in a male rat fed moldy rice (*P. islandicum*) mixed in the diet at a level of 25% for 21 days (Mallory's stain, ×100).

long-term feeding of moldy rice at moderate levels in the diet for over 9 months (Fig. 5.3). Similar cirrhosis was produced in mice fed the same moldy rice, with or without hepatocellular tumors.[7,29–31]

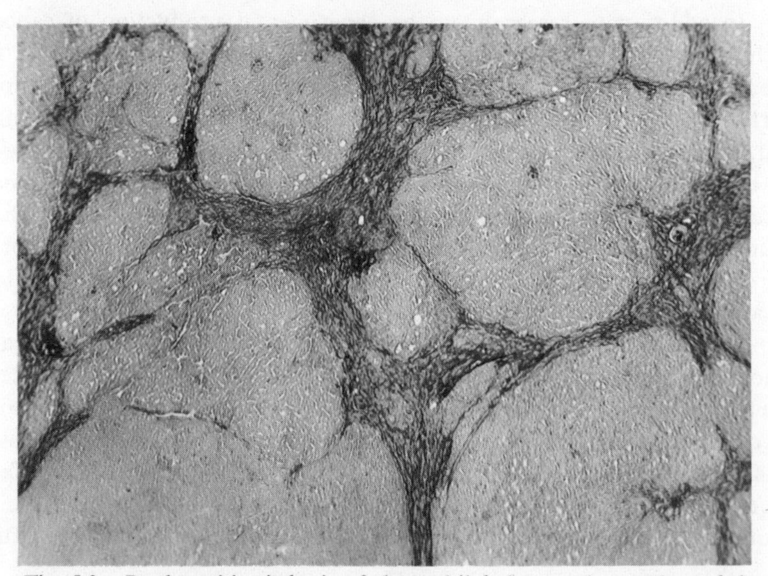

Fig. 5.3. Posthepatitic cirrhosis of the multilobular type in a male rat fed moldy rice of lower toxic potency than in Fig. 5.2 for 311 days (Mallory's tain, ×100).

Results of long-term feeding with two hepatotoxic metabolites obtained from cultures of *Penicillium* indicated that these metabolites, cyclochlorotine and luteoskyrin, have a definite hepatocarcinogenic action in mice, although they are not so potent as aflatoxin B_1. Cyclochlorotine acts as a cirrhogenic agent (Fig. 5.4). This acinus-peripheral cytotoxin produced liver fibrosis at a high frequency, but the mice developed few hepatocellular tumors[10,11](Table 5.2). Chronic liver damage induced in mice fed with luteoskyrin for over 6 months was usually characterized by necrosis, degeneration and nuclear pleomorphism of the liver cells, and relatively scant mesenchymal reaction. The interesting finding is that the morphological alterations caused by this acinus-central toxin moved away from the acinus-central area to the acinus-peripheral region during feeding experiments lasting more than 2–3 months.[32] As shown in Fig. 5.5, single cell necrosis and slight fibrosis with minimum cell infiltration are seen along the portal triads. In some cases, luteoskyrin alone can also cause cirrhotic changes with increased fibrosis, bile duct proliferation, and cell reaction as shown in Fig. 5.6. However, cirrhosis of the complete type was rarely

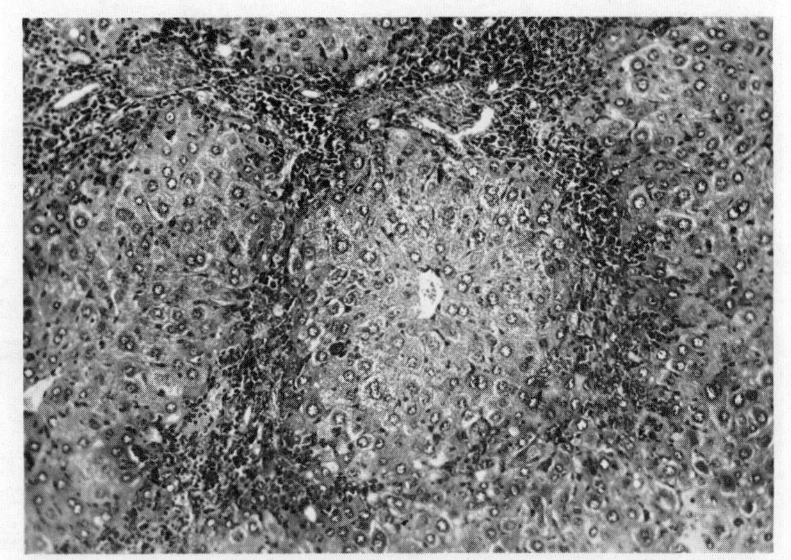

Fig. 5.4. Liver from a male mouse sacrificed after repeated administration of cyclochlorotine (20 μg/10 g, p.o.) for 72 days (H.E., ×120). Cirrhotic nodule formation with a hepatic vein in the center is seen.

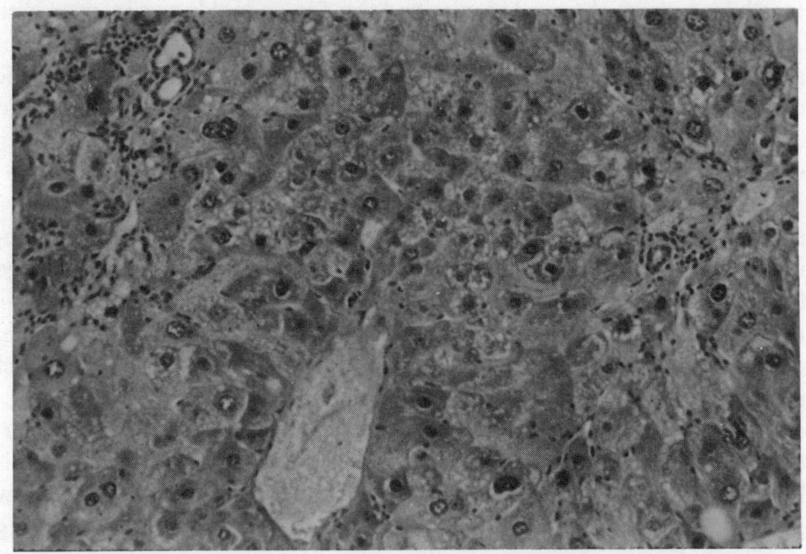

Fig. 5.5. Liver from a male mouse fed 160 μg luteoskyrin per day for 191 days (H.E., ×100). Slight fibrosis with cell infiltration is seen along the portal triads. There is no necrosis in the centrolobular area (center).

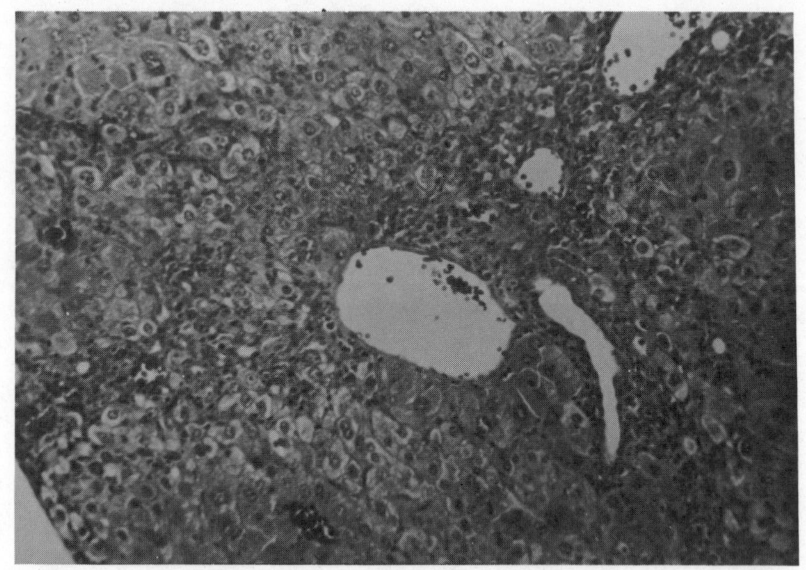

Fig. 5.6. Liver from a male mouse fed 150 μg luteoskyrin per day for 216 days, showing cirrhotic change (×100).

produced by this mycotoxin alone. The cirrhosis development with hepatocellular tumors observed in mice and rats fed on moldy rice infested with *P. islandicum* is thus attributable to the combined effects of these two hepatotoxic metabolites. This is also confirmed by the accentuated hepatic changes induced in mice after combined administration of luteoskyrin and cyclochlorotine, as compared with those in mice treated with each mycotoxin separately[33](Fig. 5.7).

There is no evidence that aflatoxins and sterigmatocystin alone induce significant cirrhosis of the liver. Torres *et al.*[34] in their study of mycotoxin-induced hepatic carcinomas, characterized the toxic non-viral liver damage found in Bantu. The findings of their biopsy series included focal cellular necrosis, ballooning degeneration and non-staining of the cytoplasm, nuclear plemorphism, the absence of inflammation, and limited bile duct proliferation with no cholestasis.

Possible factors that may induce cirrhosis in association with hepatocellular carcinomas of aflatoxin or sterigmatocystin origin, are as follows: (1) nutritional factors such as prolonged lipotrope deficiency and low protein diet, and (2) the combined effects of hepatotoxic agents including monocrotaline and lasiocarpine. These have been experimentally demonstrated.[15] In addition, (3) species differences, (4) viral hepatitis, and (5) other hepatotoxic fungal metabolites that may be related to cirrhosis induc-

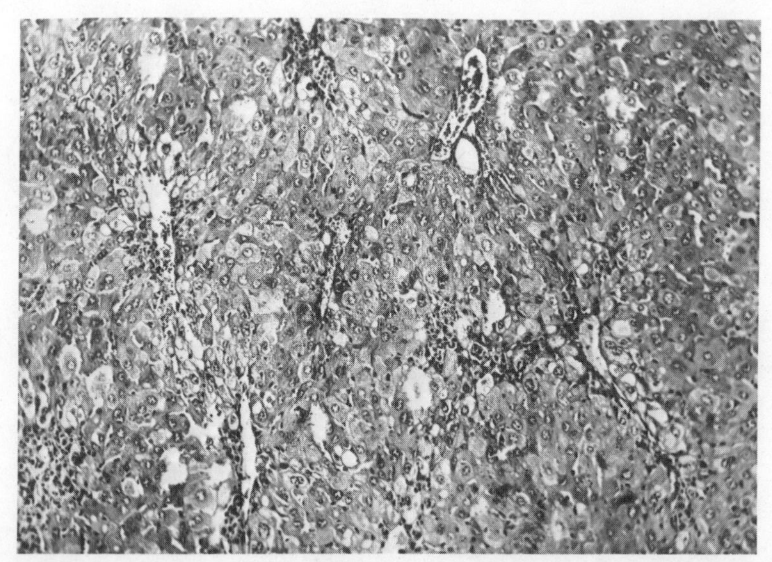

Fig. 5.7. Liver of a mouse sacrificed 7 days after combined administration of luteoskyrin (0.1 mg/10 g, p.o.) and cyclochlorotine (2 μg/10 g, s.c.) (H.E., $\times$120). Centrolobular necrosis and fatty change of the liver cells with lobular distortion are seen.

tion as precursors or cause associated changes of hepatic tumors, should be mentioned.

The following facts have been disclosed regarding the possible histogenic relationship between chronic liver lesions observed in interim studies of the liver of sacrificed mice and biopsy liver specimens of rats, and the occurrence of liver cirrhosis, hepatocellular tumors, and other effects. Many cases of liver cirrhosis were observed in a group of animals exhibiting marked acinus-central necrosis and interstitial reaction predominantly in the periportal tracts (Fig. 5.8). However, neither acinus-central necrosis nor interstitial reaction appears to be essential for the formation of hepatocellular tumors (Fig. 5.9). As shown in Fig. 5.10, all strains of mice fed on a diet containing moldy rice (at a level of 3% in the CF#1, NC and KK strains, and 1% in the $C_{57}BL$ strain) developed hepatocellular tumors at around 400 days.

Development of liver cell tumors seems to have a histogenetic relationship to nodular hyperplasia of the liver cells and to focal appearance of the atypical hepatocytes.[32] Preneoplastic changes of the former type are a usual finding induced by many known hepatocarcinogens (Fig. 5.6). The latter change is occasionally observed in the preneoplastic stage of aflatoxin or sterigmatocystin carcinogenesis. It is a scattered basophilic area

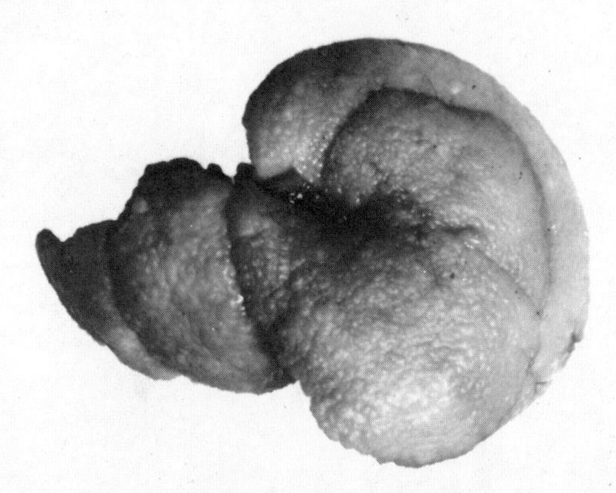

Fig. 5.8. Liver from a male mouse fed moldy rice mixed in the diet at a level of 5% for 20 weeks, showing liver cirrhosis.

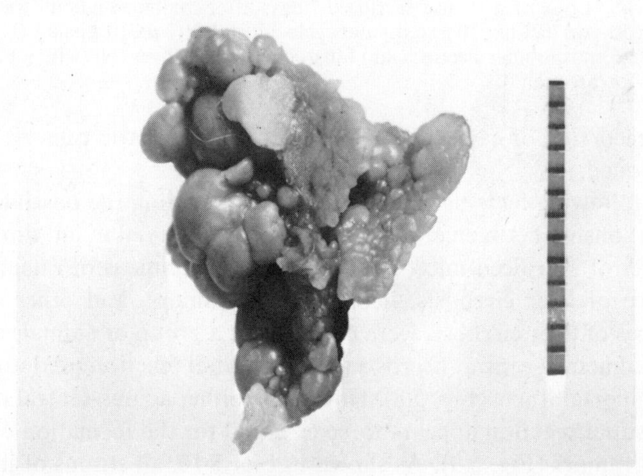

Fig. 5.9. Liver from a male mouse fed moldy rice mixed in the diet at a level of 30% for 240 days, showing multiple tumor nodules.

composed of enlarged liver cells (megalocytes) with a prominent nucleolus or big "clear" cells with enhanced glycogen storage (PAS-positive). Of significance is the demonstration in such atypical focal areas of increased activity of γ-glutamyltranspeptidase, an embryo-carcinogenic enzyme[35] (Fig. 5.11).

Proliferation of the ductular cells or cholangiofibrosis may not be essential for the later formation of an hepatocellular tumor. So-called oval

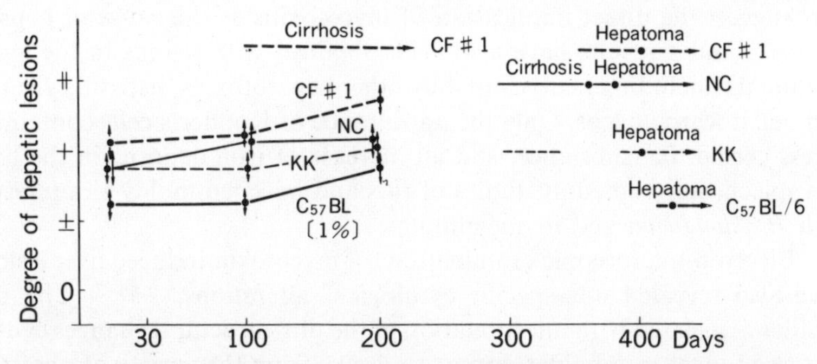

Fig. 5.10. Histological findings in the liver of mice (four strains) fed a diet containing rice cultures of *P. islandicum* (3% and 1%). (After H. Itakura *et al.*)

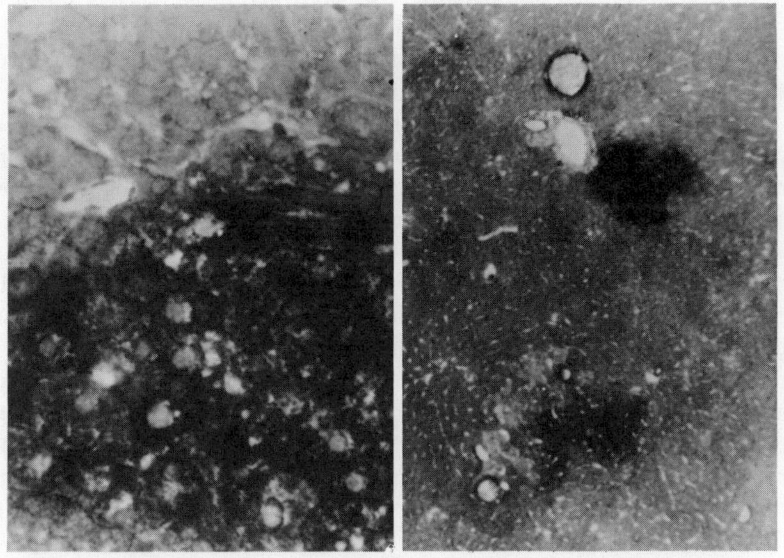

Fig. 5.11. Histochemical demonstration of γ-glutamyltranspeptidase (γ-GTP) activity in the focal area of the liver of a male rat sacrificed 36 weeks after repeated administration of aflatoxin B_1 (40 μg × 100 times, p.o.). The picture on the right (×200) is an enlargement of the γ-GTP positive area in the left picture (×40) (γ-GTP-stain).

cell proliferation was not so prominent among the mycotoxin-induced liver changes.[32)]

There have been many experiments to date on the carcinogenicity of known carcinogenic mycotoxins in a variety of animals (Table 5.1). However, they have demonstrated no specific histological or cytological changes that suggest the direct implication of mycotoxins as the cause of hepatic lesions. Similar results have also been obtained with lesions in the livers of animals (including man) caused by other hepatotoxins, hepatitis viruses, and hepatocarcinogens. Only the appearance of Kupffer's cells containing excess ceroid-like substance, and an increase of iron deposits in the liver and spleen, were noted in studies of rats and mice fed moldy rice infested with *P. islandicum* and its metabolites.

Electron-microscopic examinations of mycotoxin-induced liver lesions have also revealed non-specific cytological alterations.[5,9,36–8)] The difficulties associated with interpretation of the ultrastructural changes caused by these fungal metabolites appear to derive from the variety of observed lesions, which range from minor changes to total necrosis of hepatocytes.

In contrast to the ultrastructural findings in acutely damaged livers, no examples of liver cell nucleoli showing segregation have been noted in the liver after prolonged intoxication. However, the nuclei showed pleomorphic changes together with an increased polyploidy pattern, irregularity in chromatin distribution, increase in interchromatin granules, and occasional enlargement of the nucleoli (Fig. 5.12).

Lamellar arrangement of the rough endoplasmic reticulum (the so-called myelin-figure) and focal proliferation of the smooth endoplasmic reticulum (Fig. 5.13) were the prominent alterations. Similar results are also seen in the livers of animals after prolonged administration of synthetic chemical carcinogens or certain drugs like phenobarbital. The amount of glycogen in the hepatocytes varied considerably from one cell to the other. Some cells showed prominent glycogen deposits in the cytoplasm, revealing rosette-like aggregates intermingled with vesicular elements of the smooth endoplasmic reticulum. Liver cells of animals given prolonged administration of luteoskyrin or moldy rice infested with *P. islandicum* occasionally showed irregular arrangement of the rough endoplasmic reticulum, dilatation of its cistern, and the appearance of deformed mitochondria[38)] (Fig. 5.14). Microbodies, autophagic vacuoles, small fat droplets and lysosomes increased in many cells. Pathological changes in the fine structures of the bile canaliculi were not prominent, other than moderate dilatation between the damaged liver cells (Fig. 5.15).

5.2.3. Some Observations on Hepatitis B Antigen (HBAg) Carriers and Human Hepatocellular Carcinomas

Recent extensive investigations of virus-induced hepatic diseases in man, originating with the discovery of HBAg, have disclosed a higher percentage of HBAg-carriers among residents of areas with a high inci-

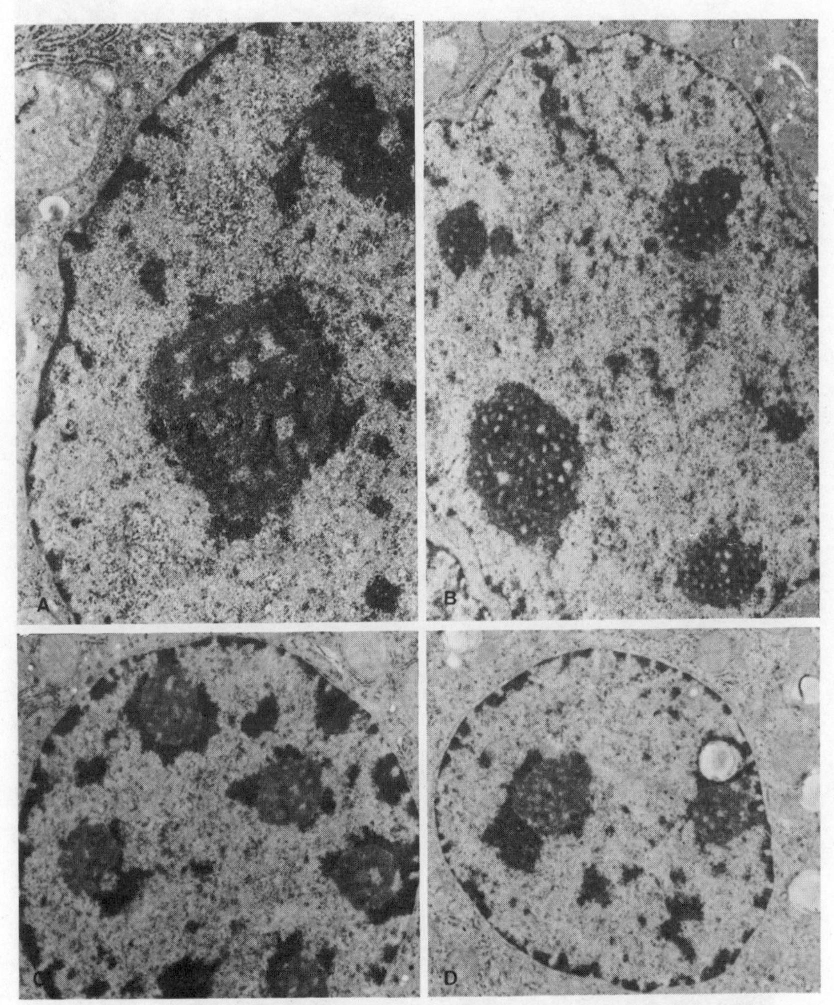

Fig. 5.12. A variety of nuclear changes found in liver parenchymal cells of mice after prolonged administration of 35 or 70 ppm luteoskyrin for 10–20 weeks: increase in interchromatin granules (A, C), appearance of many nucleoli associated with perinucleolar chromatin (B, D), or without it (C).

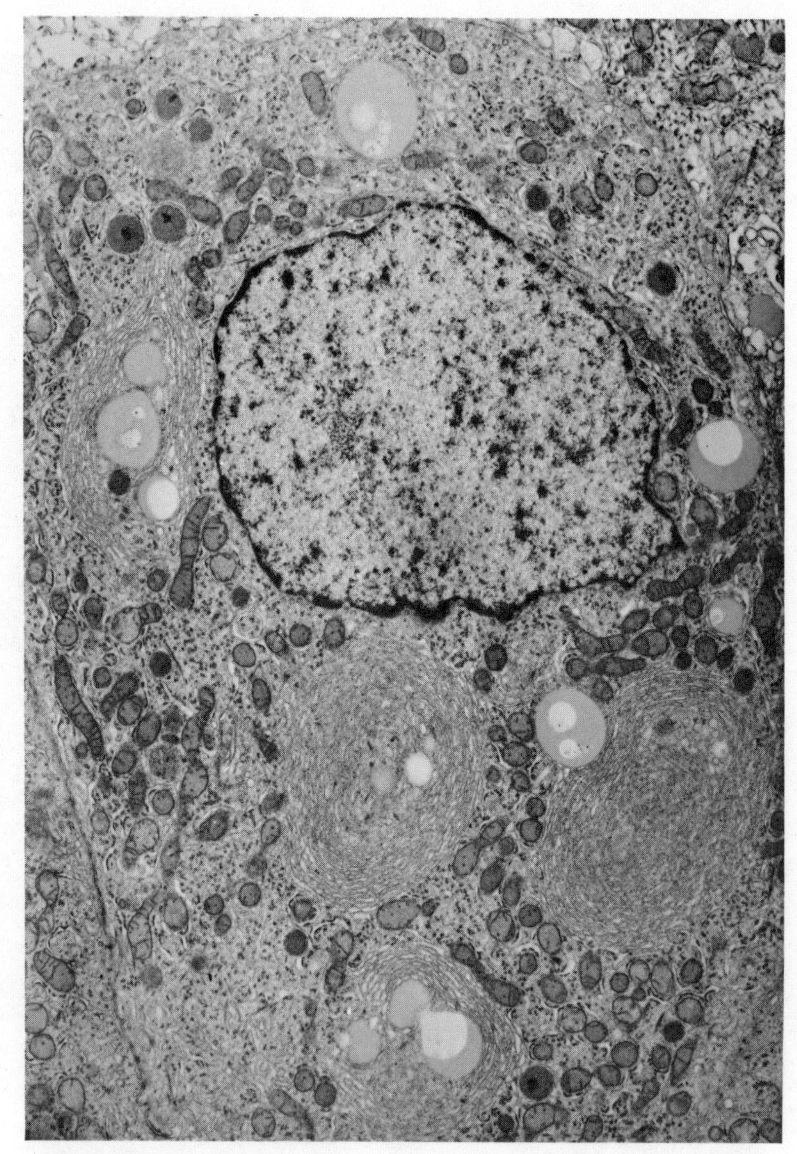

Fig. 5.13. Marked proliferation of the smooth endoplasmic reticulum of a liver parenchymal cell from male rat after feeding with 10 ppm *O*-acetylsterigmatocystin for 52 weeks.[9] (By courtesy of Dr. K. Terao, Research Institute for Chemobiodynamics, Narashino, Japan.)

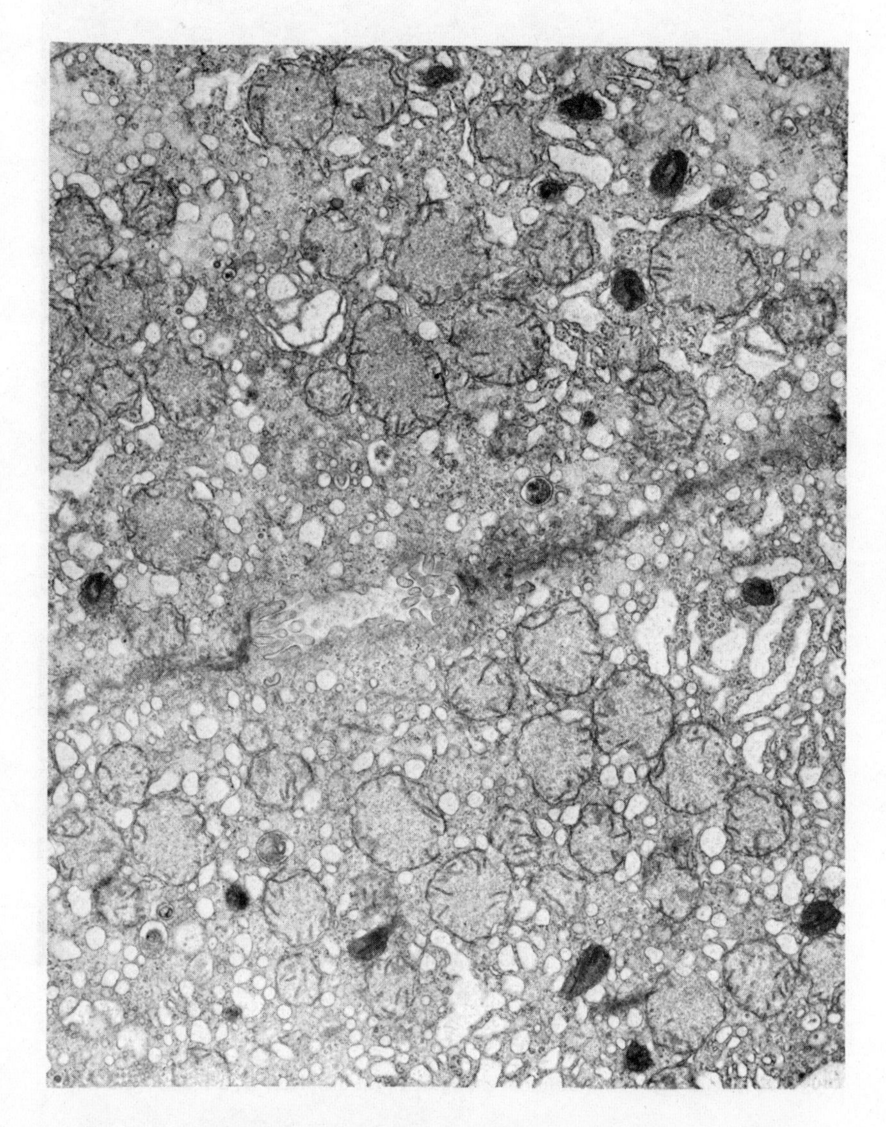

Fig. 5.14. Deformed mitochondria and increased appearance of microbodies in the liver parenchymal cells of a mouse fed 35 ppm luteoskyrin for 20 weeks. No remarkable changes are visible in bile capillaries.

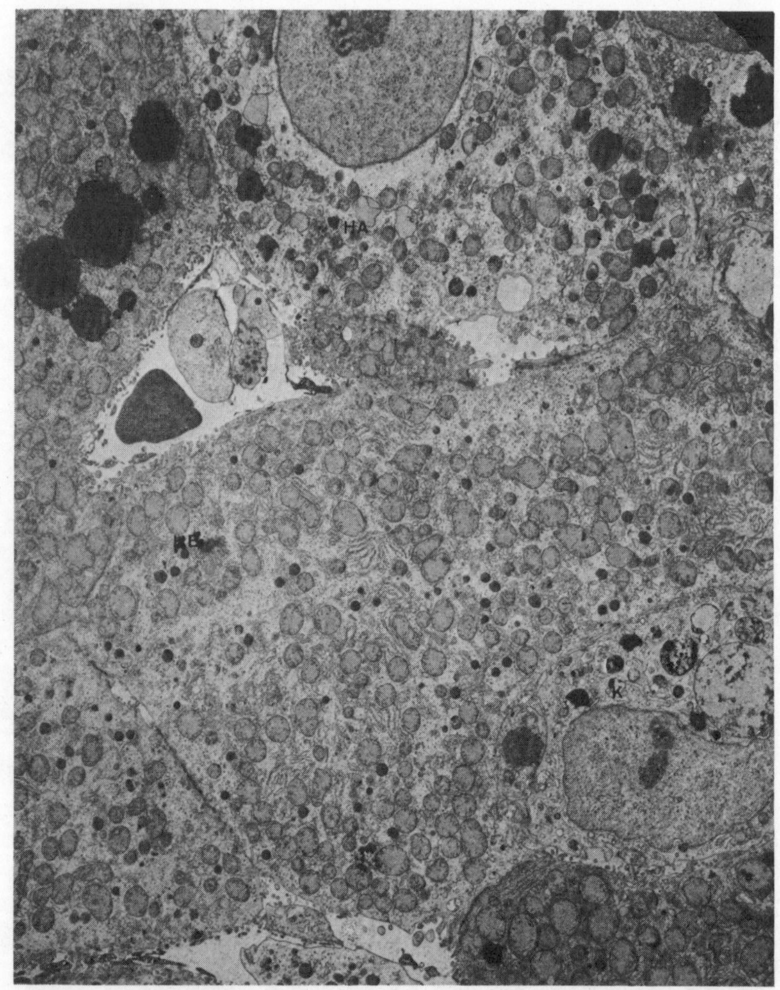

Fig. 5.15. Electron micrograph of liver parenchymal cells from a rat fed a diet containing a rice culture of *P. islandicum* (5%) for 25 weeks, showing persistent toxic damage as seen in the case of acutely damaged liver with luteoskyrin or moldy rice. HA, Hepatocyte showing marked degenerative changes; HB, hepatocyte with minimal changes; F, fat droplets; K, a Kupffer's cell.

dence of hepatocellular carcinomas. The high percentage of HBAg (HBs and HBc)- or antibody-positive cases in patients with acute hepatitis, chronic hepatitis, liver cirrhosis, and especially hepatocellular carcinomas, also strengthens the implication of HB as the cause of human hepatocellular carcinoma. However, further evidence is needed to evaluate hepatitis viruses or mycotoxins as causative agents of human liver cancer. There remains the possibility of mere association of such agents in the liver, especially in liver cells displaying altered metabolic activities. Whether they play a significance, active role in injury and neoplastic transformation of the liver cells is still not clear.

Pathologists have provided easy explanations of the morphological changes, including ground-glass hepatocytes, hyaline or globular inclusions, filamentous structure, and virus-like inclusions or particles, which are observed in the human liver in association with HB. However, no definitive explanations yet exist for the similar changes which appear in the livers of animals treated with mycotoxins and synthetic chemical carcinogens, or for the virus-like particles seen in the control animal livers.

5.3. Future Study of Carcinogenic Mycotoxins

Experimental evidence on the production of cancer in animals has been obtained in the past 20 yr with several fungal toxins, including the strongest known hepatocarcinogen, aflatoxin B_1. Information has been collected on the actual contamination of a variety of human foodstuffs, especially such staple foods as rice, maize and peanuts, by these toxins. A possibility of indirect spoilage of human foods, i.e. via the milk, eggs or meat of livestock given feed contaminated with fungal toxins, has also been suggested (Fig. 5.16). However, solutions to a number of problems are still needed to reach the final goal of understanding the exact etiologic role that fungal toxins play in the human body after long-term exposure.

Carcinogenic mycotoxins, as low-molecular compounds, show a close similarity in their biological effects to synthetic chemical carcinogens. However, as the secondary metabolites of fungi, they are easily influenced by the many conditions that govern fungal growth and mycotoxin production. It is also important to clarify the susceptibility of man to them and to determine their metabolic fate, including their metabolic activation, in the human body. Research into the morphological features of mycotoxin-induced hepatic lesions in animals and man, if these occur, will help to elucidate the histogenesis of liver cirrhosis and hepatocellular tumors.

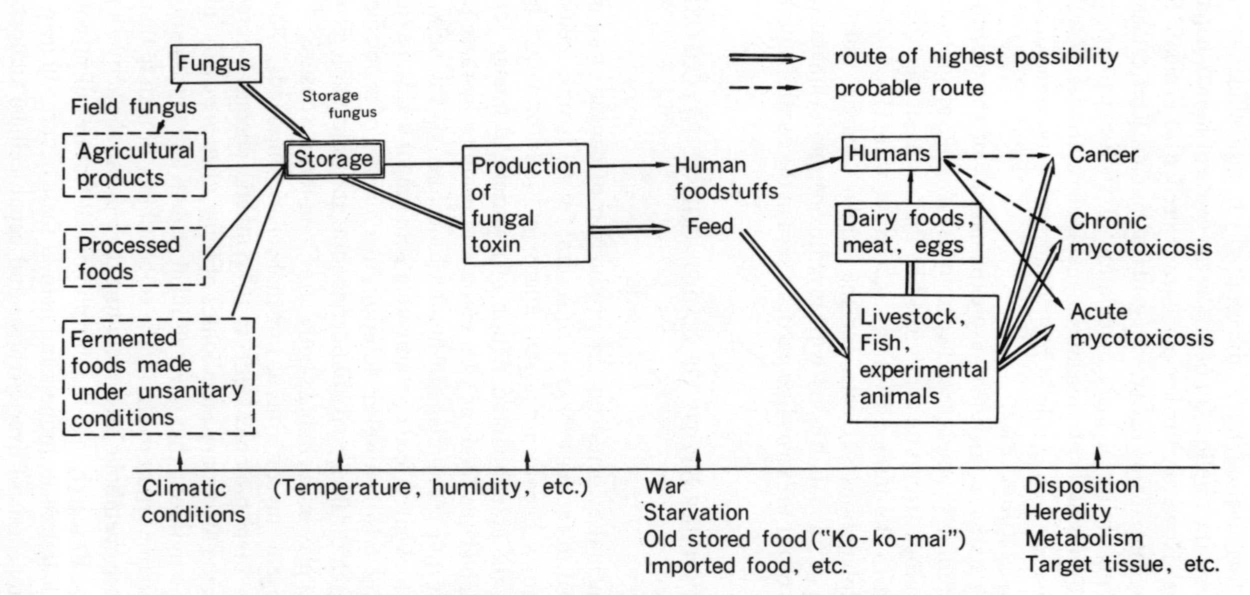

Fig. 5.16. Mycotoxin-induced diseases of man and animals: related conditions and causal relationships are shown. (After M. Enomoto, In: *Cancer* (ed. Y. Yamamura and T. Sugimura) (Japanese), vol. 15, Iwanami Shoten, 1976.)

References

1. E. W. Hurst and G. E. Paget, *Brit. J. Derm.*, **75**, 105 (1963).
2. S. S. Epstein, J. Andrea, S. Joshi and N. Mantel, *Cancer Res.*, **27**, 1900 (1967).
3. R. W. Detroy, E. B. Ciegler and A. Ciegler, *Microbial Toxins* (ed. A. Ciegler, S. Kadis and S. J. Ajl), vol VI, p. 3, Academic Press (1971).
4. D. M. Newberne and W. H. Butler, *Cancer Res.*, **29**, 236 (1969).
5. W. H. Butler, *Mycotoxins* (ed. I. F. H. Purchase), p. 1, Elsevier (1974).
6. J. J. Van der Watt, *ibid.*, p. 369, Elsevier (1974).
7. M. Saito, M. Enomoto, T. Tatsuno and K. Uraguchi, *Microbial Toxins*(ed. A. Ciegler, S. Kadis and S. J. Ajl), vol. VI, p. 299, Academic Press (1971).
8. Y. Ueno, I. Ueno, N. Sato, Y. Iitoi, M. Saito, M. Enomoto and H. Tsunoda, *Japan. J. Exptl. Med.* **41**, 177 (1971).
9. K. Terao and M. Yamazaki, Proc. Japan. Cancer Ass., 34th Ann. Mtg. Osaka (Japanese), p. 19 (1975).
10. K. Uraguchi, M. Saito, Y. Noguchi, K. Takahashi, M. Enomoto and T. Tatsuno, *Fd. Cosmet. Toxicol.*, **10**, 193 (1972).
11. I. Ueno, M. Saito M. Enomoto and K. Uraguchi, Proc. Japan. Cancer Ass., 32nd Ann. Mtg. Tokyo (Japanese), p. 187 (1973).
12. S. D. Vesselinovitch, N. Mihailovichi, G. N. Wogan, L. S. Lombard and K. V. N. Rao, *Cancer Res.*, **32**, 2289 (1972).
13. K. Fujii, H. Kurata, Y. Hatsuta and N. Odashima, Proc. Japan. Cancer Ass., 34th Ann. Mtg. Osaka (Japanese), p. 45 (1975).
14. T. Tanaka, *Proc. Japan. Ass. Mycotoxicol.* (Japanese), **1**, 6 (1975).
15. P. M. Newberne, *Environmental Health Perspectives*, **9**, 1 (1974).
16. H. Masuda, S. Naoe, M. Mabuchi, M. Enomoto, O. Hayashi, N. Matsui, M. Ichinoe and H. Kurata, Proc. Japan. Cancer Ass., 34th Ann. Mtg. Osaka (Japanese), p. 37 (1975).
17. P. M. Newberne and A. E. Rogers, *J. Natl. Cancer Inst.*, **50**, 439 (1973).
18. Y. Shinohara, M. Arai, S. Sugihara, K. Hirao, K. Nakanishi, H. Tsunoda and N. Ito, *Gann*, **67**, 147 (1976).
19. M. Saito, M. Enomoto and T. Tatsuno, *Gann*, **60**, 599 (1969).
20. E. B. Smalley and F. M. Strong, *Mycotoxins* (ed. I. F. H. Purchase), p. 199, Elsevier (1974).
21. M. Saito and K. Ohtsubo, *ibid.*, p. 263, Elsevier (1974).
22. M. Saito, K. Ohtsubo, M. Umeda, M. Enomoto, H. Kurata, S. Udagawa, F. Sakabe and M. Ichinoe, *Japan. J. Exptl. Med.*, **41**, 1 (1971).
23. M. Saito, T. Ishiko, M. Enomoto, K. Ohtsubo, M. Umeda, H. Kurata, S. Udagawa, S. Taniguchi and S. Sekita, *ibid.*, **44**, 63 (1974).
24. H. Toyokawa, T. Nuguishi, M. Hirohata and T. Sakaki, *Japan. J. Pub. Health*, **15**, 789 (1968).
25. C. H. Bourgeois, L. Olson, D. Comer, H. Evans, N. Evans, N. Keschamras, R. Cotton, R. Grossman and T. Smith, *Am. J. Clin. Path.*, **56**, 558 (1971).
26. C. H. Bourgeois, R. C. Shank, R. A. Grossman, D. O. Johnson, W. L. Wooding and P. Chandavimol, *Lab. Invest.*, **24**, 206 (1971).
27. S. J. Van Rensburg, J. J. Van der Watt, I. F. H. Purchase, L. Pereira Coutinho and R. Markham, *S. African Med. J.*, **48**, 2508a (1974).
28. K. A. V. R. Krishnamachari, R. V. Baht, V. Nagarajan and T. B. G. Tilak, *Lancet*, May 10, 1061 (1975).
29. M. Miyake, M. Saito, M. Enomoto, T. Shikata, T. Ishiko, K. Uraguchi, F. Sakai, T. Tatsuno, M. Tsukioka and Y. Sakai, *Acta. Path. Japon.*, **10**, 75 (1960).
30. M. Saito, *ibid.*, **9**, 785 (1959).
31. M. Enomoto, *ibid.*, **9**, 189 (1959).
32. M. Enomoto and M. Saito, *ibid.*, **23**, 655 (1973).
33. M. Enomoto and M. Saito, *Ann. Rev. Microbiol.*, **26**, 279 (1972).

34. F. O. Torres, I. F. H. Purchase and J. J. Van der Watt, *J. Path.*, **102**, 163 (1970).
35. M. Harada, K. Okabe, K. Shibata, H. Masuda, K. Miyata, and M. Enomoto, *Acta Histochem. Cytochem.*, **8**, 80 (1975).
36. D. Svoboda, H. J. Grady and J. Higginson, *Am. J. Path.*, **49**, 1023 (1966).
37. W. H. Bulter, *Chem. Biol. Interactions*, **4**, 49 (1971).
38. M. Enomoto, Proc. 1st Int. Congr. IAMS Tokyo, vol 4, p. 380, Science Council of Japan (1974).

CHAPTER 6

Aflatoxin: Investigations on Traditional Foods and Imported Foodstuffs in Japan

Kageaki Aibara
National Institute of Health, Kamiosaki, Shingawa-ku, Tokyo 141, Japan

6.1. Introduction

Mycotoxins have become an important problem both in public health and agriculture in recent years. Basic studies and multidisciplinary research embracing mycology, biology, chemistry, biochemistry, pathology, pharmacology, toxicology and other fields, are currently being conducted throughout the world. Particularly with reference to aflatoxin, which is believed to possess the highest carcinogenetic potency among the known substances, both basic research and surveys of various agricultural products and foods for aflatoxin contamination are being conducted in the advanced countries. Several epidemiological surveys are also being undertaken in those developing nations situated in the tropical zone to determine the hepatoma incidence among local inhabitants and the distribution of aflatoxin in foods and feedstuffs. On the basis of the results, practical countermeasures and administrative steps have already been systematically implemented in many countries of the world.

6.2. IMPORTANCE OF THE AFLATOXIN PROBLEM

The importance of aflatoxin from the standpoint of public health and food protection may be summarized as follows.

(1) Of all the substances known today, both natural and artificially synthesized, aflatoxin displays the highest oral carcinogenicity in a wide range of animals.[1-3] In 1967, Wogan and Newberne[4] reported that purified aflatoxin B_1 at levels of 15 ppb added to a semi-synthetic diet induced liver-cell carcinomas in 25/25 Fischer rats surviving for 68 weeks (male) to 80 weeks (female). The dosage corresponded to an intake of about 0.2 μg/day/rat and the total amount of aflatoxin B_1 for carcinoma incidence can be estimated at < 100 μg. Seven years later, they also reported[5] that tumors were induced in 2/22 male Fischer rats fed 1 ppb aflatoxin B_1 under conditions essentially the same as in the earlier experiment.

(2) Aflatoxin is a metabolic product derived from a living organism (fungus) which exists in the natural ecological system. It is therefore impossible to eradicate aflatoxin-producing fungi completely from the system.

(3) Staple agricultural products, particularly rice, wheat, corn, etc., are well known to be the best substrates for the experimental production of aflatoxin.[6]

(4) Aflatoxin-producing fungi are widely distributed in tropical and subtropical areas. This indicates that the population in these areas faces a great danger of contamination from aflatoxin through daily meals. The results of some epidemiological studies explained later demonstrate this possibility.

(5) It is also well known that aflatoxin-producing fungi are widely distributed over much of the U.S.A.,[7] which represents the world's largest granary today, and on the S. American Continent which may emerge as a major food supplier for the remainder of the world in the future. When we consider the worldwide food supply of the future, with signs already showing of shortages, how to protect essential foodstuffs from aflatoxin contamination becomes an important question for scientific solution, together with the problem of aflatoxin contamination that threatens the developing nations.

(6) Reports indicate that aflatoxin not only shows the highest potency for carcinogenicity as a chronic effect but also exerts acute fatal effects on infants and small children.[8]

(7) It is reported that aflatoxin B_1 and its metabolite M_1 occur in the breast milk of mothers who have eaten aflatoxin-contaminated foods.[9,10]

(8) Results of feeding experiments with aflatoxin-containing diets using farm animals have indicated that aflatoxin B_1 transmitted from the feed and aflatoxin M_1 metabolized in the animal body can be detected in animal products such as cow's milk, eggs, and meat.[11–5] The amounts of aflatoxins in the products depend on the amount of aflatoxin B_1 present in the diet. Both aflatoxin B_1 and M_1 are heat stable; the decomposition temperature of B_1 is 268–9°C and that of M_1 is 299–300°C. It should be noted therefore that both B_1 and M_1 are difficult to destroy at the range of temperatures usually employed in the processing or cooking of foods.

(9) The existing food processing techniques, with the exception of the technique for refining edible oil,[16] are insufficiently sophisticated to destroy or remove contaminated aflatoxin from agricultural products, dairy products or foods without spoiling their marketable or nutritive value.[17]

(10) The strongest other reason to take a serious view of aflatoxin from the standpoint of public health and food protection, in addition to the items mentioned above, can be found in the results of the several epidemiological studies conducted to clarify the relationship between the aflatoxin content of contaminanted foods and the incidence of hepatomas in man.[18–21] Unlike the case of experiments with animals whose feeding conditions can be effectively controlled, the causes of cancer incidence in man are extremely complicated. It is, therefore, extremely difficult to conclude that a specific substance is the single cancer-inducing agent. Particularly in advanced countries where demographic movements are frequent and where the quality of eating habits constantly changes, it is virtually impossible to prove epidemiologically the relationship between foods, food-contaminating substances, or food additives on the one hand, and cancer incidence on the other. In fact, no such attempts have so far proved successful. However, the epidemiological survey of aflatoxin represents a different story.

Results of epidemiological surveys in the tropical zone where eating habits have not changed qualitatively for more than half a century and where population movements have been limited indicate that aflatoxin ingested by man in his food has some relationship to the incidence of human hepatomas. The findings of well-designed epidemiological studies on a large scale conducted in tropical S. Asia, Thailand and Hong Kong, and reported by Shank *et al.* in 1972[18] are famous. Other reports known in this connection are that of Peers and Linsell[21] on a survey in Kenya, that on a survey in Uganda by Alpert *et al.*,[20] and that of Keen and Martin[19] on a survey in Swaziland. Campbell and Stoloff[22] have summarized these epidemiological investigations in their outstanding review.

It is thus evident that, among the numerous mycotoxins, aflatoxin holds an extremely important position in relationship to the health of human beings.

6.3. TOLERANCE LIMITS OF AFLATOXIN CONTAMINATION

Many countries have set up either a guideline or tolerance level for aflatoxin contamination of foods and feedstuffs or are now making preparations to institute them. Table 6.1 shows the guidelines and tolerance levels approved by various countries.[23]

Moreover, the WHO/FAO Joint Expert Committee, in view of the fact that agricultural products are distributed throughout the world, recommended in 1968 that the total amount of aflatoxin B_1, B_2, G_1 and G_2 in agricultural products should not exceed 30 ppb.

6.4. ANALYTICAL METHODS FOR AFLATOXIN DETERMINATION

Thin-layer chromatography is commonly employed for the chemical analysis of aflatoxin. However, the chemical or biochemical properties of agricultural commodities and individual foodstuffs to be analyzed vary widely, each substrate presenting problems characteristic to itself. Accordingly, slightly different pre-analytical procedures or clean-up procedures have been devised for each type of food, foodstuff or feed. Each country has adopted methods for the analysis of aflatoxin which are best suited to its own circumstances. The method explained in Chapter 26 of the Official Methods of Analysis of the Association of Official Analytical Chemists (AOAC), 12th Ed. (1975) and the methods recommended by the International Union of Pure and Applied Chemistry (IUPAC) can be regarded as internationally acknowledged methods for the analysis of aflatoxin.

Many scientists have pointed out that, in the analysis of aflatoxin, the problem of sampling is equally important to the need to develop analytical

TABLE 6.1. Aflatoxin control in various countries (after L. Stoloff, FDA)

Country	Legal control	Food(s)	Feed	Aflatoxin limit (μg/kg)	Remarks
Belgium	Royal decree		All feedstuffs	0 B_1†[1]	Analytical method sets 40 B_1
Brazil	Ministerial decree		Peanut meal	50 B_1	Export control
Canada	Administrative guideline	Consumer nuts and nut products		15 Total†[2]	Control under hazards to health
Denmark	Statutory order	All foods		0	Analytical method sets 5 B_1
			Feed containing peanut products	0	Analytical method sets 100 B_1
France	Decree		Feed ingredients	700 B_1	
			Finished feed	50 B_1	For sheep, goats, adult cattle
			Finished feed	20 B_1	For dairy cattle, adult pigs, poultry except ducks
			Finished feed	0	For others; analytical method sets 10 B_1
Germany	Regulation		Mixed feed	0–200 B_1	Dependent on animal
India	Indian Standards Institute suggestion	Peanut meal		30	Based on WHO/PAG recommendation
Israel	Stored Products Research Lab. recommendation		All feedstuffs	20 B_1	
Italy	Ministry of Health circular	Peanuts and peanut products	Peanuts and peanut products	50 B_1	
Japan	Regulation	All foods		0	Analytical method sets 10 B_1
			Peanut meal	1000 B_1	Use in feed limited: 0%—chicks, calves, baby pigs; 2%—dairy cows; 4%—other livestock

TABLE 6.1—*Continued*

Country	Legal control	Food(s)	Feed	Aflatoxin limi t(μg/kg)	Remarks
Malawi	Export regulation	Peanuts		5 B_1	
Malaysia	Food code	All foods		0	Set by analytical method
Norway	Ministry of Agriculture regulation		Oilseed meals	600 B_1	Limited in feed concentration less than 8%
Poland	Import control by Sanitary Authorities	All foods		0	Set by analytical method
			All feedstuffs	0	Set by analytical method
Rhodesia	Voluntary code	Peanuts		25 B_1	
			Mixed feeds	50–400 B_1	Dependent on animal
Sweden	Accepted practice	All foods		0	Analytical method sets 5–10 B_1
	Advisory Standard		Peanut meal	600 B_1	Dairy feed limit, 15%
United Kungdom	Tariff regulations	Peanuts		50 B_1	
	Voluntary code		Peanut meal	0–500 B_1	Dependent on animal
U.S.A.	Regulated tolerance	Consumer peanut products		15 Total	
	Administrative guideline	All other foods	All feedstuffs	20 Total	Except raw shelled peanuts —25 Total

†[1]B_1 = aflatoxin B_1.
†[2]Total = aflatoxin B_1, B_2, G_1 and G_2.

methods best suited to each kind of agricultural product,[24–6] dairy product, or other item. This applies also to the analysis of other mycotoxins. How to collect appropriate and representative samples from huge amounts of grain weighing scores or even hundreds of tons is always a difficult question. If the stored grains are sufficiently large in size and possess a relatively high market value such as peanuts or other nuts, it is possible to remove the infected pieces using an automatic sorter or a selector, or to pick them up by hand if necessary. In fact, peanuts for human consumption that are suspected of aflatoxin contamination are actually being removed in these ways.[27,28] In many cases, nuts are served as light snacks after simple treatment such as roasting. It should be noted, however, that aflatoxin is not destroyed completely under these conditions, and that due to its size, a single nut could contain sufficient aflatoxin to cause acute aflatoxicosis in man.

6.5. Protection of Food from Invasion by Aflatoxin

To state the conclusion first, the surest and best way to prevent the hazard of aflatoxicosis is to prevent the production of aflatoxin in agricultural products.

It is now known that aflatoxin-producing fungi require certain appropriate conditions to produce aflatoxin as a secondary metabolite. In other words, aflatoxin is produced only when certain environmental conditions, such as the temperature, humidity, level of oxygen, physiological properties and/or chemical and biochemical characteristics of the agricultural products which serve as the substrate for aflatoxin production, are satisfied.[29] Not only the problem of aflatoxin contamination of stored commodities but also the possibility that pre-harvest corn exposed to the attack of insects may be most liable to initiate invasions of aflatoxin-producing fungi and production of aflatoxin, have been pointed out recently in the U.S.A.[30] Prevention of the growth of aflatoxin-producing fungi is thus fundamental in protecting agricultural products from contamination by aflatoxin and ultimately in protecting human health from the hazards of aflatoxin.

One important step for safeguarding man's health would be to measure the amounts of aflatoxin contained in all foodstuffs destined for human consumption, and to designate the foodstuffs as edible or not based on the tolerance level mentioned above. This step, however, is a last resort and is essentially passive. In order to avert the dangers of aflatoxin more positively, aflatoxin production in foods and foodstuffs must be prevented. The route by which aflatoxin enters foods naturally depends on the kind and type of food or foodstuffs—farm products, animal products, dairy products, fishery products, etc. Aflatoxin contamination can be traced back, in principle, to farm products. Rodricks and Stoloff[15)] have attempted to calculate average ratios of feed levels of aflatoxin B_1 to edible tissue levels of aflatoxin B_1 and M_1 from available data of more than a dozen experimental feeding trials on farm animals. According to their tentative conclusions, the average feed to tissue ratios of aflatoxin B_1 in the liver of beef cattle, broilers and swine were 14,000, 1200 and 800, respectively. The ratio in the case of eggs was 2200, and that for aflatoxin M_1 in the case of milk was 300. They suggested, however, that further investigations into the metabolic fate of aflatoxin in food-producing animals might necessitate a later reassessment.

In practice, it is virtually impossible to eliminate aflatoxin-producing fungi from agricultural products produced on land where the fungi exist and grow naturally all the time. It is not possible to go so far as exterminating the fungi by destroying the ecosystem in aflatoxin-affected areas. There is no alternative but to restrain the growth of aflatoxin-producing fungi in the course of harvesting, storage and processing of agricultural products so as to minimize either the contamination or the production of aflatoxin. The fact that agricultural products are naturally contaminated more or less by fungi has to be recognized. It is well known that correct handling of agricultural products at the time of harvest represents an important key. Care must be taken so that agricultural products are not injured at the time of harvesting, since injured parts are susceptible to invasion by aflatoxin-producing fungi. More important still, the agricultural products should be dried as quickly as possible immediately after harvesting in order to ensure a condition most detrimental to the growth of the fungi. In practice however, mechanical drying machines and equipment are not always available, particularly in tropical areas where the distribution of aflatoxin-producing fungi is wide and the aflatoxin hazard is high. In some cases, quick mechanical drying does not have a favorable effect on the nutritional, chemical or marketable value of the products, such as by affecting taste, flavor, color and/or texture. In order to resolve the aflatoxin contamination problem faced by developing countries situated in the high-temperature, high-humidity zone where the possibilities for growth of aflatoxin-producing fungi are very high, renewed efforts for controlling the fungi as well as for improving agricultural technology must be made with international cooperation.

Low temperature storage represents another way to combat the fungi. This method has already been used for some agricultural products in advanced industrialized countries, but the energy. requirements and costs for maintaining the low-temperature warehouses are enormous.

6.6. Aflatoxin and Traditional Fermented Foods in Japan

When the International Symposium on Oil-Seed Protein was held in Japan in May, 1964, many scientists from Europe and the U.S.A. presented papers on aflatoxin. This was the first occasion for Japanese scientists to engage in detailed international discussions on the aflatoxin problem. Some of them warned of possible or suspected aflatoxin contamination of traditional Japanese fermented foods. Responding to this warning, extensive research on aflatoxin was initiated in Japan immediately after the symposium by K. Miyaki and K. Aibara at the Department of Biomedical Research on Foods, National Institute of Health. As a first study under this program, chemical (TLC, UV absorption analysis, fluorescence analysis) and biological tests were made on more than 180 strains of koji-molds (*Aspergillus oryzae* group) used in the Japanese fermentation industry for the production of traditional fermentation foods. The purpose was to determine whether or not they exhibited any aflatoxin productivity. The results were entirely negative.[31)]

Following this first study, similar tests were made by several research institutions in succession on the koji-molds used in the production of Japanese fermented foods. These studies included the work of H. Murakami *et al.*, National Research Institute of Brewing, on more than 610 strains of koji-molds used for "sake" production,[32)] that of M. Manabe and S. Matsuura, National Food Research Institute, on more than 200 strains of koji-molds used for producing "miso" (soybean paste) and "shoyu" (soy sauce),[33)] and that of T. Yokotsuka *et al.*, Central Research Institute of Kikkoman Shoyu Co. Ltd., on more than 70 strains of koji-molds used in their own products.[34)] The results clearly showed that aflatoxin productivity is not present in the koji-molds used for industrial production of Japanese traditional fermentation foods. A research group of the National Food Research Institute also attempted experimental production of "miso" with an aflatoxin-producing fungus instead of the common koji-mold, but found that it was impossible. This suggests that the enzymatic system with characteristic proteinase activity that is indispensable in the final processing of fermented products with koji-molds, was deficient in the tested aflatoxin-producing fungus. It can be said that during the more than 2000 year history of fermented foods in Japan, the koji-molds have been selected successively to obtain those best suited for the production of improved-quality fermented foods.

In parallel with the above research which proved the lack of aflatoxin productivity in the koji-molds used in the Japanese fermentation industry, other work was undertaken on the fermented foods themselves to determine whether there was any aflatoxin contamination. M. Manabe *et al.* collected from throughout Japan, 108 samples of commercial "miso", 33 samples of home-made "miso", 28 samples of miso-koji (a solid culture of an industrial koji-mold strain inoculated on rice for making "miso"), and 39 samples of commercial soybean sauce and tested them for detectable aflatoxin contamination. They obtained negative results from chemical, thin-layer chromatographic analysis of all samples and biological tests using the yolk sac inoculation method.[35)]

6.7. AFLATOXIN AND RICE IN JAPAN

M. Manabe and S. Matsuura, National Food Research Institute, have reported results for the aflatoxin analysis of stored rice in warehouses controlled by the Japanese Government. When it is considered that autoclaved polished rice constitutes the best substrate for experimental aflatoxin production, it is of maximum concern to the rice-eating Japanese that the relationship between stored rice and aflatoxin contamination be known clearly.[36)] These two investigators collected from 46 areas throughout Japan samples of brown rice stored in rice warehouses for periods of 6 to 18 months and conducted aflatoxin analysis on them. They also collected from 440 warehouses brown rice which had been stored for 3–4 yr under well controlled conditions, and carried out aflatoxin analysis as well as ochratoxin A and sterigmatocystin analysis. They found no mycotoxin of any kind in these well stored samples. However, among samples of brown rice stored accidentally for longer periods of 5–7 yr or more under poor storage conditions, and showing visible mold contamination, there were cases in which sterigmatocystin and ochratoxin A contamination was detected, although no aflatoxin was found. These and related studies are still continuing at the National Food Research Institute. The above results suggest that, although rice represents a good substrate for experimental mycotoxin production, it can be safely stored for several years if the conditions of storage are adequately controlled.

6.8. Inspection of Imported Peanuts and Peanut Products in Japan

In November, 1970, K. Aibara *et al.*, National Institute of Health, collected commercially sold peanut butter, peanut paste and some other types of peanut products from the ordinary Japanese market and tested them for aflatoxin contamination.[37] The collected peanut products consisted of 102 lots from 22 different manufacturers. The total number of samples tested was 207. It was discovered that 61 of these samples (from 33 lots of 11 different manufacturers) showed aflatoxin B_1 contamination ranging from a minimum of 3 ppb to a maximum of 82 ppb. Concerning the products of three of the above 11 manufacturers which showed a particularly high degree of contamination, follow-up investigations were conducted on the source of aflatoxin contamination. It was found that among peanuts which were being imported as raw materials for peanut butter, those produced in Brazil, Nigeria and Tanzania possessed high degrees of aflatoxin contamination. In the case of bulk peanut butter for use in making confectionery, some of the tested samples showed aflatoxin B_1 levels of 280–320 ppb. Aflatoxin contamination was not detected in domestically produced and harvested peanuts or in peanuts and peanut butter imported from the northern part of continental China. Similar trends were observed in subsequent investigations, suggesting that aflatoxin-producing fungi may not occur in the soil of regions located in relatively high latitudes, such as northern China and Japan, or if they do, that they are very thinly dispersed. This conclusion is supported by the results on soil fungal distribution obtained by O. Tsuruta *et al.*, National Food Research Institute.[38] In other words the climate in these regions may not be conducive to the maintenance of a constant population of aflatoxin-producing fungi.

Investigations conducted by the Department of Biomedical Research on Foods, National Institute of Health, have established, as explained above, that aflatoxin contamination exists in peanuts and peanut products distributed in the general market of Japan and that the source of contamination has been imported peanuts. In accordance with a suggestion made by the Food Sanitation Investigation Council of Japan which reviewed the survey results, the Ministry of Health and Welfare decided to carry out import inspections of all peanuts and peanut products brought into Japan from throughout the world after April, 1971. The Council set the guideline level for aflatoxin B_1 at less than 10 ppb for aflatoxin contamination of foods.

In the first year after import inspection was instituted (April, 1971 through March, 1972—the 1971 fiscal year), 4.6% of all imported peanuts and peanut products were rejected due to high levels of aflatoxin B_1 above the guideline. In September, 1973, the Mycotoxin Research Association was established to specialize in mycotoxin inspection and analysis of imported foods and foodstuffs under the supervision of the Ministry of Health and Welfare, Japan. Table 6.2 shows the results of import inspections of peanuts and peanut products made during fiscal 1974.[39] In that year, about 60,000 tons of raw peanuts were imported into Japan.

Of 773 peanut samples tested, which represented each imported lot, 615 (79.6%) were of the Spanish type, and 158 (20.4%) were of the Virginian and other types. Aflatoxin contamination was detected in 40 (5.2%) of the samples tested, and all were Spanish type peanuts. Of these, 19 samples (2.5% of the total) failed to meet the guideline, having more than 10 ppb aflatoxin B_1. The average contamination was 447 ppb (range 11.0 to 1696 ppb). The average concentration in the other contaminated samples was 2.1 ppb (0.2–9.2 ppb). Most were below 1.5 ppb.

As for peanut butter, 183 representative samples from all lots of imported peanut butter were inspected in fiscal 1974. Aflatoxin contamination was detected in 39.9% (73) of the samples. Those which exceeded the guideline level and were disqualified amounted to only 1.1% of the total, i.e. 2 samples (10.5 and 20.8 ppb).

Inspections were conducted on 102 samples of peanut products other than peanut butter. Aflatoxin contamination was detected in 9 out of 12 (75%) samples of peanut cream, 9 out of 9 (100%) samples of peanut chocolate, 6 out of 23 (26.1%) samples of peanut cookies and cakes, and 2 out of 10 (20%) samples of peanut oil. Cases of contamination exceeding the B_1 guideline amounted to 5 out of 12 (41.7%) of the tested samples of peanut cream. The average aflatoxin B_1 concentration was 25.6 ppb (23.5–27.8 ppb).

Of 308 test samples including peanut butter and other peanut products in 1974, 30.8% showed aflatoxin contamination below the guideline level. Most of the aflatoxin-positive samples showed a concentration of less than 5 ppb, although a few had fairly high contamination (6–8 ppb). In summary, of the total samples (308), 11% were below 2 ppb, 10.4% at 2.1–4.0 ppb, and 3.2% at 4.1–5.0 ppb. Those below 5 ppb amounted to 24.7% of the total. The test results thus indicated that about 1/4 of the peanut products on the commercial market had positive aflatoxin contamination of 5 ppb or less. Clearly, it is very important to control aflatoxin contamination in the raw peanuts used as raw materials for processed peanut products.

TABLE 6.2. Results of aflatoxin examinations of imported peanuts in Japan between April, 1974 and March, 1975 (Fiscal 1974)†

Exporting countries	Samples examined representing each lot		aflatoxin B_1			% with detected aflatoxin B_1		% with no detected aflatoxin B_1
	No.	% of total	10 ppb	Up to 10 ppb	None detected	10 ppb	up to 10 ppb	
U.S.A.	280	36.2	0	1	279	0	0.4	99.6
Indonesia	134	17.3	9	4	121	6.7	3.0	90.3
China	105	13.6	1	0	104	1.0	0	99.0
India	85	11.0	3	2	80	3.5	2.4	94.1
S. Africa	77	10.0	0	0	77	0	0	100.0
Mozambique	28	3.6	0	0	28	0	0	100.0
Sudan	15	1.9	0	7	8	0	46.7	53.3
Brazil	15	1.9	2	3	10	13.3	20.0	66.7
Australia	12	1.6	0	0	12	0	0	100.0
Thailand	10	1.3	3	1	6	30.0	10.0	60.0
Unknown	9	1.2	1	3	5	33.3	11.1	55.6
Ethiopia	2	0.3	0	0	2	0	0	100.0
Vietnam	1	0.1	0	0	1	0	0	100.0
Total	773	100.0	19	21	733	2.5	2.7	94.8
Peanut butter								
U.S.A.	113	61.8	1	51	61	0.9	45.1	54.0
Unknown	29	15.8	1	4	24	3.4	13.8	82.8
China	24	13.1	0	8	16	0	33.3	66.7
S. Africa	10	5.5	0	3	7	0	30.0	70.0
Indonesia	7	3.8	0	5	2	0	71.4	28.6
Total	183	100.0	2	71	110	1.1	38.8	60.1

†Examining Laboratory: Mycotoxin Research Association, supervised by the National Institute of Health, Japan.

6.9. REMOVAL OF AFLATOXIN FROM CONTAMINATED COMMODITIES

The removal or inactivation of aflatoxin from contaminated agricultural commodities represents one of the most challenging problems, and many possibilities have been investigated. Goldblatt and his research group lead in this field, and the ammoniation procedure for animal feed was established through their efforts.[17] Toxicological studies for safety evaluation of the ammoniated products have been carried out in the U.S.A.

An entirely new approach for the removal of aflatoxin was recently proposed by K. Aibara and N. Yano.[40] The principle of their patented method is organic solvent extraction using methoxymethane (dimethyl ether) which had a low boiling point (−24.2°C) at atmospheric pressure. The method overcomes several problems that have been rather difficult to solve with conventional oil extraction or the usual aflatoxin removal procedures so far proposed. The advantages are as follows. Both aflatoxin removal and oil extraction can be effected by a monoprocess at room temperatures. The solvent can be easily and thoroughly removed from the defatted products due to its low boiling point. The defatted products are still native and intact under the mild extraction conditions and desolventization process. The solvent can be recycled in the new process, so that there may be a saving on operating costs. It may be necessary, however, to undertake further research on practical applications on a large scale, toxicological studies for safety evaluation, including the possibility of trace residues in the final products, and nutritional studies on the final. treated materials.

REFERENCES

1. M. C. Lancaster, *Cancer Res.*, **28**, 2288 (1968).
2. P. M. Newberne and W. H. Butler, *ibid.*, **29**, 236 (1969).
3. R. O. Sinnhuber, D. J. Lee, J. H. Wales, M. K. Landers and A. C. Keyl, *J. Natl. Cancer Inst.*, **53**, 1285 (1974).
4. G. N. Wogan and P. M. Newberne, *Cancer Res.*, **27**, 2370 (1967).
5. G. N. Wogan, S. Paglialunga and P. M. Newberne, *Fd. Cosmet. Toxicol.*, **12**, 681 (1974).
6. U. L. Diener and N. D. Davis, *Aflatoxin* (ed. L. A. Goldblatt), p. 26, Academic Press (1969).
7. C. W. Hesseltine, *Mycopath. Mycol. Appl.*, **53**, 141 (1974).
8. R. C. Shank, C. H. Bourgeois, N. Keschamras and P. Chandavimol, *Fd. Cosmet. Toxicol.*, **9**, 501 (1971).

9. T. C. Campbell, J. P. Caedo, J. Bulatao-Jayme, L. Salamat and R. W. Engel, *Nature,* **227**, 403 (1970).
10. P. Robinson, *Clin. Pediat.*, **6**, 57 (1967).
11. A. C. Keyl, A. N. Booth, M. S. Masri, M. R. Gaubmann and W. E. Gagne, Proc. 1st U.S.-Japan Conf. Toxic Microorganisms (ed. M. Herzberg), LCC No. 77–604719, p. 72, Govt. Printing Office, Washington D. C. (1970).
12. A. C. Keyl and A. N. Booth, *J. Am. Oil Chem. Soc.*, **48**, 599 (1971).
13. W. C. Jacobson and H. G. Wiseman, *Poultry Sci.*, **53**, 1743 (1974).
14. M. S. Mabee and J. R. Chipley, *Appl. Microbiol.*, **25**, 763 (1973).
15. J. V. Rodricks and L. Stoloff, Proc. Int. Symp. Mycotoxins in Human and Animal Health, Univ. of Maryland (1976), *to be published.*
16. W. A. Parker and D. Melnick, *J. Am. Oil Chem. Soc.*, **43**, 635 (1966).
17. L. A. Goldblatt and F. G. Dollear, Proc. Int. Symp. Mycotoxins in Human and Animal Health, Univ. of Maryland (1976), *to be published.*
18. R. C. Schank, G. N. Wogan and J. B. Gibson, *Fd. Cosmet. Toxicol.*, **10**, 51 (1972); R. C. Shank, G. N. Wogan, J. B. Gibson and A. Nondasuta, *ibid.*, **10**, 61 (1972); R. C. Shank, J. E. Gordon, G. N. Wogan, A. Nondasuta and B. Subhamani, *ibid.*, **10**, 71 (1972); R. C. Shank, N. Bhamarapravati, J. E. Gordon and G. N. Wogan, *ibid.*, **10**, 171 (1972); R. C. Shank, P. Siddhichai, B. Subhamani, N. Bhamarapravati, J. E. Gordon and G. N. Wogan, *ibid.*, **10**, 181 (1972).
19. P. Keen and P. Martin, *Trop. Geogr. Med.*, **23**, 35, 44 (1971).
20. M. E. Alpert, M. S. R. Hutt and C. S. Davidson, *Lancet,* 1266 (1968); M. E. Alpert, M. S. R. Hutt, G. N. Wogan and C. S. Daviclson, *Cancer,* **28**, 253 (1971).
21. F. G. Peers and C. A. Linsell, *Brit. J. Cancer,* **27**, 473 (1973).
22. T. C. Campbell and L. Stoloff, *J. Agr. Fd. Chem.*, **22**, 1006 (1974).
23. L. Stoloff, Proc. Int. Symp. Mycotoxins in Human and Animal Health, Univ. of Maryland (1976), *to be published.*
24. *Technical Reports,* no. 10, Information Bulletin, Intl. Union Pure Appl. Chem. (1974).
25. FDA, *Federal Register,* vol. 40, no. 136 (Tuesday, July 15, 1975), Marketing Agreement 146, Peanut–1975 Crop, p. 29739.
26. *Voluntary Code of Good Practices for Peanut Product Manufacturers,* 11th ed., The Research Committee of the National Peanut Council, McLean, Virginia (1976).
27. O. L. Brekke, A. J. Peplinski and E. L. Griffen, Jr., *Cereal Chem.*, **52**, 205 (1975); O. L. Brekke, A. J. Peplinski, G. E. N. Nelson and E. L. Griffen, Jr., *ibid.*, **52**, 205 (1975).
28. J. W. Dickens and T. B. Whitaker, *Peanuts Sci.*, **2**, 45 (1975).
29. As ref. 6, p. 13–54.
30. E. B. Lillehoj, W. F. Kwolek, D. I. Fennel and M. S. Milburn, *Cereal Chem.*, **52**, 603 (1975); E. B. Lillehoj and C. W. Hesseltine, Proc. Int. Symp. Mycotoxins in Human and Animal Health, Univ of Maryland (1976), *to be published.*
31. K. Aibara and K. Miyaki, Ann. Mtg. Summary, J. Agr. Chem. Soc. Japan, p. 86 (1965).
32. H. Murakami, S. Takase and T. Ishi, *J. Gen. Appl. Microbiol.*, **13**, 323 (1967); **14**, 97, 251 (1968).
33. M. Manabe, S. Matsuura and M. Nakano, *J. Fd. Sci. Technol. Japan,* **15**, 341 (1968).
34. T. Yokotsuka, M. Sasaki, T. Kikuchi, Y. Asao and A. Nobuhara, *J. Agr. Chem. Soc. Japan,* **41**, 32 (1967); *Biochemistry of Some Foodborne Microbial Toxins* (ed. R. I. Mateles and G. N. Wogan), p. 131, MIT Press (1967).
35. M. Manabe, S. Ohnuma and S. Matsuura, *J. Fd. Sci. Technol. Japan,* **19**, 76 (1972).
36. M. Manabe and S. Matsuura, *ibid.*, **19**, 268 (1972); *J. Agr. Chem. Soc. Japan,* **47**, 209 (1973); *Trans. Mycol. Soc. Japan,* **16**, 399 (1975).
37. K. Aibara and K. Miyaki, *unpublished data* (Report to the Food Sanitation Investigation Council, Ministry of Health and Welfare, Japan) (1970).
38. O. Tsuruta and M. Manabe, *Trans. Mycol. Soc. Japan,* **15**, 401 (1974); **16**, 190 (1975).

39. K. Aibara and K. Maeda, *unpublished data* (Report to the Ministry of Health and Welfare, Japan (1974).
40. K. Aibara and N. Yano, Proc. Int. Symp. Mycotoxins in Human and Animal Health, Univ. of Maryland (1976), *to be published*.

Subject Index

Q, R

S

T

Index of Organisms